Chemistry and Biology
of Nucleosides and Nucleotides

Chemistry and Biology
of Nucleosides and Nucleotides

Edited by

Robert E. Harmon
Department of Chemistry
Western Michigan University
Kalamazoo, Michigan

Roland K. Robins
Department of Chemistry
Brigham Young University
Provo, Utah

Leroy B. Townsend
Department of Chemistry
University of Utah
Salt Lake City, Utah

ACADEMIC PRESS   NEW YORK   SAN FRANCISCO   LONDON   1978

A Subsidiary of Harcourt Brace Jovanovich, Publishers

C ++on

ACADEMIC PRESS, INC.
111 Fifth Avenue, New York, New York 10003

United Kingdom Edition published by
ACADEMIC PRESS, INC. (LONDON) LTD.
24/28 Oval Road, London NW1 7DX

ISBN 0-12-326140-6

PRINTED IN THE UNITED STATES OF AMERICA

78 79 80 81 82   9 8 7 6 5 4 3 2 1

*Dedicated to my wife, Marilyn,
and to my children, Greg, Jeff, Eric, and Heidi.*

Contents

List of Contributors *xi*

Preface *xix*

Total Synthesis of a Bacterial Gene II: The Total DNA
Including the Promotor and Its Transcription *Ramamoorthy
Belagaje, Eugene L. Brown, Hans-Joachim Fritz,
Michael J. Gait, Robert G. Lees, Kjeld Norris, Takao
Sekiya, Tatsuo Takeya, Roland Contreras, Hans
Küpper, Michael J. Ryan, and H. Gobind Khorana* 1

Total Synthesis of Bacterial Gene I: Chemical Synthesis of
Oligonucleotides and Use of High-Pressure Liquid
Chromatography *Hans-Joachim Fritz, Ramamoorthy
Belagaje, Eugene L. Brown, Rebecca H. Frita, Roger A.
Jones, Robert G. Lees, and H. Gobind Khorana* 21

The Use of Microparticulate Reversed-Phase Packing in
High-Pressure Liquid Chromatography of Nucleosides *John
A. Montgomery, H. Jeanette Thomas, Robert D. Elliott,
and Anita T. Shortnacy* 37

Chemical and Enzymatic Methods in the Synthesis of
Modified Polynucleotides *Thomas J. Bardos and
Yau-Kwan Ho* 55

Simple Models of Nucleic Acid Interactions *J. Zemlicka,
M. Murata, and J. Owens* 69

Synthesis of Nucleotides which Inactivate Enzymes
Utilizing Adenine Nucleotides *Alexander Hampton* 85

Synthesis, Antiviral Activity, and Mechanism of Action of a
Novel Series of Pyrimidine Nucleoside Analogs   *William H.
Prusoff, Tai-Shun Lin, Ming-Shen Chen, George T. Shiau,
and David C. Ward*   97

The Chemistry and Biological Activity of C-Nucleosides
Related to Pseudouridine   *Dean S. Wise, Robert A. Earl,
and Leroy B. Townsend*   109

The Chemistry and Biological Activity of Certain 4-Substituted
and 3,4-Disubstituted Pyrazolo (3,4-d) Pyrimidine
Nucleosides   *Raymond P. Panzica, Ganapati A. Bhat,
Robert A. Earl, Jean-Louis G. Montero, Linda W.
Roti Roti, and Leroy B. Townsend*   121

The Chemistry and Biology of Some New Nucleoside
Analogs Active against Tumor Cells   *Miroslav Bobek and
Alexander Bloch*   135

Potentiation of the Chemotherapeutic Action of Antineoplastic
Agents by Nucleosides   *H. Osswald*   149

Coformycin and Deoxycoformycin: Tight-Binding
Inhibitors of Adenoxine Deaminase   *R. P. Agarwal,
Sungman Cha, G. W. Crabtree, and R. E. Parks, Jr.*   159

Synthesis and Chemistry of a New Antiviral–Antibiotic
Nucleoside in Anhydro Form: Application of a Novel
1,3-Electrophilic (C-N-C) Annulation Reagent   *Wendell
Wierenga and John A. Woltersom*   199

Synthesis of 1-(2-Deoxy-β-D-Ribofuranosyl)-5,6-Dihydro-
5-Methyl-s-Triazin-2,4(1H,3H)Dione   *H. I. Skulnick*   211

An Approach to the Synthesis of 1,9-Diribosyl Purine
Bernard Rayner, Claude Tapiero, and Jean-Louis Imbach   229

An Initial Study on the Selectivity and Stereospecificity of
the Ribosylation Reaction   *Jean-Louis Barascut and
Jean-Louis Imbach*   239

On the Mechanism of Nucleoside Synthesis *Helmut F. Vorbrüggen, Ulrich Niedballa, Konrad Krolikiewicz, Bärbel Bennua, and Gerhard Höfle*   251

Synthesis and Chemistry of Certain Azole Nucleosides *Joseph T. Witkowski and Roland K. Robins*   267

The Synthesis of Heterocyclic Analogs of Purine Nucleosides and Nucleotides Containing a Bridgehead Nitrogen Atom *Ganapathi R. Revankar and Roland K. Robins*   287

Determination of the Site of Glycosylation in Various Nucleosides by Carbon-13 NMR Spectroscopy *Phoebe Dea and Roland K. Robins*   301

A "Geometry-Only" ^{1}H NMR Method for Determination of the Anomeric Configuration of Ribofuranosyl Compounds *Morris J. Robins and Malcolm Maccoss*   311

Investigation of Nucleoside Alkylphosphonates *M. N. Preobrazhenskaya, S. Ya. Melnik, T. P. Nedorezova, I. D. Shingarova, and D. M. Oleinik*   329

The *Tert*-Butyldimethylsilyl Group as a Protecting Group in Oligonucleotide Synthesis *Wilfried Köhler, Wilhelm Schlosser, Geeta Charubala, and Wolfgang Pfleiderer*   347

Advances in the Synthesis of *C*-Glycosyl Nucleosides *John G. Moffatt, Hans P. Albrecht, Gordon H. Jones, David B. Repke, Günter Trummlitz, Hiroshi Ohrui, and Chhitar M. Gupta*   359

Synthesis of C-Glycosyl Thiazoles *M. Fuertes, M. T. García-López, G. García-Muñoz, and M. Stud*   381

A New Method for the Synthesis of 2'-Deoxy-2'-Substituted Purine Nucleosides. Synthesis of the Antibiotic 2'-Deoxy-2'-Aminoguanosine *Morio Ikehara, Tokumi Maruyama, and Hiroko Miki*   387

Synthesis and Reaction of Nucleosides Containing Sulfur
Functions in the Sugar Portion   *Tohru Ueda, Akira Matsuda,
Tamotsu Asano, and Hideo Inoue*   397

New C-Nucleoside Isosteres of Some Nucleoside Antibiotics
*Jack J. Fox, Kyoichi A. Watanabe, Robert S. Klein,
Chung K. Chu, Steve Y-K. Tam, Uri Reichman,
Kosaku Hirota, J.-S. Hwang, Federico G. De Las Heras, and
Iris Wempen*   415

Index   *441*

List of Contributors

Numbers in parentheses indicate the pages on which the authors' contributions begin.

R. P. Agarwal (159), Section of Biochemical Pharmacology, Division of Biology and Medicine, Brown University, Providence, Rhode Island

Hans P. Albrecht (359), Institute of Molecular Biology, Syntex Research, Palo Alto, California

Tamotsu Asano (397), Faculty of Pharmaceutical Sciences, Hokkaido University, Sapporo, Japan

Jean-Louis Barascut (239), Université des Sciences et Techniques du Languedoc, Laboratoire de Chimie Bio-Organique, Place E. Bataillon, 34060 Montpellier Cedex, France

Thomas J. Bardos (55), Department of Medicinal Chemistry, State University of New York at Buffalo, New York 14214

Ramamoorthy Belagaje (1, 21), Departments of Biology and Chemistry, Massachusetts Institute of Technology, Cambridge, Massachusetts

Bärbel Bennua (251), Research Laboratories of Schering, A.G., Berlin/Bergkamen-, Germany

Ganapati A. Bhat (121), Division of Medicinal Chemistry, Department of Biopharmaceutical Science and Department of Chemistry, University of Utah, Salt Lake City, Utah

Alexander Bloch, (135), Department of Experimental Therapeutics, Grace Cancer Drug Center, Roswell Park Memorial Institute, Buffalo, New York

Miroslav Bobek (135), Department of Experimental Therapeutics, Grace Cancer Drug Center, Roswell Park Memorial Institute, Buffalo, New York

Eugene L. Brown (1, 21), Departments of Biology and Chemistry, Massachusetts Institute of Technology, Cambridge, Massachusetts 02139

Sungman Cha (159), Section of Biochemical Pharmacology, Division of Biology and Medicine, Brown University, Providence, Rhode Island

Geeta Charubala (347), Fachbereich Chemie, Universität Konstanz, Konstanz, West Germany

Ming-Shen Chen (97), Department of Pharmacology, Yale University School of Medicine, New Haven, Connecticut 06510

Chung K. Chu (415), Laboratory of Organic Chemistry, Memorial Sloan–Kettering Cancer Center, Sloan–Kettering Institute, Sloan–Kettering Division of Graduate School of Medical Sciences, Cornell University, Rye, New York

Roland Contreras (1), Laboratorium voor Moleculaire Biologie, Geur, Belgium

G. W. Crabtree (159), Section of Biochemical Pharmacology, Division of Biology and Medicine, Brown University, Providence, Rhode Island

Federico G. De Las Heras (415), Laboratory of Organic Chemistry, Memorial Sloan–Kettering Cancer Center, Sloan–Kettering Institute, Sloan–Kettering Division of Graduate School of Medical Sciences, Cornell University, Rye, New York

Phoebe Dea (301), Department of Chemistry, California State University at Los Angeles, Los Angeles, California 90032

Robert A. Earl (109, 121), Division of Medicinal Chemistry, Department of Biopharmaceutical Sciences and Department of Chemistry, University of Utah, Salt Lake City, Utah

Robert D. Elliott (37), Kettering-Meyer Laboratory, Southern Research Institute, Birmingham, Alabama

Jack J. Fox (415), Laboratory of Organic Chemistry, Memorial Sloan–Kettering Cancer Center, Sloan–Kettering Institute, Sloan–Kettering Division of Graduate School of Medical Sciences, Cornell University, Rye, New York

Rebecca H. Frita (21), Departments of Biology and Chemistry, Massachusetts Institute of Technology, Cambridge, Massachusetts

Hans-Joachim Fritz (1, 21), Departments of Biology and Chemistry, Massachusetts Institute of Technology, Cambridge, Massachusetts 02139

M. Fuertes (381), Institute de Qúimica Médica, Juan de la Cierva, Madrid, Spain

Michael J. Gait (1), MRC Laboratory of Molecular Biology, Hills Road, Cambridge, England

M. T. García-López (381), Instituto de Química Médica, Juan de la Cierva, Madrid, Spain

G. García-Muñoz (381), Institute de Química Médica, Juan de la Cierva, Madrid, Spain

Chhitar M. Gupta (359), Institute of Molecular Biology, Syntex Research, Palo Alto, California

Alexander Hampton (85), The Institute of Cancer Research, The Fox Chase Cancer Center, Philadelphia, Pennsylvania

Kosaku Hirota (415), Laboratory of Organic Chemistry, Memorial Sloan–Kettering Cancer Center, Sloan–Kettering Institute, Sloan–Kettering Division of Graduate School of Medical Sciences, Cornell University, Rye, New York

Yau-Kwan Ho (55), Department of Medicinal Chemistry, State University of New York at Buffalo, Buffalo, New York 14214

Gerhard Höfle (251), Department of Organic Chemistry, Technical University, Berlin 12, Germany

J-S. Hwang (415), Laboratory of Organic Chemistry, Memorial Sloan–Kettering Cancer Center, Sloan–Kettering Institute, Sloan–Kettering Division of Graduate School of Medical Sciences, Cornell University, Rye, New York

Morio Ikehara (387), Faculty of Pharmaceutical Sciences, Osaka University, Suita, Osaka, Japan

Jean-Louis Imbach (229, 239), Université des Sciences et Techniques du Languedoc, Laboratoire de Chemie Bio-Organique, 34060 Montpellier Cedex, France

Hideo Inoue (397), Faculty of Pharmaceutical Sciences, Hokkaido University, Sapporo, Japan

Gordon H. Jones (359), Institute of Molecular Biology, Syntex Research, Palo Alto, California

Roger A. Jones (21), Departments of Biology and Chemistry, Massachusetts Institute of Technology, Cambridge, Massachusetts 02139

H. Gobind Khorana (1, 21), Departments of Biology and Chemistry, Massachusetts Institute of Technology, Cambridge, Massachusetts 02139

Robert S. Klein (415), Laboratory of Organic Chemistry, Memorial Sloan–Kettering Cancer Center, Sloan–Kettering Institute, Sloan–Kettering Division of Graduate School of Medical Sciences, Cornell University, Rye, New York

Wilfried Köhler (347), Fachbereich Chemie, Universität Konstanz, West Germany

Konrad Krolikiewicz (251), Research Laboratories of Schering A. G., Berlin/Bergkamen, Germany

Hans Küpper (1), Molekulare Genetik Der Universitat, Heidelberg, West Germany

Robert G. Lees (1, 21), Department of Chemistry, Fordham University, Bronx, New York

Tai-Shun Lin (97), Department of Pharmacology Yale University School of Medicine, New Haven, Connecticut 06510

Malcolm Maccoss (311),[1] Argonne National Laboratory, Argonne, Illinois

Tokumi Maruyama (387), Faculty of Pharmaceutical Sciences, Osaka University, Suita, Osaka, Japan

Akira Matsuda (397), Faculty of Pharmaceutical Sciences, Hakkaido University, Sapporo, Japan

S. Ya. Melnik (329), Cancer Research Center of the USSR, Academy of Medical Sciences, 115478 Moscow, USSR

Hiroko Miki (387), Faculty of Pharmaceutical Sciences, Osaka University, Suita, Osaka, Japan

John G. Moffatt (359), Institute of Molecular Biology, Syntex Research, Palo Alto, California

Jean-Louis G. Montero (121), Division of Medicinal Chemistry, Department of Biopharmaceutical Sciences and Department of Chemistry, University of Utah, Salt Lake City, Utah

John A. Montgomery (37), Kettering-Meyer Laboratory, Southern Research Institute, Birmingham, Alabama

M. Murata (69), Michigan Cancer Foundation and Department of Oncology, Wayne State University School of Medicine, Detroit, Michigan

T. P. Nedorezova (329), Cancer Research Center of the USSR Academy of Medical Sciences, 115478 Moscow, USSR

Ulrich Niedballa (251), Research Laboratories of Schering A.G., Berlin/Bergkamen, Germany

Kjeld Norris (1), The Danish Institute of Protein Chemistry, Hørsholm, Denmark

Hiroshi Ohrui (359), Institute of Molecular Biology, Syntex Research, Palo Alto, California

D. M. Oleinik (329), Cancer Research Center of the Academy of Medical Sciences, 115478, Moscow, USSR

H. Osswald (149), Institute of Toxicology and Chemotherapy, German Cancer Research Center, Heidelberg, Federal Republic of Germany

[1]Present address: Department of Chemistry, The University of Alberta, Edmonton, Alberta, Canada T6G2G2

J. Owens (69), Michigan Cancer Foundation and Department of Oncology, Wayne State University School of Medicine, Detroit, Michigan

Raymond P. Panzica (121), Department of Medicinal Chemistry, University of Rhode Island, Kingston, Rhode Island

R. E. Parks, Jr. (159), Section of Biochemical Pharmacology, Division of Biology and Medicine, Brown University, Providence, Rhode Island

Wolfgang Pfleiderer (347), Fachereich Chemie, Universitat Konstanz, Konstanz, West Germany

M. N. Preobrazhenskaya (329), Cancer Research Center of the USSR Academy of Medical Sciences, 115478, Moscow, USSR

William H. Prusoff (97), Department of Pharmacology, Yale University School of Medicine, New Haven, Connecticut 06510

Bernard Rayner (229), Université des Sciences et Techniques du Languedoc, Laboratoire de Chemie Bio-Organique, Montpellier, Cedex, France

Uri Reichman (415), Laboratory of Organic Chemistry, Memorial Sloan–Kettering Cancer Center, Sloan–Kettering Institute, Sloan–Kettering Division of Graduate School of Medical Sciences, Cornell University, Rye, New York

David B. Repke (359), Institute of Molecular Biology, Syntex Research, Palo Alto, California

Ganapathi R. Revankar (287), ICN Pharmaceuticals, Inc., Nucleic Acid Research Institute, Irvine, California

Morris J. Robins (311), Department of Chemistry, The University of Alberta, Edmonton, Alberta, Canada T6G2G2

Roland K. Robins (267, 287, 301),[2] ICN Pharmaceuticals, Inc., Nucleic Acid Research Institute, Irvine, California

Linda W. Roti Roti, (121), Division of Medicinal Chemistry, Department of Biopharmaceutical Sciences and Department of Chemistry, University of Utah, Salt Lake City, Utah

Michael J. Ryan (1), Departments of Biology and Chemistry, Massachusetts Institute of Technology, Cambridge, Massachusetts 02139

Wilhelm Schlosser (347), Fachbereich Chemie Universitat Konstanz, Konstanz, West Germany

Takao Sekiya (1), Microbial Chemistry Research Foundation, Institute of Microbial Chemistry, Tokyo, Japan

[2]Department of Chemistry, Brigham Young University, Provo, Utah

George T. Shiau (97), Department of Pharmacology, Yale University School of Medicine, New Haven, Connecticut 06510

I. D. Shingarova (329), Cancer Research Center of the USSR Academy of Medical Sciences, 115478 Moscow, USSR

Anita T. Shortnacy (37), Kettering–Meyer Laboratory, Southern Research Institute, Birmingham, Alabama

H. I. Skulnick (211), The Upjohn Company, Kalamazoo, Michigan

M. Stud (381), Instituto de Química Médica, Juan de la Cierva, Madrid, Spain

Tatsuo Takeya (1), Institute for Chemical Research, Kyoto University, Kyoto, Japan

Steve Y-K. Tam (415), Laboratory of Organic Chemistry, Memorial Sloan–Kettering Cancer Center, Sloan–Kettering Institute, Sloan–Kettering Division of Graduate School of Medical Sciences, Cornell University, Rye, New York

Claude Tapiero (229), Université des Sciences et Techniques du Languedoc, Laboratoire de Chimie Bio-Organique, Montpellier Cedex, France

H. Jeanette Thomas (37), Kettering–Meyer Laboratory, Southern Research Institute, Birmingham, Alabama

Leroy B. Townsend (109, 121), Division of Medicinal Chemistry, Department of Biopharmaceutical Sciences and Department of Chemistry, University of Utah, Salt Lake City, Utah

Günter Trummlitz (359), Institute of Molecular Biology, Syntex Research, Palo Alto, California

Tohru Ueda (397), Faculty of Pharmaceutical Sciences, Hakkaido University, Sapporo, Japan

Helmut F. Vorbrüggen (251), Research Laboratories of Schering A.G., Berlin/Bergkamen, Germany

David C. Ward (97), Departments of Human Genetics and Molecular Biophysics and Biochemistry, Yale University School of Medicine, New Haven, Connecticut 06510

Kyoichi A. Watanabe (415), Laboratory of Organic Chemistry, Memorial Sloan–Kettering Cancer Center, Sloan–Kettering Institute, Sloan–Kettering Division of Graduate School of Medical Sciences, Cornell University, Rye, New York

Iris Wempen (415), Laboratory of Organic Chemistry, Memorial Sloan–Kettering Cancer Center, Sloan–Kettering Institute, Sloan–Kettering Division of Graduate School of Medical Sciences, Cornell University, Rye, New York

Wendell Wierenga (199), Experimental Chemistry Research, The Upjohn Company, Kalamazoo, Michigan 49001

Dean S. Wise (109), Division of Medicinal Chemistry, Department of Biopharmaceutical Sciences and Department of Chemistry, University of Utah, Salt Lake City, Utah

Joseph T. Witkowski (267), ICN Pharmaceuticals, Inc., Nucleic Acid Research Institute, Irvine, California

John A. Woltersom (199), Experimental Chemistry Research, The Upjohn Company, Kalamazoo, Michigan 49001

J. Zemlicka (69), Michigan Cancer Foundation and Department of Oncology, Wayne State University School of Medicine, Detroit, Michigan

Preface

This volume is a permanent record of the symposium on the Chemistry and Biology of Nucleosides and Nucleotides held August 30–September 1, 1976, as part of the San Francisco Centennial Meeting of the Carbohydrate Division of the American Chemical Society. This symposium was partially supported by Grant 1 R13 CA-20090-1 from the National Cancer Institute and was also designated as the Second International Round Table on Nucleosides and Nucleotides. The First International Round Table was held in 1974 in Montpellier, France, and was organized by Jean-Louie Imbach, Laboratoire de Chemie Bioorganic, Université des Science et Techniques du Languedoc, Montpellier, France.

The chemistry and biology of nucleosides and nucleotides was deemed to be a very appropriate topic for a symposium to be held in conjunction with a meeting celebrating the hundredth anniversary of the founding of the American Chemical Society, since about 100 years ago, a young Swiss physician, Friedrich Miescher, published the first paper on "nuclein" (or nucleohistone, using current nomenclature) and thus launched chemical research on nucleic acids.

It was also deemed appropriate that the first paper of the symposium should describe the synthesis of a gene. The second paper described the introduction of this synthetic gene into a biological system where it proved to be functional. Nearly 25 years ago nucleic acids were identified as the physical basis of genes, and the first papers presented at the symposium represent the first reported synthesis of a functional gene.

Knowledge of nucleosides and nucleotides is being used extensively in the areas of cell biology, microbiology, virology, oncology, genetic therapy, genetic engineering, and chemotherapy. The papers in this symposium describe many new chemical and instrumental techniques used in the synthesis of nucleosides, nucleotides, and nucleic acids.

Several of the papers describe the synthesis of nucleosides and nucleotides with anticancer and antiviral activity. The rational for the design and synthesis of anticancer and antiviral compounds is also discussed. The study of transition state analog inhibition of enzymatic reactions (e.g., adenosine deaminase) is reported. A transition state analog may not have any anticancer or antiviral activity itself but because of its ability to essentially irreversibly bind to certain enzymes the half-life of an antiviral or anticancer agent may be greatly increased thereby greatly increasing the latter's efficacy. An example would be the increase in antiviral activity of "Ara-A" (9-β-D-arabino-furanosyladenine) brought out by the tight-binding adenosine deaminase inhibitors coformycin and deoxycoformycin. The latter compounds inhibit the conversion of Ara-A to Ara-H (9-β-D-arabinofuranosylhypoxanthine). The antiviral activity of Ara-H is much lower than that for Ara-A.

The synthesis of modified nucleic acids as new types of macromolecular antimetabolites is done by altering the intact nucleic acids using techniques that will be of interest to the synthetic medicinal chemist as well as biomedical researchers. The use of high pressure liquid chromatography to separate complex synthetic mixtures of oligionucleotides into pure oligonucleotides represents a significant advance in this area. This breakthrough made it possible to synthesize a gene.

With the advent of ^{13}C nuclear magnetic resonance, the unambiguous assignment of structure to ribosidated bicylic nitrogen-containing heterocyclics is now possible. The use of proton magnetic resonance on certain derivatives of nucleosides (e.g., 2',3'-di-0-isopropylidene, 3',5'-cyclic phosphate) allows one to make an unambiguous assignment of configuration at the anomeric carbon on the ribose portion of the nucleoside or nucleotide.

The description of new selective protecting groups and their applications to nucleoside and nucleotide chemistry will also be of great interest to the synthetic medicinal chemists, as well as others. New routes to the synthesis of C-nucleosides and nucleotides, which may function as metabolites as well as antimetabolites, were reported in several papers. The synthesis of 5',8'-cyclic nucleosides using photochemical techniques is also discussed.

The clinical use of naturally occurring nucleosides such as thymidine and inosine to potentiate the anticancer activity of structurally unrelated carcinostatic agents reveals a new approach to combination chemotherapy.

Hopefully, this volume will serve as reference and source material for many workers in biomedical research as teaching material for instructors of advanced science courses.

I would like to thank Drs. Harry B. Wood, Roland K. Robins, Leroy B. Townsend, and Tai-shun Lin for serving as chairmen of the various sessions of this symposium.

I would like to express my sincere appreciation and thanks to my wife, Marilyn K. Harmon, and to Mrs. Barbara Hodsdon, Mrs. Chris Harmon, and Mr. A. L. Pence of the American Chemical Society staff for handling the on-site management of the finances, meeting room arrangements, hosting of the foreign participants at the symposium, etc. I would also like to thank Marilyn K. Harmon for preparing the index for this volume. Without the help of all these people, the symposium and this volume would not have been possible.

Lastly, I would like to commend all the contributors to the symposium on their well-prepared talks and thank them for the prompt submission of their manuscripts.

Robert E. Harmon

Executive Editor

TOTAL SYNTHESIS OF A BACTERIAL GENE II:
THE TOTAL DNA INCLUDING THE PROMOTER AND
ITS TRANSCRIPTION*

*RAMAMOORTHY BELAGAJE, EUGENE L. BROWN, HANS-JOACHIM
FRITZ, MICHAEL J. GAIT[a], ROBERT G. LEES[b],
KJELD NORRIS[c], TAKAO SEKIYA[d],
TATSUO TAKEYA[e], ROLAND CONTRERAS[f],
HANS KÜPPER[g], MICHAEL J. RYAN, AND H. GOBIND KHORANA*

*Departments of Biology and Chemistry,
Massachusetts Institute of Technology,
Cambridge, Massachusetts*

[a]MRC Laboratory of Molecular Biology, Hills Road, Cambridge,
England.

[b]Department of Chemistry, Fordham University, Bronx, New
York.

[c]The Danish Institute of Protein Chemistry, Hørsholm, Denmark.

[d]Microbial Chemistry Research Foundation, Institute of Microbial Chemistry, Tokyo, Japan.

[e]Institute for Chemical Research, Kyoto University, Japan.

[f]Laboratorium voor Moleculaire Biologie, Geur, Belgium

[g]Molekulare Genetik Der Universitat, Heidelberg, West Germany.

ABSTRACT

The total synthesis of a 207 nucleotide base-pairs long DNA duplex corresponding to the *Escherichia coli* tyrosine suppressor
transfer RNA gene has been completed. The total DNA consists
of (a) a 126 nucleotide-long DNA duplex corresponding to the
structural gene; (b) a DNA duplex, 25 nucleotides long, repre-
senting the region adjoining the C-C-A end; and (c) a 56 nucleotide-long duplex corresponding to the promoter region (prein-
itiation of transcription region.) *In vitro* transcription of
this DNA duplex containing the promoter as well as the DNA duplex
(149 nucleotide base pairs) lacking the promoter has been accomplished. The products have been fully characterized and are
consistent with the expected transfer RNA. The synthetic gene
is also shown to be fully functional *in vivo* by demonstrating
its suppressor activity.

An *E. coli* tyrosine tRNA precursor of 126 nucleotide chain length
was discovered by Altman and Smith (1) and this was then processed
by various enzymes to give a functional tRNA of 85 nucleotide
chain length. The primary sequence and a possible secondary
structure of this precursor is shown in Fig. 1. One of the most
important aspects of this structure to be noted is that this precursor carries a 5'-triphosphate group at the 5'-terminus. This
immediately tells us very firmly that this is the point where the
transcription by the RNA polymerase started. Knowing the precise
initiation point for the precursor, we can present a model for
the structural components or elements of the gene, as shown in
Fig. 2. Shown here are the regions of the DNA that specify the
tRNA itself and its precursor. Presumably, the promoter region
or start signal where the binding of the RNA polymerase occurs is
to the left of this precursor; similarly, to the right of the pre-

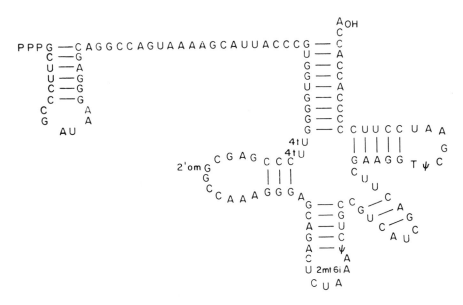

FIGURE 1. *The primary nucleotide sequence of an E.coli tyrosine tRNA precursor.*

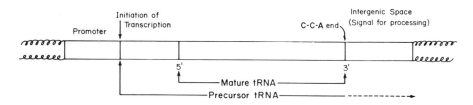

FIGURE 2. *Diagrammatic representation of the linear arrangements of transcriptional control elements and the structural gene for tyrosine transfer RNA.*

cursor gene, there is a special sequence that contains a recognition site for a processing enzyme. Indeed, it has recently been shown (2) that the primary product of transcription of this tyrosine tRNA gene is about 190 nucleotides long. In order to study in detail the mechanism of transcription and structure function relationships of this tRNA, we have undertaken the total synthesis of the bihelical DNA corresponding to the full length of this gene with different sections, as pointed out in the diagram. While the sequence of the "structural gene" specifying the 126-nucleotides long precursor could be simply derived by the

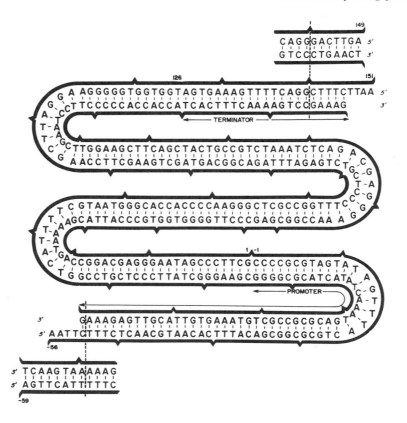

FIGURE 3. *The totally synthetic DNA corresponding to the tyrosine suppressor transfer RNA gene.*

base pairing rule, the sequences of the promoter region and the region adjoining the C-C-A end had to be deduced, and this was successfully done in our laboratory, as documented in earlier papers (3-7). The total sequences are included in the total DNA corresponding to the gene shown in Fig. 3. Shown here are the sequences of 126 nucleotides corresponding to the precursor gene, the 25 nucleotides adjoining its C-C-A end containing a recognition site for a processing enzyme, and the 56 nucleotides corresponding to the promoter region. It should be noted that this structure includes EcoR1 restriction sequences at the opposite ends. This is done by exchanging a few base pairs of the natural sequence (shown on either side of the diagram) in such a way that the total chemical work is reduced and at the same time it should allow its transformation into a suitable vector for biological multiplication of the synthetic gene.

In this paper, we review briefly the methods we have used for the total synthesis of this gene and then describe our work on the transcription of the gene and stepwise processing of the corresponding RNA product to give the mature tRNA.

The general methodology for the step-by step construction of the bihelical DNA involves three steps:

1. Chemical synthesis of polydeoxyribonucleotide segments of chain length in the range of 8 - 12 units with 3'- and 5'-hydroxyl end-groups free; the segments should correspond to the entire two strands of the intended DNA, and those belonging to the complementary strands should have an overlap of four to five nucleotides.

2. The phosphorylation of the 5'-hydroxyl group with $(\gamma-^{32}P)$-ATP using T_4 polynucleotide kinase.

3. The head-to-tail joining of the appropriate segments when they are aligned to form short bihelical complexes using T_4 polynucleotide ligase.

The synthetic plan that we have adopted for the total synthesis of the gene is shown in Fig. 4. Following extensive experimentation, the chemically synthesized segments were grouped into a total of eight duplexes as shown in this diagram. Thus, duplexes (I) - (IV) correspond to the structural gene, duplex (Va) (natural sequence) or duplex (Vb) (modified sequence) corresponds to the terminator region, and duplexes (P_{1-3}) and (P_{4-10}). correspond to the promoter region (the subscript indicates the number of oligonucleotide segments contained in the duplex). The carets show the points of enzymatic joinings performed by ligase, while distances between neighboring carets show the sequences of oligonucleotides that are chemically synthesized. Thus, it involves a total of 39 chemically synthesized segments.

The task of chemical synthesis of this magnitude was clearly the most demanding of the whole project and required sustained group effort over a period of several years (8). However, we must also add that a large amount of work is necessary to determine the combinations of segments that would give optimum yield in the overall joining reactions. One might be tempted to ask "how is this plan deduced?" Some of the factors influencing our decision are the following:

1. There is, of course, the basic requirement of the overlap of four or more base pairs.

2. Are there any sequences that occur more than once? The answer is yes. Multiple use of common blocks or parts of segments would certainly reduce the total chemical work. In this way, a complete search using a computer program was carried out

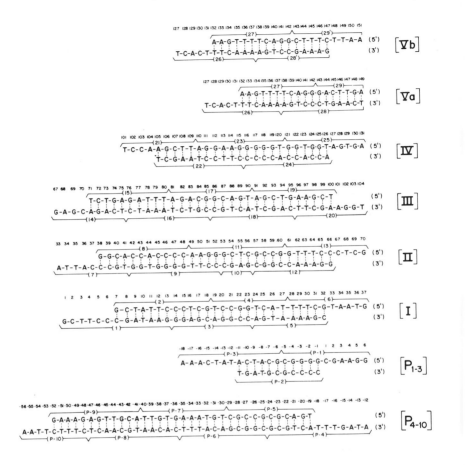

FIGURE 4. Plan for the total synthesis of the tyrosine suppressor transfer RNA gene showing chemically synthesized dioxyribopolynucleotide segments, which constitute the duplex (I) to (V), (P_{1-3}) and (P_{4-10}).

for the occurrence of common sequences. Thus, the hexanucleotide sequences C-C-C-C-A-C-C-A-C (segments 8 and 24) and A-A-T-T-C-T-T-T-C (segments P-10, 29') occur twice. Arguments of this kind can be applied for parts of quite a few other segments.

3. The third consideration is that of partial or total complementarity in six single-stranded regions of duplex. Such features lead to wrong and unwanted joinings and should therefore be totally avoided.

4. Having prepared six duplexes as shown in the diagram, the next step is to quantitatively phosphorylate the terminal 5'-hydroxyl groups. Experiments with a number of defined duplexes showed that the extent and rate of phosphorylation of these 5'-hydroxyl groups at the termini is greatly influenced by the structure of the duplex. Therefore, to ensure complete and facile phosphorylation, it is essential to have terminal 5'-hydroxyl groups at the protruding single-stranded ends of these duplexes.

Having briefly justified our synthetic plan, we now give the general methodology used for the characterization of these duplexes. Three basic concepts are being used.

1. As the duplex is formed from the corresponding oligonucleotide segments contain radioactively labeled phosphate groups (^{32}P or ^{33}P at different levels of specific activity, we can easily follow the progress of the reaction by measuring the kinetics in this way.
3. The isolated duplex is further characterized by nearest-neighbor analysis after degradation to 3'- and 5'-mononucleotides. The duplex is degraded either to 3'-mononucleotides using micrococcal nuclease and spleen phosphodiesterase or to 5'-mononucleotides using pancreatic deoxyribonuclease and snake venom phosphodiesterase. The degraded mononucleotides are separated by paper chromatography using suitable solvents.

In this way, all the duplexes were characterized. Only brief descriptions of the synthetic results can now be given to illustrate the kinds of results that are obtained in the joining reactions.

I. TOTAL SYNTHESIS OF THE PROMOTER, DUPLEX (P)

Because of the remarkable symmetry in the structure (palindromes) of the promoter region, it is expected that some segments can occupy more than one place in the duplex. For example, segments P-2 and P-5 can easily substitute for each other if they are present in the same reaction mixture. Following extensive experimentation using various combinations of three, four, or five segments at a time, the desired duplex was prepared using the following steps.

8   *Ramamoorthy Belagaje et al.*

*1. STEP 1 PREPARATION OF SINGLE-STRANDED DNA CONTAINING SEGMENTS
P-1 AND P-3*

This was prepared, as shown in Fig. 5, by joining phosphorylated segments P-1 to P-3 (^{32}P at the 5'-end) in the presence of
phosphorylated segments P-2 (^{32}P at the 5'-end) and then separating the two strands under denaturing conditions. The reaction
was performed for 23 hr at 5°C, after which the desired product
was isolated by electrophoresis on 24% polyacrylamide gel containing 7 *M* urea.

2. STEP 2 PREPARATION OF THE DUPLEX (P_{4-7})

This was prepared by using four phosphorylated segments
(P-4, P-5, P-6, and P-7) in one reaction mixture in 85% yield.
The kinetics of joining and the isolation of duplex (P_{4-7}) on
Sephadex G-100 column are shown in Fig. 6. The isolated duplex
was then fully characterized by measuring the resistance to the
alkaline phosphatase followed by nearest-neighbor analysis after
degradation to 3'- and 5'-mononucleotides as explained earlier.

*3. STEP 3 PREPARATION OF THE DUPLEX CONTAINING SEGMENTS P-4
THROUGH P-10*

To the performed duplex (P_{4-7}) was then added phosphorylated
segments P-8 and P-9 (^{32}P at the 5'-end) and an excess of unphosphorylated segment P-10. The kinetics of joining and isolation
of the duplex on 15% polyacrylamide gel are shown in Fig. 7.

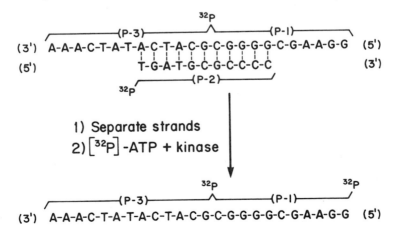

*FIGURE 5. Preparation of the single-stranded DNA containing
segments P-1 joined to segment P-3 (promoter region, Fig. 4).*

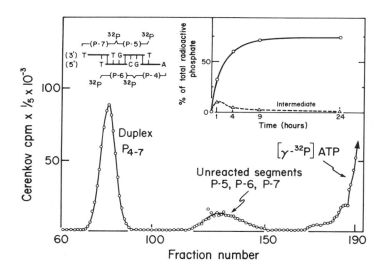

FIGURE 6. *Joining of segments P-4, P-5, P-6, and P-7 to form the duplex (P$_{4-7}$) (Promoter region, Fig. 4). The kinetics of joining are in the inset. The separation on a Sephadex G-100 column is shown. The first peak is duplex (P$_{4-7}$).*

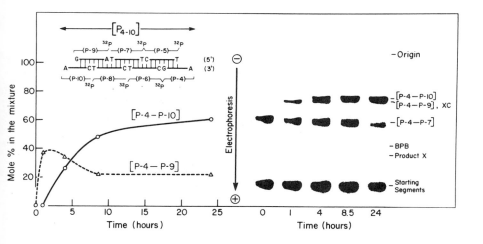

FIGURE 7. *Addition of Segments P-8, P-9, and P-10 to the duplex (P$_{4-7}$) (promoter region, Fig. 4). The kinetics of joining are shown (left). Time aliquots were taken and subjected to electrophoresis on 15% polyacrylamide gel (right).*

4. *STEP 4   THE TOTAL PROMOTER DUPLEX* (P_{1-10})

To the above preformed duplex containing segments P-4 through
P-10 was added preformed single-stranded DNA P1-3 (^{32}P at the
5'-end) and phosphorylated segment P-2 (^{32}P at the 5'-end.)   The
kinetics of joining and isolation of the duplex in 15% polyacrylamide gel are shown in Fig. 8.   The duplex was prepared in large
amounts and fully characterized by the general methods described
earlier.

II. SYNTHESIS OF DUPLEX (I)

Duplex (I) was prepared by using five phosphorylated segments (^{32}P segments 2 - 5, ^{33}P segment 6) and unphosphorylated
segment 1 (2-to 2.5- fold excess, relative to other segments).
The kinetics of joining (as measured by the resistance to bacterial alkaline phosphatase) and isolation of the duplex on Agarose column (0.5 M) are shown in Fig. 9.   Peak B contained the
desired duplex (I), while peak A contained its dimer, as estimated by its size from electrophoretic mobility on polyacrylamide
gel.

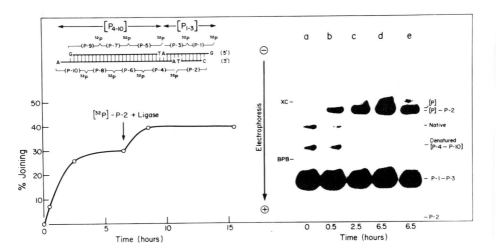

FIGURE 8.   *Preparation of the total promoter, duplex (P)*
(Fig. 4).   The kinetics of joining are shown (left).   At the time
indicated, aliquots of the reaction mixture were subjected to
electrophoresis on 15% polyacrylamide gel (right).

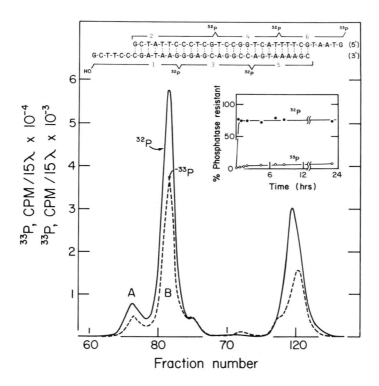

FIGURE 9. The synthesis and purification of the duplex (I)
(Fig. 4). The kinetics of the reaction are given in the inset.
The separation on an Agarose 0.5 M column (150 X 1cm) is shown.
Peak B contained the desired duplex (I) while peak A contained
its dimer.

III.   SYNTHESIS OF DUPLEX (II)

This section ultimately turned out to contain segments 7
through 13, as shown in Fig. 10. Duplex (II) has remarkably high
G-C content and also involved a tetranucleotide piece (segment
10). Many experiments were necessary to determine the optimal
conditions for successful joining. While the yield was still not
high, as seen in Fig. 10, this duplex has been prepared and fully
characterized. An improvement in the enzymatic joinings in this
part is still under investigation by undertaking the chemical synthesis of the tetradecanucleotide that combines the segment 10
with segment 12.

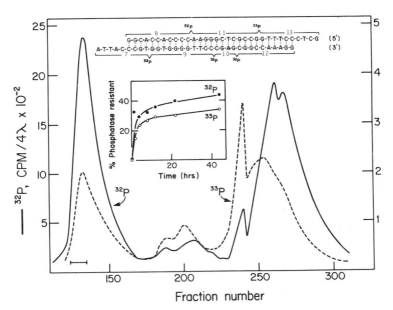

FIGURE 10. *Synthesis and purification of the duplex (II)*
(Fig. 4).   The kinetics of joining are given in the inset.   The
separation on a Sephadex G-100 column (130 X 1cm) is shown.   The
first peak contained the desired duplex (II) (segments 7 through
13).

IV.   SYNTHESIS OF DUPLEX (III)

The scheme adopted for preparing this duplex is shown in
Fig. 11. How was the maximum number of segments contained in
the duplex (II) determined? The fact that segments 20 and 21
have a common sequence at their 5'-ends precludes their simultan-
eous use in one reaction mixture. Therefore duplex (III) was
terminated at segment 20 and duplex (IV) was begun at segment 21.
Systematic experiments further showed that all of the segments up
to segment 20 could be used in a one-step joining reaction. The
isolation of duplex (III) and the kinetics of joinings are shown
in Fig. 11. This duplex was prepared in large amounts and has
been fully characterized.

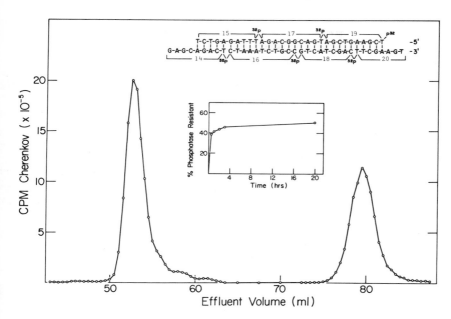

FIGURE 11. *Synthesis and purification of the duplex (III) (Fig. 4). The dinetics of joining are given in the inset. The separation on a 0.5 M Biogel column (1 X 150 cm) is shown. The first peak contained the joined product (segments 14 through 20).*

V.   SYNTHESIS OF DUPLEXES (IV) AND (VB)

These duplexes were prepared as above, in a one-step reaction, in large amounts, and fully characterized (9). The analytical data for all these duplexes were quite satisfactory. It should be noted that the internal phosphate groupings containing ^{32}P will, of course, decay in radioactivity, but the termini that are to be linked to other polynucleotide chains are phosphorylated using [$-^{32}P$]-ATP of high specific radioactivity whenever desired. This concept has been used in joining work as far as possible.

VI.   JOINING OF THE VARIOUS DUPLEXES TO GIVE THE TOTAL DNA

Once the single stranded pieces have been covalently joined to form suitable duplexes such as the duplex (P) and duplex (I)

to duplex (V), their subsequent joining through their complementary single-stranded ends has invariably proved to be rapid
and efficient. Now we show the joining reactions performed between these duplexes. For example, duplex (I) and duplex (II)
undergo a joining reaction as shown in Fig. 12.   Figure 13 shows
the joining of duplex (P), 56 nucleotides long, to the preformed
duplex (I + II), 70 nucleotides long, to give a duplex of 126
nucleotides. The other part of the total DNA *viz* duplex (III +
IV + Vb), 81 nucleotides long , was similarly prepared by joining of the duplex (Vb) to the preformed duplex (III + IV). Finally, the joining of the duplex (P + I + II), 126 nucleotide units long, to the duplex (III + IV + Vb), 81 nucleotide units long
to form the total DNA, is shown in Fig. 14.   The analytical data
as well as 3'- and 5'-nucleotide analysis were quite satisfactory
for the expected total DNA structure 207 nucleotides long shown
in Fig. 3.

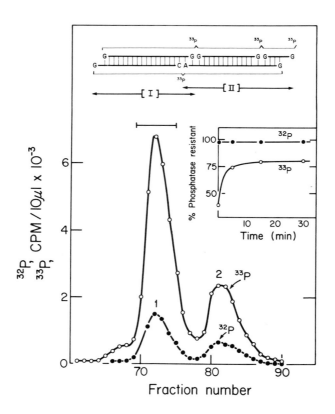

*FIGURE 12.   The joining of duplex (I) to duplex (II) and
separation of the products on an Agarose column (0.5 M).   The
kinetics of joining are given in the inset.*

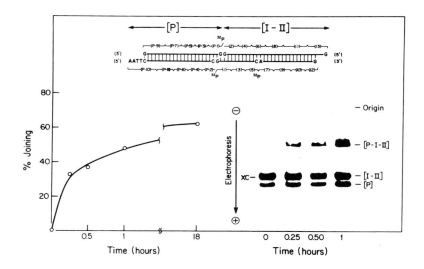

FIGURE 13. Joining of duplex (P) to duplex (I-II). Left: the kinetics of joining are shown. Right: polyacrylamide gel (10%) electrophoresis of the reaction mixture at different time intervals.

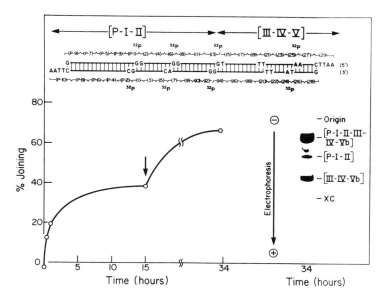

FIGURE 14. The joining of the duplex (P-I-II) to duplex (III-IV-Vb) to form the total DNA (Fig. 3). The time course of the joining is shown (left). (Data obtained by gel electrophoretic analyses.) After a total of 36-hr incubation, the reaction mixture was subjected to electrophoresis on a 10% polyacrylamide gel (right).

VII.   IN VITRO TRANSCRIPTION OF SYNTHETIC GENES

Now we present the results obtained for the transcription
of the synthetic gene and processing of the RNA product to give
functional tRNA.   While the chemical synthesis of oligonucleotides
corresponding to the promoter region were in progress, we performed transcription experiments on the synthetic precursor gene
(see Fig. 15) containing 149 nucleotide base pairs but lacking
the promoter region.   In this case, transcription by the RNA polymerase core enzyme was achieved using the ribotetranucleotide
C-C-C-G as a primer.   It gave a copy of the template, and the
structure of the corresponding RNA product is shown in Fig. 16.
This primary transcript, when exposed to a 100,000 x g) supernatant of an *E. coli* (S-100) extract was evidently processed by
multiple enzymes to give a 4S transfer RNA.   The structure of the
primary transcript as well as the processed RNA was fully characterized by enzymatic degradation with T_1 RNase and pancreatic
RNase followed by two-dimensional fingerprinting procedure using
electrophoresis on cellulose acetate, pH 3.5, in one direction
and TLC on PEI cellulose plate (2 *M* pyridinium formate, pH 3.5)
in the second direction.   The fingerprints are shown in Figs. 17
and 18 and are completely consistent with the expected products
(the primary transcript and the tRNA, respectively).

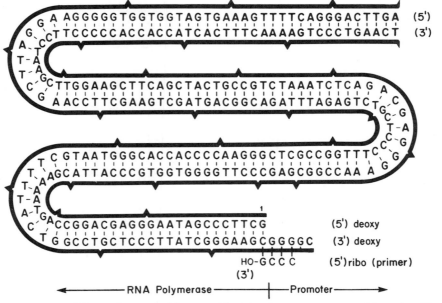

*FIGURE 15.   The totally synthetic DNA corresponding to the
tyrosine suppressor transfer RNA gene lacking the promoter.*

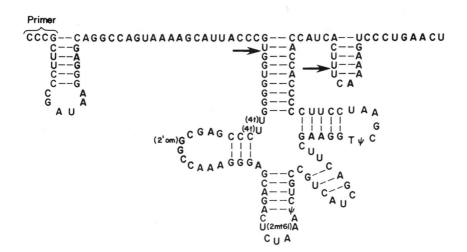

FIGURE 16. Synthetic tyrosine transfer RNA gene: the primary transcript and processing to tRNA.

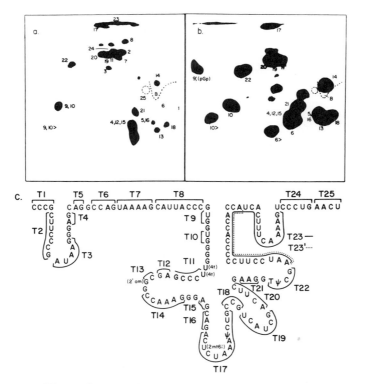

FIGURE 17. Fingerprints of the primary transcript and processed tRNA (T_1-RNase).

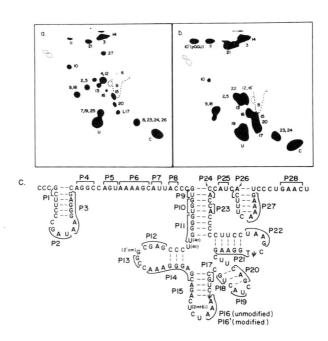

FIGURE 18. *Fingerprints of the primary transcript and processed tRNA (panc-RNase).*

The transcription of the synthetic gene (207 nucleotide units long) containing the promoter region was also similarly accomplished. In this case, it gave complete promoter-dependent transcription, and analysis of the primary transcript showed that the transcription started specifically from the expected initiation site. The primary transcript, when exposed to S-100 extract, was processed by multiple enzymes to give a 4S transfer RNA as above. The fingerprints were completely consistent with the expected products (the primary transcript and the tRNA, respectively). The biochemical activity of this processed tRNA in a protein synthesizing system remains to be demonstrated.

One of the ways to prove biological function of the synthesized tyrosine tRNA gene is to show suppressor activity *in vivo*. The gene product--the tyrosine transfer RNA--has the ability to suppress certain genetic defects (called amber mutations) that result in the production of incomplete, nonfunctional proteins. This activity was tested by incorporating the complete synthetic gene into a bacterial virus (λ phage, constructed by Fred Blattner at the University of Wisconsin, containing two ECoRI restriction enzyme sites, two amber mutations (*A* am 32 and *B* am 1), and a deletion in CI gene) that normally does not grow. When this virus carrying the synthetic gene was allowed to infect a strain of *E. coli*, it was found to grow normally; without the synthetic gene it was unable to grow. This was taken as the indication that the necessary functional proteins were being made as a result of the properly functioning suppressor gene. This accomplishes for the first time the total synthesis of a gene that is biologically functional in a living cell.

VIII   CONCLUDING REMARKS

The present successful synthesis of a 207 nucleotides-long DNA duplex shows that the methodology is now available for the stwp-by-step synthesis of long bihelical DNA. The first phase consists of the chemical synthesis of short segments of single-stranded deoxyribopolynucleotides corresponding to the total length of the double-stranded DNA. The lengths required at this stage are such that the segments can be unambiguously characterized. The second step involves the joining of a few to several single-stranded segments to form short duplexes, which can again be thoroughly characterized. The final step is the joining of the short duplexes to form the total duplex. Since chemical synthesis continues to be the most demanding phase of the work, efforts toward increasing efficiency and rapidity are still necessary. As reported in Chapter 2, high-pressure liquid chromatography has now been developed as a fast and sensitive method for analysis as well as for the efficient and preparative separation of reaction mixtures.

The yields in the different enzymatic joining reactions have varied widely, and the results have generally been unpredictable. Therefore, a very large amount of empirical but systematic work is necessary before the subgrouping can be determined. Nevertheless, the present work clearly demonstrated that unambiguous synthesis of genes or genetic materials in general are feasible in the laboratory.

Finally, it is hoped that this work would mark the beginning of extensive structure-function studies in the areas of gene function, of the promoter and terminator functions, and in general of the nucleic acid-protein interactions.

REFERENCES

1. S. Altman and J. D. Smith, *Nature (London), New Biol. 233,* 35 (1971).
2. H. Küpper, R. Contreras, A. Landy, and H. G. Khorana, *Proc. Natl. Acad. Sci. USA 72,* 4754 (1975).
3. R. Belagaje, R. G. Lees, D. G. Kleid, and H. G. Khorana, *J. Biol. Chem. 251,* 676 (1976).
4. T. Sekiya, M. J. Gait, K. Norris, R. Belagaje, and H. G. Khorana, *J. Biol. Chem. 251,* 4481 (1976).
5. T. Sekiya, H. Vanormondt, and H. G. Khorana, *J. Biol. Chem. 250,* 1087 (1975).
6. T. Sekiya and H. G. Khorana, *Proc. Natl. Acad. Sci. USA, 71,* 2978 (1974).
7. P. C. Loewen, T. Sekiya, and H. G. Khorana, *J. Biol. Chem. 249,* 217 (1974).
8. H. G. Khorana, K. L. Agarwal, P. Besmer, H. Buchi, M. H. Caruthers, P. J. Cashion, M. Fridkin, E. Jay, K. Kleppe, R. Kleppe, A. Kumar, P. C. Loewen, R. C. Miller, K. Minamoto, A. Panet, U. L. Raj Bhandary, R. Belagaje, T. Sekiya, T. Takeya, and J. Van De Sande. *J. Biol. Chem. 251,* 565 (1976) and accompanying papers.
9. T. Sekiya, P. Besmer, T. Takeya, and H. G. Khorana, *J. Biol. Chem. 251,* 634 (1976).

TOTAL SYNTHESIS OF A BACTERIAL GENE I:
CHEMICAL SYNTHESIS OF OLIGONUCLEOTIDES AND
USE OF HIGH-PRESSURE LIQUID CHROMATOGRAPHY[1]

*HANS-JOACHIM FRITZ,[2] RAMAMOORTHY BELAGAJE,
EUGENE L. BROWN, REBECCA H. FRITA, ROGER A. JONES,[3]
ROBERT G. LEES,[4] AND H. GOBIND KHORANA*

*Departments of Biology and Chemistry,
Massachusetts Institute of Technology,
Cambridge, Massachusetts*

ABSTRACT

Total synthesis of an *Escherichia coli* tyrosine suppressor tRNA gene including signals for initiation of transcription and post-transcriptional processing has been completed. Some basic principles of chemical oligonucleotide synthesis are outlined. Use of high-pressure liquid chromatography for analytical and preparative separations of nucleic acid components and intermediates in polynucleotide synthesis is described. The advantages of protecting 3'-hydroxyl groups of (oligo) nucleotides by phenylacetyl- or *n*-butyldiphenylsilyl groups are discussed.

[1]This work has been supported by grants from the National Institutes of Health, U.S. Public Health Service (Grant No. CA11981), the National Science Foundation, Washington, D.C. (Grant No. BMS73-06757); E. L. Brown and R. A. Jones were supported by Postdoctoral Fellowships awarded by the National Institutes of Health (Fellowship Nos. CA01599 and GM05388, respectively)

[2]Recipient of a fellowship from Deutsche Forschungsgemeinschaft. Present address: Institut für Genetik der Universität du Köln, 5 Köln 41, Weyertal 121, Federal Republic of Germany.

[3]Present address: Department of Chemistry, Rutgers University, Douglass College, New Brunswick, New Jersey.

[4]Present address: Department of Chemistry, Fordham University, Bronx, New York.

I.   GENERAL INTRODUCTION AND REVIEW OF ESTABLISHED METHODOLOGY FOR POLYNUCLEOTIDE SYNTHESIS

Total synthesis of an *Escherichia coli* tyrosine suppressor tRNA gene including signals for the initiation of transcription and post-transcriptional processing has been completed (1-3).   The entire DNA duplex is shown in Fig. 1: It has an overall length of 207 nucleotide base pairs.

In the total synthesis of informational DNA two principal phases can be distinguished.

First, sequence-specific synthesis of oligonucleotides of a chain length typically around 10-12.   This is done using the methodology of organic chemistry *exclusively*.   In Fig. 1 all of the oligonucleotide segments thus prepared are shown between successive carets.   The complete gene is made up of 39 of these segments.

The second phase in the total synthesis is the enzymatic joining of these segments to form the entire duplex.   This is also accomplished in two steps.   In the first step the chemically

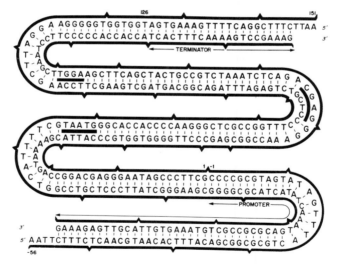

Figure 1.   Structure of synthetic Escherichia coli *tyrosine suppressor tRNA gene.*

synthesized segments are grouped so as to form duplexes represent-
ing different parts of the gene. The short duplexes are then
joined to form the entire gene (see next chapter). This chapter
deals with the first phase.

Commercially available 2'-deoxyribonucleoside-5'-monophos-
phates (Fig. 2) are the starting materials in the diester approach to polynucleotide synthesis.

The central synthetic problem is how to link in a sequence-
specific way the 5'-phosphate of one unit to the 3'-hydroxyl of
the next. In the course of such a reaction, a phosphomonoester
is converted to a phosphodiester. This is not a spontaneous process and is only brought about by the action of energy-rich phosphate activating agents.

These condensation reactions have to be done in a carefully
controlled manner to make sure that only the *desired* nucleotide
sequence is built up. The synthesis is therefore carried out
step by step (one phosphodiester linkage at a time) and a set of
blocking groups is used to permit only the formation of the desired covalent linkages.

The general synthetic strategy has been dealt with in detail
in numerous publications from this laboratory (see, for example,
Ref. 1,4-6). Some of the most basic essentials are summarized
in Fig. 3: 5'-hydroxyl groups are blocked by monomethoxytrityl.
This protecting group is removed by mild acid. The classical
blocking group for 3'-hydroxyls is acetyl. Selective removal
is achieved by very brief treatment with 1 *N* sodium hydroxide.
Amino groups of heterocyclic bases in nucleotides also have to
be protected. For this purpose the following acyl groups are
being used: benzoyl on adenine, anisoyl on cytosine, and isobu-
tyryl on guanine. Amino groups are deacylated by concentrated
aqueous ammonia.

Condensing agents are: dicyclohexylcarbodiimide (DCC), mes-
itylenesulfonyl chloride (MS), and 2,4,6-triisopropyl-benzenesul-
fonylchloride (TPS). TPS is the one most often used.

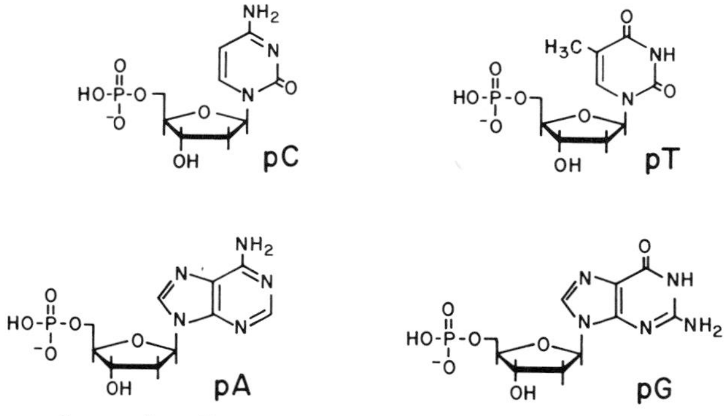

*Figure 2. The four common 2'-deoxyribonucleoside-5'-mono-
phosphates.*

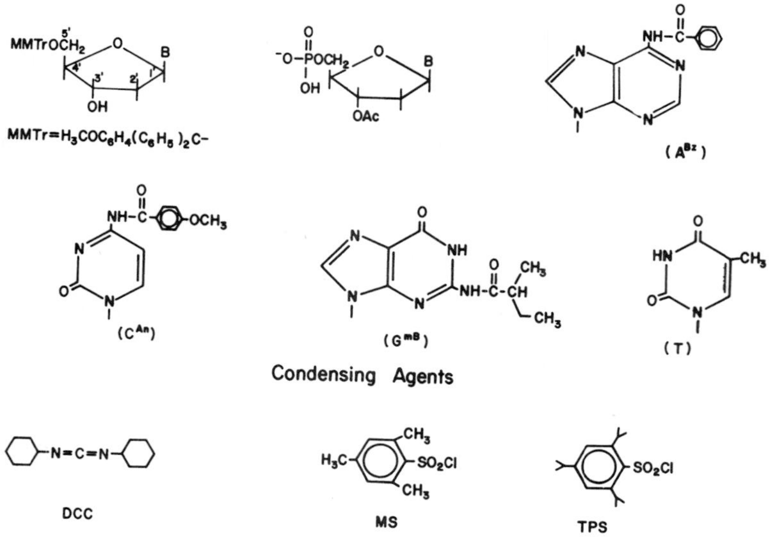

MMTr=H₃COC₆H₄(C₆H₅)₂C-

$MMTr = H_3COC_6H_4(C_6H_5)_2C-$

(A^Bz)

(C^An)   (G^mB)   (T)

Condensing Agents

DCC   MS   TPS

Figure 3. Protecting groups and condensing agents.

Figure 4 shows a typical flow chart for the preparation of an undecanucleotide. The sequence of the desired polynucleotide in this case is 5'-GGAAGCYYAAC-3'.

Synthesis starts at the 5'-end of the chain, with a nucleoside carrying a 5'-monomethoxytrityl group. TPS-mediated condensation with the nucleotide pG^ib-OAc yields the dinucleotide.

The 3'-*O*-acetyl group on the incoming nucleotide prevents self-condensation of the latter. Treatment with 1 *N* NaOH frees the 3'-hydroxyl end group without attacking any other protecting group. This 3'-hydroxyl group again provides the site for con-

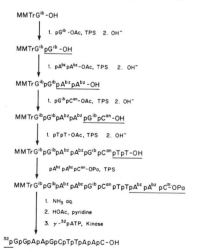

Figure 4. Typical flow chart for the chemical synthesis of an oligonucleotide.

densation with the phosphomonoester function of a second nucleotide in the same controlled fashion.

In the next step the incoming block is pA^{bz}pA^{bz}-OAc, a preformed dinucleotide. Condensation yields the tetranucleotide,
base treatment again liberates the 3'-terminal hydroxyl group,
and so on.

The reasons why we use di-, tri-, and occasionally tetranucleotide blocks are twofold.

First, using these blocks, the growing chain is subjected
to fewer elongation steps, thus keeping complications by side
reactions minimal.

Second, separation of reaction mixtures is facilitated because in the case of block condensations a more pronounced change
in physical properties (e.g., net charge) of the growing chain
is achieved.

In this stepwise manner, proceeding from the 5'-to the 3'-
end, the desired undecanucleotide is built up and, finally, all
the protecting groups are removed in two subsequent treatments.
After final purification by one or more methods, the 5'-hydroxyl
group is phosphorylated with the help of an enzyme (polynucleotide kinase).

The short oligonucleotide building blocks required are prepared as outlined Fig. 5.

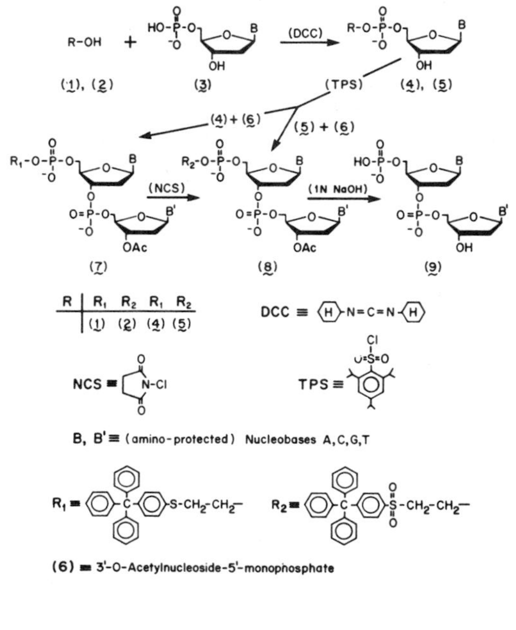

*Figure 5. Synthesis of an oligonucleotide building block
carrying 5'-phosphate groups.*

As in all condensations, the partners involve one that
carries a free hydroxyl group and a second that carries a free
phosphate group. Unlike chain elongation steps, the hydroxyl
group partner also carries a 5'-phosphate group, which has to be
blocked. This is done by introducing either a *p*-tritylphenylthio-
ethyl (TPTE) or a *p*-tritylphenylsulfonylethyl (TPSE) group.
The phosphate reaction partner is again protected at the 3'-hy-
droxyl group by acetylation. After condensation (and in the case
of TPTE subsequent oxidation with *N*-chlorosuccinimide) *both* end
protecting groups are removed in one base treatment step. In an
alkaline medium TPSE derivatives of phosphates undergo β elimination.
     As described in detail in a recent publication (6), one of
the main advantages of TPTE and TPSE groups is that all intermediates carrying these groups can be isolated and purified by rapid
solvent extraction procedures.
     In the case of growing oligonucleotide chains, the 5'-meth-
oxytrityl group makes possible solvent-solvent extraction up to
about a pentanucleotide. Beyond this point, separations have to
be accomplished by chromatography.
     The classical separation method in our synthetic approach
has been anion-exchange chromatography on DEAE-cellulose (cellulose having diethylaminoethyl groups linked to some of its hy-
droxylgroups). At a pH around 7, the stationary phase is partly
protonated and separation is largely determined by the differences
in the number of negative charges and thus the chain length of
the olignucleotides. The method has a reasonably high resolving
power but is very time consuming. It may take as long as two
weeks or more to complete a complicated separation on a DEAE-
cellulose column. Fractions to be pooled have to be checked
carefully for the identity and purity of the compounds they contain. Aliquots of selected fractions of a peak have to be investigated, classically by a combination of thin-layer chromatography, paper chormatography and uv spectroscopy.
     This thorough purity examination is absolutely essential to
the controlled stepwise synthesis. A large proportion of the
overall time in polynucleotide synthesis used to be spent on these
tedious characterizations.

II.   RECENT DEVELOPMENTS IN METHODOLOGY

     In an attempt to simplify and speed up purification and
characterization of synthetic intermediates, new types of liquid
chromatography systems and their application to nucleic acid chemistry were investigated. In our hands, reversed-phase liquid
chromatography (3) proved to be very powerful on both analytical

and preparative levels in terms of resolution, sensitivity, speed, recovery, reproducibility, and capacity.

The stationary phase in analytical work consists of fully porous silica particles of 10 μm diameter that have saturated C_{18} hydrocarbon chains covalently bound to their surface, thus giving them strongly lipophilic characteristics.

One can expect that compounds will elute from the column in the order of increasing lipophilicity. A given compound will elute faster with increasing lipophilicity of the mobile phase. Figure 6 shows the separation of a mixture of the four deoxynucleoside-5'-monophosphates on such an analytical reversed-phase column. The parameters of the run are as follows: column size, 0.4 x 31 cm; mobile phase, 1% acetonitrile in 0.1 M ammonium acetate; slowrate, 2 ml/min, building up a back pressure of about 2,000 psi; sample size, about 0.1 0.D.$_{260}$ units each; elution is monitored with a uv detector, operating in this experiment at 254 nm.

Phosphomonoesters, being very hydrophilic, require only very little acetonitrile (acting as a lipophilic additive to the buffer) to give the unprotected nucleotides a high mobility. Purines have a stronger affinity to the stationary phase than pyrimidines.

Figure 7 shows the separation of the four common ribonucleosides. Now in the absence of the phosphate group, the differences in affinity to the C_{18} surface are so great that a gradient from 1 to 25% acetonitrile must be applied in order to elute all four components within a reasonable time.

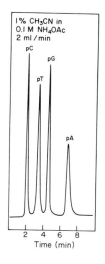

Figure 6. Separation of the four common 2'-deoxyribonucleoside-5'-monophosphates.

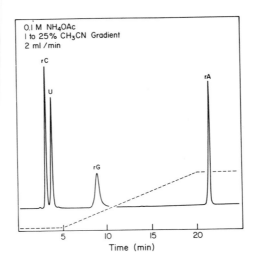

Figure 7. Separation of the four common ribonucleosides.

That the retention behavior of a compound can indeed be correlated with lipophilic contributions made by various functional groups present in the molecule is illustrated in Fig. 8.

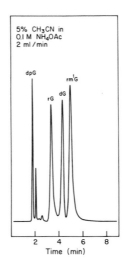

Figure 8. Separation of various derivatives of the purine base guanine.

This figure shows a chromatogram obtained from a mixture of various derivatives of the base guanine, 2'-deoxyguanosine-5'-monophosphate, and three nucleosides. The nucleotide elutes first at 2% acetonitrile. Riboguanosine elutes earlier than 2'-deoxyriboguanosine because of the hydrophilic contribution made by the 2'-hydroxyl group. Ribo-1-methylguanosine, a natural component of certain transfer RNAs, elutes much later than its unmethylated precursor.

It seems that this type of chromatography could be very useful in the isolation and characterization of modified nucleosides from tRNAs. Many of these modified nucleosides have a more or less high degree of alkylation as their main unusual structural feature.

Figure 9. illustrates an effect that is interesting in connection with intermediates in oligonucleotide synthesis: absence or presence of a 3'-O-acetyl group has a pronounced influence on the retention behavior. Base-protecting groups make an even greater difference.

As mentioned earlier, growing chains in oligonucleotide synthesis carry 5'-monomethoxytrityl groups. The behavior of this class of compounds is schematically illustrated in Fig. 10. As can be expected, the monomethoxytrityl group has an especially strong retarding effect, requiring high percentages of acetonitrile for elution. Moreover, there is a critical acetonitrile concentration range, usually around 35%, where the retention time of tritylated material is extremely sensitive to even small changes in solvent composition. This effect can be utilized in two ways. First, on an analytical scale one can very easily distin-

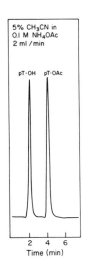

Figure 9. Separation of thymidine-5'-monophosphate from its 3'-acetate.

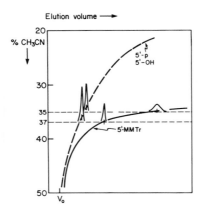

Figure 10. Schematic illustration of the retention behavior of oligonucleotides carrying 5'-monomethoxytrityl groups.

guish tritylated from nontritylated compounds by injecting the same mixture at two different concentrations, i.e., 37 and 35% acetonitrile. Only tritylated compounds will show a sharp increase in retention time. Second, this behavior can be exploited for preparative separations: this is the topic to be discussed next.

Figure 11 shows liquid chromatography tracings of the preparation of an octanucleotide from a hexanucleotide. The starting material, $MMTrG^{ib}pG^{ib}pA^{bz}pA^{bz}pG^{ib}pC^{an}$-OH (panel 1) elutes from the analytical column after about 6 min at 35% acetonitrile concentration. This compound was condensed with a large excess of pTpT-OAc to ensure a high turnover of the starting hexanucleotide. TPS was removed by extraction and the products in the reaction mixture were precipitated from ether. Liquid chromatography analysis of this mixture (panel 2) showed that some hexanucleotide was still present. The large peak corresponds to the desired octanucleotide. The material eluting very fast is the excess of the dinucleotide, its pyrophosphate, and pyridine.

Since thymine in the above dinucleotide block makes no significant lipophilic contribution, the elongated chain elutes faster than the hexanucleotide starting material, due mostly to the presence of two more phosphate groups. This shift to shorter retention time can be further increased by deacetylation (performed on the total reaction mixture). Even with an octanucltotide, an acetyl group has a significant influence, as shown in panel 3 (Fig. 11). The main peak at longer retention time now is the octanucleotide with a free 3'-hydroxyl group. Preparative separation (panel 4 Fig. 11) is accomplished on a 0.7 x 183 cm. column packed with 60 μm diameter fully porous C_{18} silica particles.

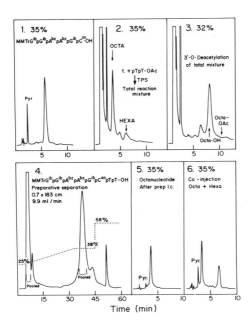

Figure 11. High-pressure liquid chromatography tracings from the synthesis of an octanucleotide.

The mobile phase in preparative separations is 0.1 *M* triethylammonium acetate with varying percentages of acetonitrile. We operated this larger column at a flowrate of 9.9 ml/min, creating a back pressure of about 2,000 psi. In typical runs about 250 mg of a mixture can be separated in 45 to 90 min. In the example shown, the following solvent program is used: start at 25% acetonitrile to elute quickly any excess of the dinucleotide block and its pyrophosphate. The tritylated material elutes after a linear gradient (30 min) from 25 to 38% of acetonitrile is applied.

Before reusing the column, a step gradient to 58% acetonitrile is applied to wash off tightly bound impurities. The small peak appearing at about 43 min contains the starting hexanucleotide. The main peak was cut as indicated and the product checked by analytical liquid chromatography (panel 5, Fig. 11). Isolated yield in this preparation was 33%. Absence of the starting material was confirmed by co-injection of the hexa- and the isolated octanucleotide (panel 6 Fig. 11).

The above serves as an example where the pure elongation product can be isolated by preparative liquid chromatography alone. However, it is not always the case that the longer chains elute distinctly faster than the shorter starting materials.

The usual base-protecting groups have a fairly pronounced influence. Thus, depending on the base composition of the incoming block, the reaction product can elute anywhere, i.e., before, together with, or following the starting material. Attempts were therefore made to retard the elongated products consistently on the reversed phase packings by using more lipophilic protecting groups on the 3'-OH end groups. Some results are shown in Fig. 12. The trinucleotide MMTrG^{ib}pTpA^{bz}-OH (panel 1). was condensed with 6 eq. of pTpA^{bz}. The 3'-OH group in the dinucleotide now carried a phenylacetyl group. Analysis of the crude reaction mixture (panel 2) indeed showed a pronounced shift to a longer retention time. Panel 3 shows a co-injection of the crude mixture with pure starting monomethoxytrityltrinucleotide.

In the preparative run (panel 4 Fig. 12) a very good separation of the excess of dinucleotide and its pyrophosphate was achieved, the latter carrying *two* lipophoilic phenylacetyl "handles". The main peak in the trityl region pooled as shown afforded pure pentanucleotide (panel 5 Fig. 12) in 56% yield. This principle of lipophilic tags can be extended by using silyl protecting groups on the 3'-hydroxyl groups. At the same time silyl groups allow new types of manipulations not possible with acyl groups.

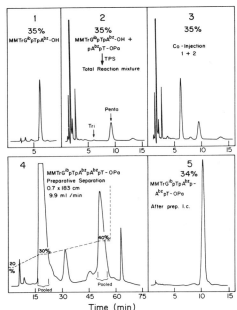

Figure 12. High-pressure liquid chromatography tracings from the synthesis of a pentanucleotide, the 3'-OH groups of the incoming dinucleotide block being protected by the phenylacetyl group.

The use of sterically hindered silanes for hydroxyl protection has been reported (7,8), including use with nucleosides and
phosphorylation to the corresponding nucleotides (9,10). We
have now found that alkyldiphenylchlorosilanes can be used with
the nucleotide itself to obtain 3'-O-silyl derivatives suitable
for use in oligonucleotide synthesis following the diester approach (3). Reaction of a nucleotide with about five equivalents
of n-butyldiphenylchlorosilane proceeds readily in pyridine and
is complete in less than two hours. The nucleotide-3'-O-n-butyl-
diphenylsilyl ether is isolated by quenching the reaction with
sodium bicarbonate, extraction into ethyl acetate/n-butanol mixtures, and precipitation from diethyl ether. These products are
slowly hydrolyzed by aqueous pyridine and therefore require careful handling. However, yields are normally greater than 90%.
Removal of the silyl group is achieved by using tetra-n-butylam-
monium fluoride in pryidine, and the product is freed from the
quaternary amine by the ion exchanger pyridinium Dowex-50.

Use of the n-butylidiphenylsilyl (nBDPS) group in the synthesis of the trinucleotide pA^bz^pA^bz^pC^an (compare Fig. 4) is
illustrated in Fig. 13.

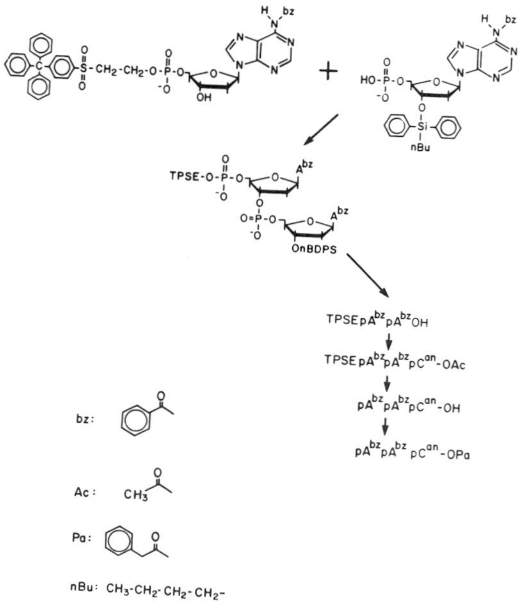

*Figure 13. Use of the n-butyldiphenylsilyl (nBDPS) group
in the synthesis of a trinucleotide with a 5'-phosphate group.*

TPSEpA^{bz}-OH was condensed with pA^{bz} carrying a nBDPS group
on the 3'-end. Treatment with 0.1 M flouride in dry pyridine
for 20 min liberated the 3'-hydroxyl terminus without attacking
the TPSE group. This type of reaction is not possible with 3'-
O-acetyl groups.

   Retaining the lipophilic TPSE handle, the dinucleotide was
elongated with pC^{an}-OAc, then treated with base and phenylacety-
lated at the 3'-end before condensing the trinucleotide to the
growing chain. Liquid chromatography tracings of this synthesis
are shown in Fig. 14. The two starting mononucleotides (panel 1)
behaved rather similarly at 45% acetonitrile. The condensation
product required 55% acetonitrile for elution from the analytical
column after 9 min (panel 2). This is the most lipophilic nucleotide derivative in our experience.

   After removal of the silyl group, the dinucleotide eluted
at 40% acetonitrile concentration after 9 min. Conversion of the
di- to the trinucleotide could again be monitored by analytical
liquid chromatography (shown in panels 4-6 Fig. 14) at 38% acetonitrile concentration, to enhance the only slight difference
in retention. The three chromatograms in panels 4-6 show: panel
4, the dinucleotide TPSEpA^{bz}pA^{bz}-OH; panel 5, the condensation

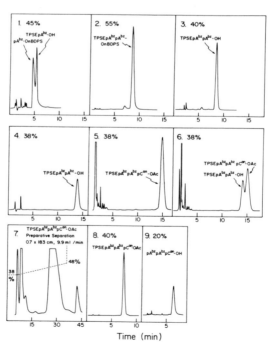

*Figure 14. High-pressure liquid chromatography tracings
from the synthesis of pA^{bz}pA^{bz}pC^{an}-OH, as illustrated in Fig. 13.*

reaction mixture, prepurified by extraction into methylene chloride containing 10% *n*-butanol, i.e., crude TPSEpA^{bz}pA^{bz}pC^{an}-OAc; and panel 6, a co-injection of the above two compounds to make sure the retention difference observed in the isolated runs was not artifactual.

Preparative liquid chromatography of the fully protected trinucleotide (panel 7, Fig. 14) takes 45 min per 250-mg injection, applied in a volume of 1.5 ml. It yields the pure product (panel 8, Fig. 14) that was deprotected at both the 5'- and 3'-ends to give the N-protected trinucleotide shown (panel 9, Fig. 14) in a run at 20% acetonitrile concentration.

III.   SUMMARY OF NEW METHODS

We have developed a set of new methods to complement the diester approach to polynucleotide synthesis. The main features are as follows.

(1) Introduction of high-pressure liquid chromatography on reversed phase as a highly effective separation tool in nucleic acid chemistry on both analytical and preparative levels.

(2) Use of highly lipophilic protecting groups on both the 5'- and 3'-ends of oligonucleotides and exploitation of this lipophilicity as an aid for rapid solvent extractions and for reversed-phase liquid chromatography.

(3) Use of the *n*-butyldiphenylsilyl group to protect 3'-hydroxyl groups. This group, while being sufficiently stable for hydrolysis during all work-up operations, can be removed specifically under nonbasic, anhydrous conditions by fluoride ion.

In the light of these recent results, requirements have become evident for new properties of the protecting groups used in polynucleotide synthesis and it seems that the entire strategy as regards protecting groups in the diester approach deserves reconsideration.

Some of the developments are still at a very early stage, but nevertheless even at this stage of development, a saving of at least 50% of the time has been effected in the synthesis of oligonucleotides.

1.   H. G. Khorana, K. L. Agarwal, P. Besmer, H. Büchi, M. H. Caruthers, P. J. Cashion, M. Fridkin, E. Jay, K. Kleppe, R. Kleppe, A. Kumar, P. C. Loewen, R. C. Miller, K. Minamoto, A. Panet, U. L. RajBhandary, B. Ramamoorthy, T. Sekiya, T. Takeya, and H. van de Sande, *J. Biol. Chem.* *251*, 565 (1976).

2.   B. Ramamoorthy, R. G. Lees, D. G. Kleid, and H. G. Khorana,
     J. Biol. Chem. 251, 676 (1976).
3.   H. G. Khorana, *et al., unpublished papers.*
4.   H. G. Khorana, K. L. Agarwal, H. Büchi, M. H. Caruthers,
     N. K. Gupta, K. Kleppe, A. Kumar, E. Ohtsuka, U. L. Raj-
     Bhandary, J. H. van de Sande, V. Sgaramella, T. Terao,
     H. Weber, and T. Yamada, *J. Mol. Biol. 72,* 209 (1972).
5.   K. L. Agarwal, A. Yamazaki, P. J. Cashion, and H. G. Khorana,
     Angew. Chem., Internat. Ed. 11, 451 (1972).
6.   K. L. Agarwal, Y. A. Berlin, H.-J. Fritz, M. J. Gait, D. G.
     Kleid, R. G. Lees, K. E. Norris, B. Ramamoorthy, and
     H. G. Khorana, *J. Am. Chem. Soc. 98,* 1065 (1976).
7.   E. J. Corey and A. Venkateswarlu, *J. Am. Chem. Soc. 94,*
     6190 (1972).
8.   S. Hanessian and P. Lavallee, *Can. J. Chem. 53,* 2975 (1975).
9.   K. K. Ogilvie, *Can. J. Chem. 51,* 3799 (1973).
10. K. K. Ogilvie, E. A. Thompson, M. A. Quilliam, and J. B.
     Westmore, *Tetrahedron Lett.* 2865 (1974).

THE USE OF MICROPARTICULATE REVERSED-PHASE PACKING IN HIGH-PRESSURE LIQUID CHROMATOGRAPHY OF NUCLEOSIDES*

JOHN A. MONTGOMERY, H. JEANETTE THOMAS, ROBERT D. ELLIOTT, AND ANITA T. SHORTNACY

Kettering-Meyer Laboratory, Southern Research Institute, Birmingham, Alabama

ABSTRACT

The utility of microparticulate reversed-phase packing in the high-pressure liquid chromatography of purine and pyrimidine nucleosides is demonstrated by four examples of difficult separations of anomers, epimers, and isomers readily carried out on a C_{18} packing using water-acetonitrile as the eluant. Two of these complex mixtures resulted from attachment of a sugar to a purine and to a pyrimidine by the fusion reaction in the first case and a Hilbert-Johnson reaction in the second. The third mixture resulted from two consecutive rearrangements: the first, the rearrangement of a triazolopyrimidine to a thiadiazolopyrimidine, and the second involving the opening of the sugar ring of a β-ribofuranose to give a mixture of α- and β- ribopentoses The last separation demonstrates the utility of this technique for the separation and identification of nucleoside metabolites from cell extracts.

*This investigation was supported by the Division of Cancer Treatment, National Cancer Institute, National Institutes of Health, Department of Health, Education, and Welfare, Contract No 1-CM-43762.

37

I.   INTRODUCTION

Chemically bonded stationary phases for use in high-pressure liquid chromatography (HPLC) were first introduced in 1970. These
chemically bonded phases are liquids or solids chemically attached
to the support material, usually silica gel, and oriented like
bristles on the column of support (1,2). Column efficiencies,
as measured by plates per foot, are much higher on the chemically
bonded phases than on the conventionally coated phases in which
liquids or solids are deposited on the support and held by physical forces. In 1974, the column efficency of these packings
was further increased by the use of supports of very small par-
ticle size--5 to 10 µm.

The separations described in this paper were carried out on
µ Bondapak C_{18}, an exceedingly nonpolar packing material produced
by the permanent bonding of a monomolecular layer of octadecyltrichlorosilane to µPorasil (10 µm) via a carbon-silicon bond (3).
A column 4-mm inside diameter and 30-cm long was employed. This
packing material has enabled the separation of such diverse organic molecules of biologic interest as the isomeric hydroxylated
metabolites of the anticancer agent N-(2-chloroethyl)-N'-cyclo-
hexyl-N-nitrosaurea (CCNU), adenosine and related purine and pyrimidine nucleosides, simple pteridines and the antifols aminop-
terin and methotrexate, and citrovorum factor and its isomers and
related compounds. This discussion will be confined to work with
purine nucleosides.

Work with µBondapak C_{18} has shown that retention times are
not always readily reproducible, and hence it is prudent to run
known compounds along with samples to be analyzed. Retention time
vary with very small changes in the percentage composition of the
eluting solvent pair and with the "age" of the column, undoubt-
edly due to an accumulation of materials not readily eluted (per-
iodic washing with N,N-dimethylformamide is helpful). Retention
times and tailing, as well as theoretical plates, also vary from
column to column, probably due to uncontrolled variables in the
packing process, making reproducible retention times from laboratory to laboratory fortuitous. For this reason, we have chosen
not to present retention times, but instead reproductions of
actual chromatograms.

All separations were carried out at room temperature and at
a flowrate of 1 ml/min with a Waters ALC-242 Liquid Chromatograph
equipped with a uv detector (254 nm) and an M-6000 pump (Waters
Associates, Inc., Milford, Massachusetts). All chromatograms
were recorded at a chart speed of 1 cm/min and an attenuation of
16 aufs (absorbance units full scale).

II. SEPARATIONS

A. *THE FUSION REACTION*

The fusion of 3-acetamido-1,2,5-(tri-O-acetyl-β-D-ribo- or
α-D-arabinofuranose) (I) with 2,6-dichloropurine (II) resulted
in anomerization at C_1 and epimerization at C_2 of the pentofur-
anoses to give a mixture of four nucleosides (III), which was
treated with ethanolic ammonia to give the 2-chloroadenine nucleosides (IV, V) (4). Figure 1A shows the separation by HPLC of a
known mixture of these four nucleosides. Such a separation could
not be achieved by TLC on silica gel G. Figure 1B, which shows
an aliquot of the reaction mixture from the arabino sugar before
purification and separation of the components, enabled the determination of relative amounts of each nucleoside present by measuring the area under each peak. These amounts agreed well with
those previously determined by PMR spectroscopy (Table I).

TABLE 1 *Nucleoside Amounts*

| Compound | Amount in reaction mixture (%) | |
	HPLC	PMR
β-IV	8	5
α-IV	4	2
α-V	62	69
β-V	26	24

B. *THE HILBERT-JOHNSON REACTION*

An example of the separation of blocked nucleosides by HPLC
is shown in Fig. 2, which is the chromatogram obtained by injection of a methanol solution of an aliquot of the reaction mixture

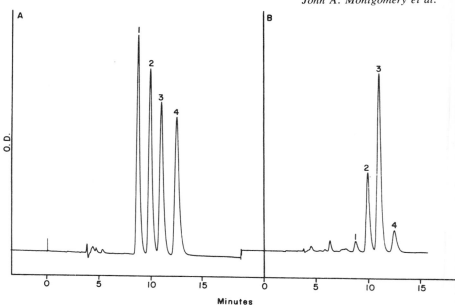

FIGURE 1. Solvent: 9:1 water-acetonitrile; attenuation: B, 32 aufs; sample weights: A--1 μg each, B-- 1 μl (10 μg) aliquot of the reaction mixture before isolation of products. (1) 9-(2-acetamido-3-deoxy-α-D-ribofuranosyl)-2-chloroadenine (α-IV), (2) 9-(3-acetamido-3-deoxy-β-D-arabinofuranosyl)-2-chloroadenine (β-V), (3) 9-(3-acetamido-3-deoxy-α-D-arabinofuranosyl)-2-chloroadenine (α-V), and (4) 9-(3-acetamido-3-deoxy-β-D-ribofuranosyl)-2-chloroadenine (β-IV).

resulting from heating a neat mixture of the chloro sugar (VII) with an excess of 5-fluoro-2,4-dimethoxypyrimidine (VI). Although the major product was 1-[3,5-*bis-O*-(4-chloro-benzoyl)-2-*O*-methyl-β-D-arabinofuranosyl]-5-fluoro-4-methoxy-2(1H)-pyrimidinone (β-IX) along with the α-anomer (α-IX) as a second product,· there was also a small amount of material (peaks 1 and 2) that was later identified by HPLC as 5-fluoro-1-(2-*O*-methyl-α- and β-D-arabinofuranosyl)-uracil (α- and β-XI). Peak 3, 5-fluoro-2,4-dimethoxypyrimidine (VI), was large because an excess of the pyrimidine was used in the reaction. The very small peak 4 yielded enough material (on work-up of the reaction mixture) to allow its identification by PMR spectroscopy as the partially hydrolyzed β-anomer (β-VIII). Peak 5 was found to be methyl 4-chlorobenzoate, probably resulting from transesterification of

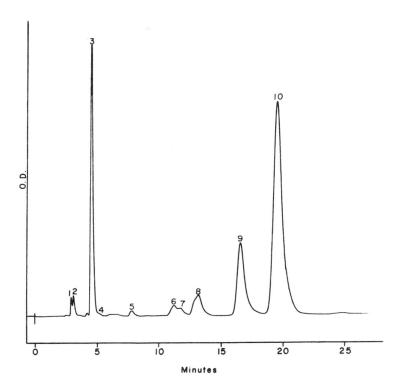

FIGURE 2. Solvent: 2:3 water-acetonitrile; sample weight:
5 µg. (1) and (2) thought to be 5-fluoro-1-(2-O-methyl-α- and β
-D-arabinofuranosyl) uracil (α- and β-XI), (3) 5-fluoro-2,4-di-
methoxypyrimidine (VI), (4) 1-(5-O-(4-chlorobenzoyl)-2-O-methyl-β-D-
arabinofuranosyl]-5-fluoro-4-methoxy-2(1H)-pyrimidinone (β-VIII),
(5) methyl 4-chlorobenzoate, (6) unidentified, (7) 1-[3,5-bis-O-
(4-chlorobenzoyl)-2-O-methyl-α-D-arabinofuranosyl]-5-fluorouracil
(α-X), (8) 1-[3,5-bis-O-(4-chlorobenzoyl)-2-O-methyl-β-D-arabino-
furanosyl]-5-fluorouracil (β-X), (9) 1-[3,5-bis-O-(4-chlorobenzoyl)
-2-O-methyl-α-D-arabinofuranosyl]-5-fluoro-4-methoxy-2(1H)-pyri-
midinone (α-IX), and (10) 1-[3,5-bis-O-(4-chlorobenzoyl)-2-O-
methyl-β-D-arabinofuranosyl]-5-fluoro-4-methoxy-2(1H)-pyrimidinone
(β-IX).

the 4-chlorobenzoyl group of IX by methanol. Peak 6 is unidentified. Peaks 7 and 8 are 1-[3,5-*bis-O*-(4-chlorobenzoyl)-2-*O*-
methyl-α- and β- and β-D-arabinofuranosyl]-5-fluorouracil (α- and
β-X), resulting from the loss of the methyl group from IX.
     Separation of α- and β-IX, accomplished by preparative TLC,
gave a β to α ratio of 4.9:1, while the ratio determined by measuring the area under the peaks (9 and 10) was 4.8:1. Figure 3A

is the chromatogram of pure α-IX (2.5 μg). The β-anomer (β-IX), shown in Fig. 3B (5 μg), was found to contain a small amount of the partially debenzoylated material (β-VIII), although β-VIII was not detected by TLC and the elemental analysis was correct for pure β-IX.

Cleavage of α- and β-IX in chloroform saturated with HCl gas gave the 5-fluorouracil derivatives α- and β-X, which were each debenzoylated with methanolic sodium methoxide to give the 2'-O-methyl-5-fluorouracil arabinosides α- and β-XI. Figure 4A is the chromatogram of a 1:2 mixture of α- and β-XI, while Fig. 4B is nearly pure α-anomer containing traces of β, and Fig. 4C is β-anomer containing traces of α. These anomers cannot be resolved by TLC on silica gel.

C.   The (1,2,3)Thiazolo(4,5)pyrimidine—(1,2,3)Thiadiazolo(5,4-d)pyrimidine—Furanose—Pyranose Rearrangement

A number of 6-substituted 8-azapurine ribonucleosides related to inosine and guanosine were recently prepared from 7-(methylthio)-3-(2,3,5-tri-O-acetyl-β-D-ribofuranosyl)-3H-[1,2,3]triazolo[4,5-d]pyrimidine (XII) (5) and 7-(benzylthio)-3-β-D-ribofuranosyl-3H-[1,2,3]triazolo[4,5-d]pyrimidin-5-amine (XIII) (6). The reaction of XII with sodium hydrosulfide followed by removal of the acetyl blocking groups with sodium methoxide gave 8-aza-6-thioinosine (3,6-dihydro-3-β-D-ribofuranosyl-7H-[1,2,3]triazolo-[4,5-d]pyrimidine-7-thione, XIV). Similar treatment of XIII with sodium hydrosulfide gave 8-aza-6-thioguanosine (5-amino-3,6-dihydro-3-β-D-ribofuranosyl-7H-[1,2,3]triazolo[4,5-d]pyrimidine-7-thione, XV). Both of these compounds (XIV and XV) were found to be unstable and to rearrange slowly at room temperature to N-β-D-ribofuranosyl[1,2,3]thiadiazolo[5,4-d]pyrimidin-7-amine (XVI) and N^7-β-D-ribofuranosyl[1,2,3]thiadiazolo[5,4-d]pyrimidine-5,7 diamine (XVII), respectively. The rearrangement product XVI was found to be about ten times more cytotoxic than XIV and also showed activity at three dose levels in the L1210 mouse leukemia

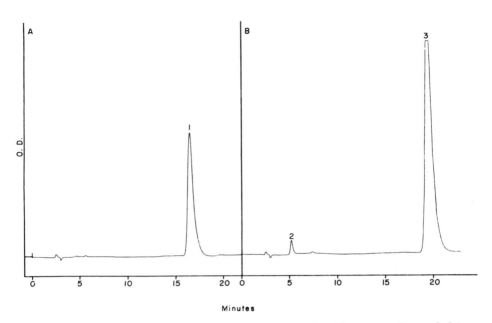

Minutes

FIGURE 3. Solvent: 2:3 water-acetonitrile; sample weights:
A, 2.5 µg, B, 5 µg. (1) 1-[3,5-bis-O-(4-chlorobenzoyl)-2-O-methyl-
α-D-arabinofuranosyl]-5-fluoro-4-methoxy-2(1H)-pyrimidinone (α-IX),
(2) 1-[5-O-(4-chlorobenzoyl)-2-O-methyl-β-D-arabinofuranosyl]-5-
fluoro-4-methoxy-2(1H)-pyrimidinone (β-VIII), and (3) 1-[3,5-bis-
O-(4-chlorobenzoyl)-2-O-methyl-β-D-arabinofuranosyl]-5-fluoro-4-
methoxy-2(1H)-pyrimidinone (β-IX).

screen (5).   In contrast, compounds XIV, XV, and XVII showed no
significant cytotoxicity or L1210 activity.   The rearrangement of
XIV in aqueous solution at pH 7.3 was followed by HPLC (Fig. 5).
The solution of XIV containing 8% of XVI (broken line) gave, after
six days, a 41% yield of XVI and a trace of α-nucleosides.   A
chromatogram (Fig. 6) of the thiadiazolopyrimidine XVI indicated
no detectable reverse rearrangement to XIV (arrow) in pH 7.3
buffer after six days.   However, an increase in the ratio of α-
nucleosides from 2.5% at 0 time (broken line) to 16% at 6 days
was observed.   This peak further increased to 34% after 28 days.
To determine the structure of these isomerization products, a
50-mg sample of XVI was intentionally isomerized by heating in
aqueous solution at 100°C for 2.5 hr.   An expanded HPLC chromatogram (Fig. 7) of this mixture indicated the presence of four
major isomers, none of which could be visually separated by TLC

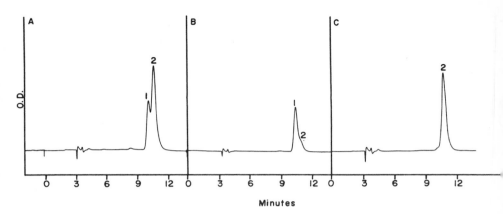

FIGURE 4. *Solvent: 24:1 water-acetonitrile; sample weights: A, 2μg; B and C, 1μg. (1) 1-(2-O-methyl-α-D-arabinofuranosyl)-5-fluorouracil (α-XI) and (2) 1-(2-O-methyl-β-D-arabinofuranosyl)-5-fluorouracil (β-XI).*

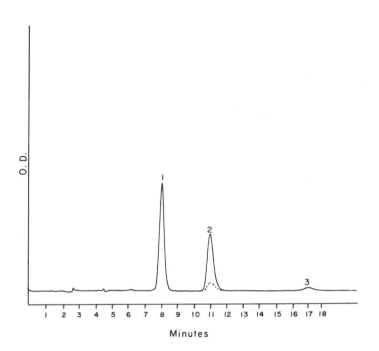

Figure 5. Solvent: 95:5 water-acetonitrile; sample weight: 3 μg. (1) 3,6-dihydro-3-β-D-ribofuranosyl-7H-1,2,3-triazolo(4, 5-d)pyrimidine-7-thione (XIV), (2) N-β-D-ribofuranosyl(1,2,3) thiodiazolo(5,4-d)pyrimidin-7-amine (XVI), and (3) α-nucleoside mixture (α-XVIII and α-XIX).

on silica gel. The estimated peak areas of the four components are 31% β-XIX, 24% β-XVIII, 26% α-XIX, and 18% α-XVIII. Although partial separation of these isomers could be achieved on the analytical column, much better separation occurred when the mixture was applied to a 7 mm (i.d.)x244 cm Bondpak C_{18}/Porasil B preparative column. The isomers were eluted with 97:3 water-acetonitrole at a flowrate of 4ml/min. The four major fractions (Fig. 8) isolated from this column were sufficiently pure to identify the four components from NMR data (see Table II) and mass spectra. The first fraction (4.7 mg), containing some XVI, was identified as N-β-D-ribopyranosyl(1,2,3)thiadiazolo(5,4-d) 9.0Hz) of the anomeric proton, which is indicative of a transdiaxial relationship of the C_1' and C_2' protons (7). The second fraction (5.6 mg), consisting of a mixture of XVI and β-XIX, was identified by comparison with authentic XVI and fraction 1. The NMR spectrum of XVI contained a triplet at 4.8 ppm assigned to the hydroxyl proton of the furanose CH_2OH. The third fraction

TABLE II PMR Spectral Data (δ in ppm)[a]

Compound	H_1' ($J_{1'2'}$)	H_2'	H_3'	H_4'	H_5'	$O_2'H$	$O_3'H$	$O_5'H$	NH	C_5H
XVI	q, 6.0	m, 4.1	m, 4.1	m, 3.8	m, 3.5	d, 5.1[b]	d, 5.0[b]	t, 4.8	d, 9.8	s, 8.7
+D_2O	d, 6.0 (5 Hz)	t, 4.2*	t, 4.6	m, 3.8	m, 3.5	--	--	--	--	s, 8.7
β-XIX	m, 5.7	-------m, 3.3 - 4.3-----------------				m, 4.9	m, 4.9	m, 4.9	9.8	s, 8.7
+D_2O	d, 5.7 (9 Hz)	-------m, 3.3 - 4.3-----------------				--	--	--	--	s, 8.7
α-XVIII	m, 6.2	---------m, 3.3 - 4.3---------------				5.6,5.7	5.6,5.7	t, 4.8	8.6	s, 8.7
+D_2O	d, 6.2 (5 Hz)	--	--	--	m, 3.5	--	--	--	--	s, 8.7
α-XIX/	m, 5.9	---------m, 3.3 - 4.3---------------				--4.9,5.4 and 5.7-6.0--		--	d, 9.4	s, 8.7
+D_2O	d, 5.9 (4 Hz)	---------m, 3.3 - 4.3---------------				--	--	--	--	s, 8.7

[a] 0.5 to 1.5% solutions in DMSO-d_6.

[b] Assigned by spin-decoupling.

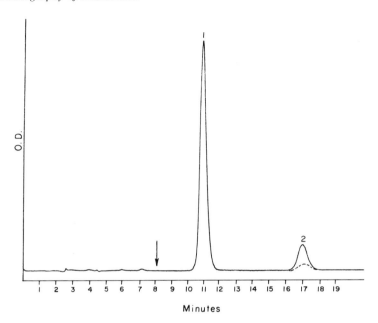

FIGURE 6. Solvent: 95:5 water-acetonitrile; sample weight: 1.5 µg. (1) N-β-D-ribofuranosyl[1,2,3]thiadiazolo[5,4-d]pyrimidin-7-amine (XVI) and (2) α-nucleoside mixture (α-XVIII and α-XIX).

(4.9 mg) was identified as N-α-D-ribopyranosyl[1,2,3]thiadiazolo-[5,4-d]pyrimidin-7-amine (α-XIX) by absence of an O_5'H triplet and the presence of the anomeric proton (H_1') at a lower field (8) then the anomeric proton of β-XIX. The fourth fraction (1.6 mg) was identified as N-α-D-ribofuranosyl[1,2,3]thiadiazolo-[5,4-d]pyrimidin-7-amine (α-XVIII) by the presence of a triplet at 4.8 ppm (CH_2OH) and the anomeric proton (H_1') at a lower field than that assigned to β-XVIII. Mass spectra of XVI, α-XVIII, α-XIX, and β-XIX exhibited mass ions of m/e 285 (M^+) consistent with the proposed structures. In addition, XVI and α-XVIII displayed mass ions of m/e 254, consistent with loss of CH_2OH. This mass ion was not observed in the mass spectrum of the pyranoside α-XIX.

D. THE METABOLISM OF 9-(α-D-ARABINOFURANOSYL)-8-AZAADENINE

9-(α-D-Arabinofuranosyl)adenine (XXI) is easily prepared by the classical mercuri procedure of Davoll and Lowy(9), but this method is less satisfactory for the preparation of 9-

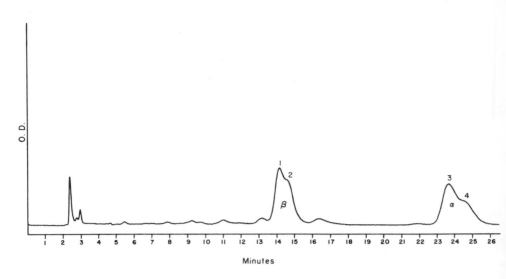

*FIGURE 7.   Solvent: 97:3 water-acetonitrile.   (1) N-β-D-
ribopyranosyl[1,2,3]thiadiazolo[5,4-d]pyrimidin-7-amine (β-XIX),
(2) N-β-D-ribofuranosyl[1,2,3]thiadiazolo[5,4-d]pyrimidin-7-amine
(XVI), (3) N-α-D-ribopyranosyl[1,2,3]thiadiazolo[5,4-d]pyrimidin-
7-amine (α-XIX), and (4) N-α-D-ribofuranosyl[1,2,3]thiadiazolo
[5,4-d]pyrimidin-7-amine (α-XVIII).*

(α-D-arabinofuranosyl)-8-azaadenine (3-α-D-arabinofuranosyl-3H-
[1,2,3]triazolo[4,5-d]pyrimidin-7-amine, α-XXVI). Originally,
this latter nucleoside was obtained along with the β-anomer from
the reaction of N-nonanoyl adenine with 2,3,5-tri-O-benzyl-α-D-
arabinofuranosyl bromide in a nonpolar solvent in the presence
of an AW-500 molecular sieve (10). It is best prepared by the
overnight reaction of 8-aza-6-(methylthio)purine (XXII) with 2,
3,5-tri-O-benzyl-D-arabinofuranosyl bromide (XXIII) with molecular sieve in toluene followed by treatment of the blocked nucleoside with methanolic ammonia (11). XXI and α-XXVI were toxic
to H. Ep. #2 cells in culture, whereas the corresponding β-anomers
were not toxic at the highest concentrations (375 μM) assayed
(Table III). α-XXVI was more cytotoxic than XXI and, therefore,
was selected for further study (12). Cell lines deficient in
adenosine kinase were resistant to inhibition by α-XXVI (Table
III). In order to study the metabolism of this nucleoside,
[2-^{14}C]-α-XXVI was prepared in five steps from sodium formate-
^{14}C (Scheme 1). Cultured H. Ep. #2 cells grown in the presence
of the labeled nucleoside contained compounds migrating, on paper chromatograms, like mono-, and di-, and triphosphates (Table
IV). Elution of the areas of paper containing the suspected

TABLE III   Inhibition of H. Ep. #2 Cells in Culture by α-Ara-8-AzaA and Some Related Compounds[a]

Compound	ED_{50} (μM)[b]		
	H. Ep. # 2/S	H. Ep. # 2/MeMPR[c]	H. Ep. # 2/FA/FAR[d]
α-XXVI	2.2	>375	>375
β-XXVI	>375	--	--
XXI	14	--	--
β-AraA	>375	--	--
6-MeMPR	1.0	>300	>300

[a]Taken from Ref. 12.

[b]The concentration required for a 50% reduction in the number of colonies from treated cells compared to controls.

[c]Cell deficient in adenosine kinase.

[d]Cell deficient in adenosine kinase and adenine phosphoribosyl transferase.

TABLE IV   Metabolism of [2-^{14}C]α-Ara-8-AzaA by H. Ep. #2 Cells[a]

Compound isolated	Cells (dpm/10^6)		Total intracellular^{14}C (%)	
	4 hr	24 hr	4 hr	24 hr
α-ara-8-azaA	754	468	7	5
α-ara-8-azaAMP	8319	6421	81	63
α-Ara-8-azaA di- and triphosphates	1229	3265	12	32

[a][2-^{14}C]α-Ara-8-azaA was added to logrithmically growing H. Ep. #2 cells at a concentration of 2 μg/ml (0.0138 μCi/ml). Cells were harvested 4 and 24 hr thereafter and extracted with hot 80% ethanol, after which the soluble fraction was analyzed by two-dimensional paper chromatography and autoradiography (see text for details). The metabolites shown here are from the same experiment from which the data of Fig. 2 were obtained (taken from Ref. 12).

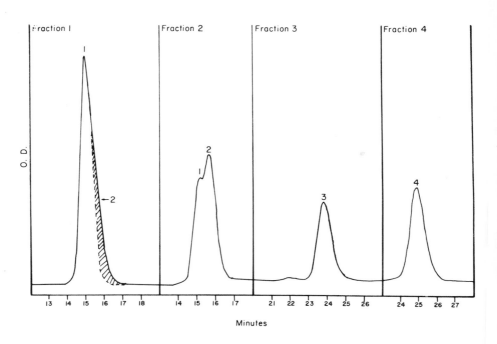

FIGURE 8.   Solvent 97:3 water-acetonitrile; sample weights: fraction 1,5 µg; fractions 2-4 were not weighed. (1) N-β-D-ribo-pyranosyl[1,2,3]thiadiazolo[5,4-d]pyrimidin-7-amine (β-XIX), (2) N-β-D-ribofuranosyl[1,2,3]thiadiazolo[5,4-d]pyrimidin-7-amine (XVI), (3) N-α-D-ribopyranosyl[1,2,3]thiadiazolo[5,4-d]pyrimidin-7-amine (α-XIX), and (4) N-α-D-ribofuranosyl[1,2,3]thiadiazolo [5,4-d]pyrimidin-7-amine (α-XVIII).

phosphates followed by treatment of the eluates with snake venom (to remove terminal phosphates) gave the expected nucleosides. The nucleosides were subjected to paper chromatography, and the radioactive areas on these chromatograms were eluted; the eluates were lyophilized; and the residues were added to 0.2 ml of an aqueous solution containing 0.5 mg each of nonradioactive α- and β-XXVI.   Up to 100 µl of these solutions were injected on the column.   The fractions under the peaks corresponding to the re-tention times of α- and β-XXVI were pooled separately.   The pro-cedure gave pure samples of α- and β-XXVI, each of which was assayed for radioactivity (Fig. 9).   More than 99% of the ^{14}C present in the metabolic samples was eluted with α-XXVI.

Subsequently, α-XXVI was shown to be a substrate for ad-enosine kinase partially purified from H. Ep. #2 cells; the K_m was 110 µM and the V_{max} was about 15% that of adenosine.   α-

XXVI was not a substrate for adenosine deaminase; several other α-nucleosides also had little or no activity as substrates for this enzyme, whereas the corresponding β-anomers were deaminated (12).

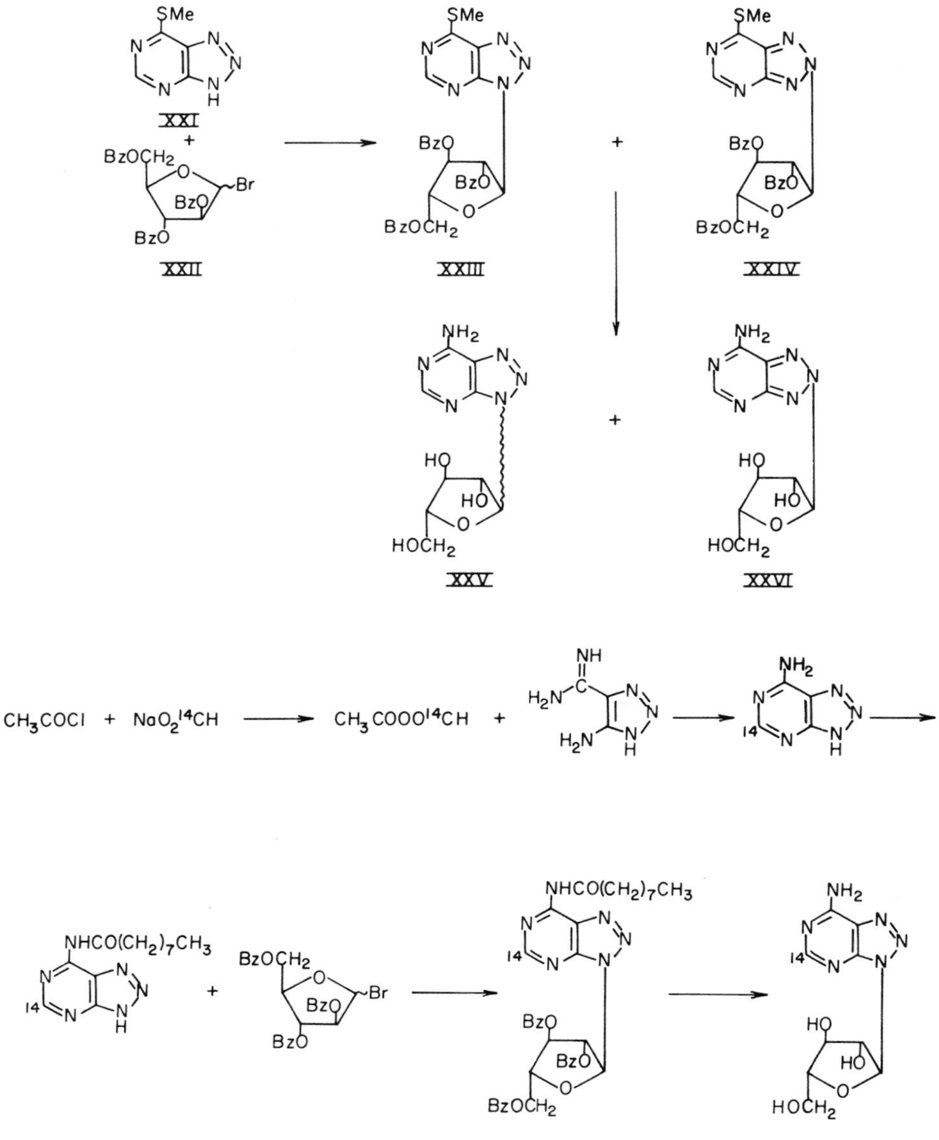

Scheme I.

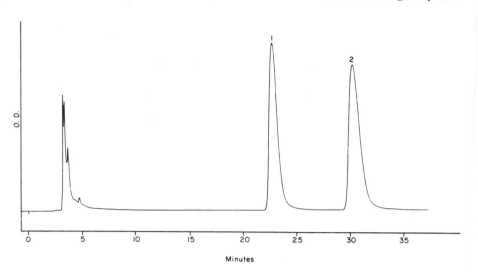

FIGURE 9.   *Solvent: 49:1 water-acetonitrile; sample weights:*
2.5 μg each.   (1) 9-α-D-arabinofuranosyl-8-azaadenine (α-XXVI)
and (2) 9-β-D-arabinofuranosyl-8-azaadenine (β-XXVI).

REFERENCES

1. J. N. Little, D. F. Horgan, and K. J. Bombaugh, Abstracts of
    the American Chemical Society Meeting, September 14-18,
    1970, Chicago, ANAL-41.
2. J. J. Kirkland, *Modern Practise of Liquid Chromatography,*
    Wiley-Interscience, New York, 1971, pp. 162 and 187.
3. Waters Associates, Inc., Technical Bulletin, February, 1974.
4. J. A. Montgomery and H. J. Thomas, *J. Carbohyd. Nucl. Nucl.*
    2, 91 (1975).
5. R. D. Elliott and J. A. Montgomery, *J. Med. Chem. 20, 116* (1977)
6. R. D. Elliot and J. A. Montgomery, *J. Med. Chem. 19,* 1186
    (1976).
7. J. A. Montgomery and H. J. Thomas, *J. Org. Chem. 36,* 1962
    (1971).
8. R. V. Lemieux and J. D. Stevens, *Can. J. Chem. 44,* 249 (1966);
    Y. H. Pau, R. K. Robins, and L. B. Townsend, *J. Heterocycl.*
    Chem. 4, 246 (1967).
9. J. Davoll and B. A. Lowry, *J. Am. Chem. Soc. 73,* 1650 (1951).
10. J. A. Montgomery and H. J. Thomas, *J. Med. Chem. 15,* 305
    (1972).

11. J. A. Montgomery, A. T. Shortnacy, G. Arnett, and W. M. Shannon, *J. Med. Chem. 20, (1977)*.
12. L. L. Bennett, Jr., P. W. Allan, D. L. Hill, H. J. Thomas, and J. W. Carpenter, *Mol. Pharmacol. 12,* 242 (1976).

CHEMICAL AND ENZYMATIC METHODS IN THE
SYNTHESIS OF MODIFIED POLYNUCLEOTIDES*

THOMAS J. BARDOS AND YAU-KWAN HO

*Department of Medicinal Chemistry, State University
of New York at Buffalo, Buffalo, New York*

ABSTRACT

"Antitemplates," i.e., inhibitory structural analogs of
functional primers and templates of DNA and RNA polymerases, have
been prepared by a novel procedure for the introduction of 5-SH
substituents into some of the uracils and/or cytosines of nucleic
acids. The partially thiolated polynucleotides were further modified by selective reactions of the ionized 5-SH groups with electrophilic reagents. S-methylation of partially thiolated poly(C)
did not abolish its antitemplate activity. Alternatively, partially thiolated polynucleotides were prepared by copolymeriza-
tion of sh^5CDP with CDP, and sh^5UDP with UDP, using polynucleotide phosphorylase, but yields were unsatisfactory. The S-methy-
lated analog sm^5UDP was a much better substrate; sm^5dUTP was also
utilized by DNA polymerases. Partially thiolated polynucleotides
are potent inhibitors of polymerases, resistant to nucleases,
and readily taken up by tumor cells.

*This work was supported by U.S.P.H.S. Grant No. CA-06695
from the National Cancer Institute, NIH, and by Grant No. CH-20C
from the American Cancer Society.

55

INTRODUCTION

Chemical modifications of either the base or the sugar moieties
of polynucleotides by treatment of the polymers with various chemical reagents have been used extensively during the past decade as
a tool in the investigation of the structures and functions of
the natural nucleic acid molecules (1). An alternative approach
to the preparation of modified polynucleotides, involving the
enzymatic incorporation of unnatural or modified nucleotide analogs into the polymeric products by polynucleotide phosphorylase
or, in a few cases, by other polymerases, has also been employed
by many investigators in studies of enzyme specificities. These
techniques have also been used in studies of the effects of structural modifications on the secondary structures of polynucleotides,
on their complex formations with complementary polynucleotide
strands, and on their stabilities to nucleases, particularly with
a view to their possible application for interferon induction (2).
     Our own interest in the synthesis of modified polynucleotides
was primarily motivated by considerations of their potantial chemotherapeutic application as a new type of "macromolecular antimetabolites" that would compete with the structurally analogous DNA
or RNA molecules of the virus, or of the cancer cells, and thus
interfere with their specific functions as templates, primers,
or messengers. For the designation of such inhibitory structural
analogs of the informational macromolecules, we are using the
term "antitemplates" (3,4).
     In order to be effective as a chemotherapeutic agent, an an-
titemplate would have to show (a) specific and much stronger "binding" to the target enzymes, e.g., polymerases, than the corresponding natural template, (b) inability to function as a template,
(c) ability to penetrate cell membranes, and (d) preferential uptake by cancer cells.
     Our previous studies with 5-mercaptouracil and 5-mercaptocy-
tosine nucleosides and nucleotides, synthesized in our laboratory
several years ago (5-7), revealed the unusual reactivity and
"binding" ability of the SH group in the 5 position of the pyrimidine ring (8,9). In the case of 5-mercaptouracil derivatives,
the SH group has a pK_a of 5.0, and in the case of the corresponding cytosines, a pK_a of 5.6; thus, in both cases, the SH group
is essentially ionized at physiologic pH and undergoes rapid au-
toxidation (10), resulting in dimeric disulfide formation as well
as in strong binding to proteins via mixed disulfide linkages.
Therefore, we thought that the introduction of this reactive
group into the uracil or cytosine bases of various polynucleotides
might produce particularly effective inhibitory template or messenger analogs.

CHEMICAL MODIFICATIONS

Based on a preliminary study of the mechanism of reaction
of 5-bromouracil derivatives with sulfur nucleophiles, we developed a novel synthetic method for the direct conversion of uracil
nucleotides into their 5-mercapto analogs under relatively mild
reaction conditions (11). This method consisted of two reaction
steps: addition of methyl hypobromite (MeOBr) in methanol and
reaction of the methyl hypobromite adduct with sodium disulfide
(Na_2S_2) in dimethylformamide, to give 70-80% of the 5-mercapto
derivative in the form of the di- and trisulfides, which are readily reduced to the free thiol with dithiothreitol (DTT). However,
in order to develop a general procedure for the modification of
both uracil and cytosine bases in polynucleotides, it was necessary to replace the Na_2S_2 in the second reaction step with sodium hydrosulfide (NaSH), although in this case only a much smaller
yield of the thiolated derivative was obtained and the major portion of the adducts were reconverted to the starting materials(12).
This general method, with some modifications, was found to
be applicable to the selective "5-thiolation" of a small percentage of the uracil and cytosine bases present in any given polynucleotide without causing any other chemical alterations or any
degradation of the polymer chain. In the standard procedure, the
polynucleotides are first converted to their organic solvent-
soluble cetyltrimethylammonium salts and then treated with MeOBr
in dry methanol at 0°C for 1-2 hr under a nitrogen atmosphere.
The cetyltrimethylammonium salts of the partially thiolated polynucleotides are reconverted to the sodium salts by precipitation
with 3 *M* NaCl, followed by dissolving in buffer and gel filtration (13,14).
The course of the reactions of the uracil and cytosine bases
in the polynucleotides is shown in Scheme 1. The percentage
yield of the conversion of the uracil and/or cytosine residues
to the corresponding 5- mercapto derivatives can be controlled by
varying the amounts of the reagents used (both MeOBr and NaSH,
correspondingly) between 0.25 and 2.5 mole equivalents, based on
the total uracil and cytosine nucleotides in the polymer. However, there appears to be a maximum yield (extent) of thiolation;
this is 10-15% in the case of polycytidylic and polyuridylic acids,
and only 2-4% in the case of various DNA and RNA isolates. Although the first reaction step, with MeOBr, proceeds nearly quantitatively (at least in the case of the homopolynucleotides),
the methoxybromide adducts of the bases are partially converted
by NaSH to the 5-mercaptoderivatives and partially reconverted to
the 5-unsubstituted pyrimidine nucleotides (as shown in Scheme 1).
Since the purine nucleotides do not react with MeOBr under these
conditions, the partial "5-thiolation" of the pyrimidine bases is
the only observed chemical change resulting from the application

of this procedure (except for the possible reactions of some of
the minor bases in the case of tRNA). This has been verified by
identification of the hydrolysis products of (^{35}S)-labeled partially thiolated polynucleotides.

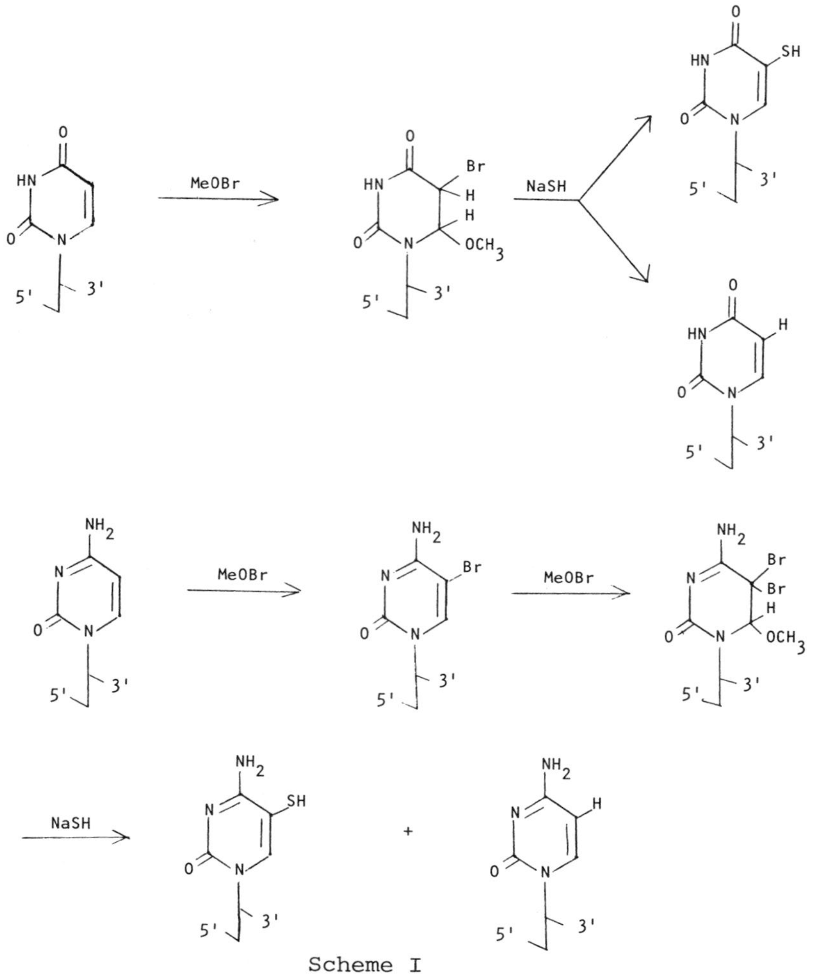

Scheme I

*Partial "5-thiolation" of uracil and cytosine residues of poly-
nucleotides.*

It is of interest that the readily ionizable 5-mercapto
groups, due to their strong nucleophilicity, can be selectively
reacted with various eletrophilic reagents, such as alkylating
agents, or organometal cations (see Scheme 2). For example, the
quantitative and selective reaction of the thiol groups with *p*-
hydroxymercuribenzoate (HMB) followed by neutron activation anal-

ysis (15) may be used for the determination of the percentage of
free SH groups (13). For the determination of the oxidized thiol
(disulfide) groups, the latter were reduced with dithiothreitol
(DDT) prior to the reaction with HMB, and after the reaction the
HMB derivative of the thiolated polynucleotide was purified from
the small molecular weight by-products by gel filtration. Following activation for 30 min in a thermal neutron flux (15), the
radioactivities (Hg[197]) of the samples were measured, and the
corresponding values for the total percentage of thiol groups
were calculated with the aid of a simultaneously irradiated Hg^{2+}
standard. In our experience, this method proved to be the most
sensitive, accurate, and reliable one for the determination of
the percentage of thiolation, and it was applicable to all types
of partially thiolated polynucleotides.

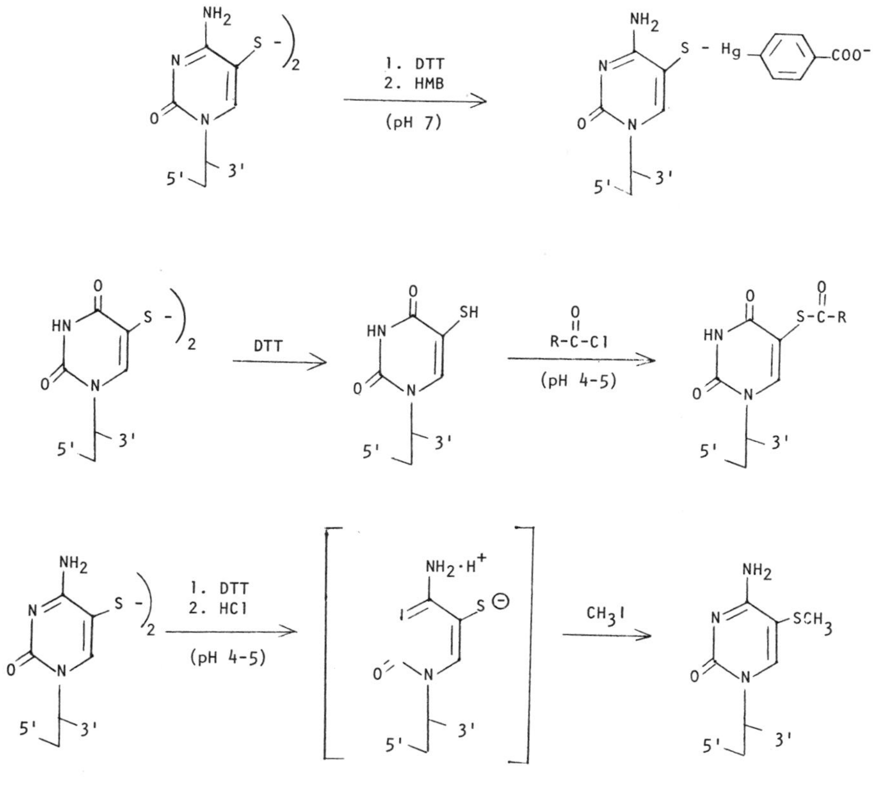

Scheme II
*Chemical synthesis of S-substituted derivatives of partially
thiolated polynucleotides.*

Further, still undeveloped applications of the HMB derivatives of partially thiolated nucleic acids may obviously present
themselves in structural studies of various natural RNA or DNA
molecules with the aid of x-ray crystallography. However, from
the point of view of chemotherapeutic application, these (or any
other heavy metal derivatives) are of little interest because of
their toxicities. In contrast, the S-acyl and S-alkyl derivatives
of the partially thiolated polynucleotides appear to be of considerable chemotherapeutic interest. The S-acyl groups, which
are more or less readily removed in the biological media via
thiolytic cleavage (16), provide temporary protection for the 5-
mercapto groups and may modify the distribution and site of action
of the partially thiolated polynucleotides; conversely, the latter
may serve as carriers for certain biologically active acyl groups.
On the other hand, the S-alkyl groups would be expected to modify
in a more permanent manner the structures and, therefore, also
the physicochemical properties and biological activities of the
partially thiolated polynucleotides. Our preliminary results with
the S-methyl derivatives (see below) are interesting and promising
for the therapeutic potential of modified polynucleotides containing alkylmercapto substituents in the 5 positions of some of the
pyrimidine residues.

For selective alkylation of the 5-mercapto groups of partially
thiolated polynucleotides, it is necessary to conduct the reaction
at an acidic pH in the range 4 to 5.5 and in the presence of mercaptoethanol (see Scheme 2). Under these conditions, all disulfide groups are reduced to the free SH, which is partially ionized
($pK_a \sim 5$) to the reactive thiolate anion, while at the same time the
amino groups and ring nitrogens of the polynucleotide bases are
protonated and protected from alkylation. Thus, a solution of a
10%-thiolated polycytidylic acid (MPC) containing mercaptoethanol
(in fivefold excess over the 5-mercapto groups) was adjusted with
hydrochloric acid to pH 4.3 and then treated with methyl iodide
(in slight excess over the total thiols present in the reaction
mixture) under vigorous stirring for 2 hr at room temperature.
The methylated polymer was purified by gel filtration through a
Sephadex G-25 column. Disappearance of the characteristic uv
absorption peak of the 5-mercapto groups at 335 nm (at pH 8, upon
addition of DTT) indicated that the methylation was complete.
However, neutron activation analysis showed that 2% of the mercapto groups remained unmethylated; thus, avoiding the use of an
excess of the methyl iodide, either slightly higher pH or longer
reaction time would be required for quantitative methylation of
the SH groups.

ENZYMATIC SYNTHESES

    In order to determine the feasibility of synthesizing par-
tially thiolated polynucleotides and their S-alkyl derivatives
by enzymatic methods, the polynucleotide phosphorylase of *Micro-*
coccus luteus was employed first, because (a) it is known to have
relatively low substrate specificity, being capable of utilizing
various modified nucleoside diphosphates as substrates, and (b)
it has no template requirement and therefore presumably would
not be inhibited by the thiolated polynucleotides produced in the
course of the enzymatic reaction.
    The 5'-diphosphates of 5-mercaptouridine and 5-mercaptocyto-
sine (sh^{5}UDP and sh^{5}CDP, respectively) were prepared from the
corresponding monophosphates by application of the morpholidate
method of Moffat and Khorana (17). The 5'-diphosphate of 5-meth-
ylmercaptouridine (sm^{5}UDP) was synthesized by reacting the 5'-
phosphoromorpholidate of 5-mercaptouridine with methyl iodide
and, subsequently, with mono(tri-N-butylammonium) dihydrogen phosphate, as shown in Scheme 3.

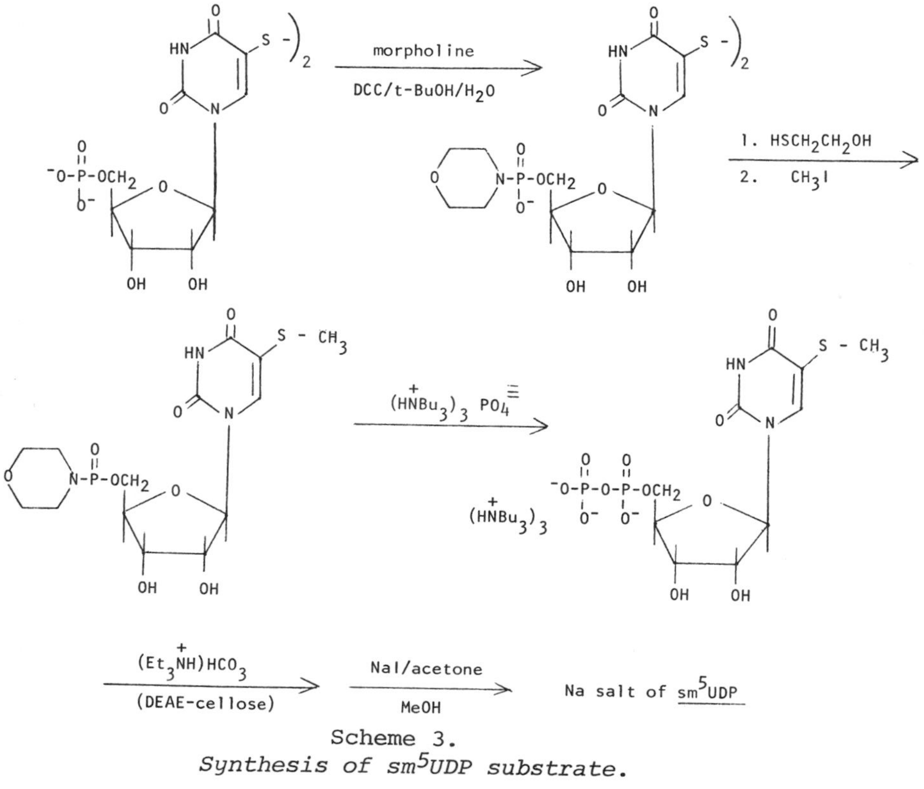

Scheme 3.
Synthesis of sm^{5}UDP substrate.

The sh[5]UDP and sh[5]CDP were purified by column chromatography on DEAE-cellulose with a linear gradient of triethylammonium bicarbonate buffer (pH 7.5) containing mercaptoethanol; the sm[5]UDP was purified in the same manner but without mercaptoethanol.

Polymerization studies with sh[5]UDP and sh[5]CDP were conducted in the presence of mercaptoethanol under conditions in Table I.

TABLE I   Effect of sh[5]UDP on the Polymerization Reaction[a]

Substrates (%)		Polymerization	sh[5]UMP in
UDP	sh[5]UDP	(%)	polynucleotide (%)
100	0	52.8	0
95	5	42.8	1.1
80	20	39.9	2.9
50	50	35.5	4.9
30	70	15.2	7.7
7	93	3.7	21.4
0	100	0	--

[a]*The reaction mixture, incubated at 37°C for 8 hr contained in a final volume of 0.2 ml: 15 mM total nucleotides (sh[5]UDP and UDP), 10 units/ml of enzyme, 100 mM Tris-HCl (pH 8.5), 5 mM $MgCl_2$, and 10 mM DTT.*

Neither of these free thiol compounds could be polymerized alone (in the absence of other substrates), but they were utilized by the enzyme in the presence of the corresponding unmodified substrates, UDP and CDP, respectively. Using a 1:1 ratio of sh[5]CDP and CDP in the substrate mixture, a very low yield of polymeric product containing only 1.6% 5-mercaptocytidylate units was obtained, which showed similar properties as a 1.6%-thiolated polycytidylic acid (MPC) prepared by chemical modification. Co-polymerization of sh[5]UDP and UDP gave somewhat better yields and a higher percentage of 5-mercapto nucleotide contents, as shown in Table 1. However, the relative utilization of sh[5]UDP by the enzyme was far below the ratio of sh[5]UDP/UDP present in the reaction mixture. This is anomalous for polynucleotide phosphorylase, which generally utilizes all nucleoside diphosphate substrates in the same ratio as they are presented. Further studies demonstrated that sh[5]UDP due to its ionized 5-SH group inhibits the polymerization reaction in a noncompetitive manner (9).

In contrast, sm[5]UDP (which has a blocked 5-mercapto group) was found to be a much better substrate for polynucleotide phosphorylase than sh[5]UDP, although it showed a qualitatively similar pattern (see Table II). It was capable of serving as the sole substrate and thus yielded a homopolymer, 5-methylmercap-

touridylic acid. However, its rate and yield of polymerization
was considerably lower than that of UDP, and when copolymerized
with the latter it was utilized less efficiently. It appears

TABLE II. Effect of sm^{5}UDP on the Polymerization Reaction[a]

Substrate (%)		Polymerization (%)	sm^{5}UMP in polynucleotide(%)
UDP	sm^{5}UDP		
100	0	51.8	0
95	5	47.4	3.41
80	20	39.7	10.3
50	50	30.0	21.4
20	80	24.2	49.0
10	90	15.1	67.2
0	100	5.64	100

[a]*The reaction mixture, incubated at 37°C for 8 hr contained
in a final volume of 0.2 ml: 15 mM total nucleotides (sm^{5}UDP
and UDP), 10 units/ml of enzyme 100 mM Tris-HCl (pH 8.5), and
5 mM MgCl$_2$.*

that sm^{5}UDP also inhibits the polymerization reaction, but it is
certainly a much better substrate and weaker inhibitor than
sh^{5}UDP.
        We may conclude that the enzymatic synthesis of S-alkyl (or
at least, S-methyl) substitituted partially (or even fully) thiolated polynucleotides is a feasible and promising method, which
may complement the chemical method due to its different range of
applicability. However, in the case of S-unsubstituted partially
thiolated polyribonucleotides the enzymatic method using polynucleotide phosphorylase is impractical and cannot compete in feas-
ibility with the chemical thiolation procedure. Use of the terminal polydeoxyribonucleotidyl transferase for the synthesis of
partially thiolated polydeoxyribonucleotides is currently being
investigated.
        Furthermore, S-methyl-5-mercaptodeoxyuridine-5'-triphosphate
(sm^{5}dUTP) was found to be an active substrate for various DNA
polymerases including DNA polymerase α from regenerating rat liver
and DNA polymerase I of *E. coli* (18). In cultures of HSV-1 infected cells, 5-methylmercapto-2'-deoxyuridine equals or surpasses
thymidine in its rate of incorporation into the viral DNA (19).
Not surprisingly, the 5-methylmercapto derivatives appear to be
better substrates for DNA than for RNA synthesis.

BIOLOGICAL ACTIVITIES OF PARTIALLY THIOLATED POLYNUCLEOTIDES

A. *INHIBITION OF POLYMERASES*

Various partially thiolated polynucleotides were found to be
potent inhibitors of the DNA directed RNA polymerases from *M.
luteus* (13) and from Friend virus-infected mouse spleen (20).

Among the DNA directed DNA polymerases, the high-molecular-
weight DNA polymerase α, isolated from regenerating rat liver, was
the most sensitive to inhibition by partially thiolated polynucleotides. The inhibitory activity of the modified polymers was
directly related to the percentage of thiolation, but it was
found to depend also on the configuration, base composition, and
other structural properties. The inhibition was competitively reversible with the DNA template, but the "binding" of the modified
polynucleotides to the α-polymerase was much stronger than that of
the template, with K_i values of 1-3 x 10^{-6} M and $K_i/K_m \simeq 0.03$ (21)

In contrast, the low-molecular-weight DNA polymerase β, from
either regenerating (22) or normal rat liver nuclei (21), was
only partially inhibited even by much higher concentration of the
modified polynucleotides. The DNA polymerase I of *E. coli*, which
has no SH group at the active center, was not inhibited by partially thiolated polynucleotides (22).

It was shown that partially thiolated polynucleotides are
potent inhibitors of the RNA-directed DNA polymerases (reverse
transcriptases) from RNA tumor viruses. In these studies, the
various partially thiolated polynucleotides showed significant
selectivities in their inhibitory effects (22-28).

B. *STABILITY TO THE ACTION OF NUCLEASES*

From the point of view of therapeutic application, it is important that partial thiolation significantly increases the stabilities of both RNA and DNA molecules to the hydrolytic action
of nucleases (29). Thus, ribosomal RNA from Ehrlich ascites cells
(0.165 mg/0.7 ml) was digested with pancreatic RNase A (recryst.,
33 μg) at 37°C for 60 min to give 90.8% acid-soluble hydrolysis
products, based on the ultraviolet absorbance at 260 nm (A_{260}).
The same rRNA, after thiolation of 3.5% of the total uracil and
cytosine residues in the polymer, gave only 41.5% acid soluble
products under the same conditions.

Similarly, the DNA isolated from Ehrlich ascites cells
(0.165 mg/0.7 ml) was treated with pancreatic DNase I (0.08 μg;
3000 units/mg) at 37°C for 30 min to give 65% acid soluble hydrolysis products (based on (A_{260}). After thiolation of 3.9%
of the total cytosine residues, the modified DNA gave only 4.0%

acid soluble hydrolysis products under the same conditions. Thus, one 5-SH group for an average of 100 nucleotides is sufficient for almost complete protection of the DNA from degradation by DNase. This protection was not affected by the addition of mercaptoethanol to the reaction mixture (30).

C. *INTERFERON INDUCTION*

It was found that partially thiolated polycytidylic acids (MPC) readily annealed with poly(I) to form double-stranded 1:1 complexes (9) that induced antiviral activity in both human and murine cells (31). A difference was found in the induction of interferon by the modified MPC·poly(I) when compared with unmodified poly (C)·poly(I), which may prove practical in circumventing the hyporesponsiveness or resistance to repeated stimulation with interferon inducers (31).

D. *UPTAKE BY TUMOR CELLS*

The uptake of partially thiolated DNA by Ehrlich ascites cells proceeds much more readily than that of the corresponding unmodified DNA; furthermore, the major portion of the cell-associated [^{35}S]]-thiolated DNA was recovered from the isolated nuclei (32). Excess Ca^{2+} ions during incubation dramatically increased the uptake of the thiolated DNA by the cells and its transport into the nuclear fraction. Under these conditions, about 70% of the total [^{35}S]-DNA added to the incubation media was recovered from the isolated cell nuclei; this amount corresponded to 15-20% of the total endogenous nuclear DNA of the ascites cells. Studies with [^{35}S]-labeled partially thiolated polycytidylic acid (MPC) showed considerable differences in the rates and extent of uptake between various cell types (33). Differences of uptake between normal and malignant cells are currently being investigated.

E. *CHEMOTHERAPEUTIC ACTIVITY*

To date, only some preliminary results (34, 35) have been reported relating to the chemotherapeutic activity of partially thiolated polycytidylic acid (MPC). Further studies are being conducted in several laboratories for the purpose of evaluating the chemotherapeutic potential of MPC and, particularly, of other partially thiolated polynucleotides having more specific structural features.

BIOLOGICAL ACTIVITIES OF S-SUBSTITUTED PARTIALLY THIOLATED POLY-
NUCLEOTIDES

Preliminary studies indicated that S-methyl-MPC also inhibited the DNA polymerase-α of mammalian cells although it was less
active than the MPC before alkylation (36). This result indicates that the S-methyl compound may also bind to the polymerase,
even though it cannot form mixed disulfide linkages with the enzymic sulfhydryl groups. One possibility is that the neighboring
Me-$\overset{..}{\underset{..}{S}}$-($C_5$) and H_2N-(C_4) groups may bind to an essential metal ion
(presumably, Zn^{2+}) present at the active center of the polymerase
via chelation, and thus inhibit its catalytic activity. However,
further studies are required to clarify this question and to establish the apparent "antitemplate" activities and range of selectivities of the S-alkylated partially thiolated polynucleotides.

REFERENCES

1.   Kochetkov, N. K. and Budowsky, E. I., *Prog. Nucl. Acid Res.*
        Mol. Biol. 9, 403 (1969); Hayatsu, H. *Prog. Nucl. Acid Res*
        Mol. Biol. 16, 75 (1976); Budowsky, E. I. *Prog. Nucl. Acid*
        Res. Mol. Biol. 16, 125 (1976).
2.   DeClercq, E. in *Topics in Current Chemistry, 52,* Springer Ver
        lag, Berlin, 1974, p. 173.
3.   Bardos, T. J., Baranski, K., Chakrabarti, P., Kalman, T. I.,
        and Mikulski, A. J. *Proc Am. Assoc. Cancer Res. 13,* 359
        (1972).
4.   Bardos, T. J. in *Topics in Current Chemistry, 52,* Springer
        Verlag, Berlin, 1974, p. 92.
5.   Kotick, M. P., Szantay, C., and Bardos, T. J. *J. Org. Chem.*
        34, 3806 (1969).
6.   Szekeres, G. L. and Bardos, T. J., *J. Med. Chem. 13,* 708 (197
7.   Szekeres, G. L., Ph. D. thesis, State University of New York
        at Buffalo, 1970.
8.   Kalman, T. I. and Bardos, T. J. *Mol. Pharmacol. 6,* 621 (1970).
9.   Bardos, T. J., Aradi, J., Ho, Y. K., and Kalman, T. I. *Ann.*
        N. Y. Acad. Sci. 255, 522 (1975)
10. Kalman, T. I. and Bardos, T. J. *J. Am. Chem. Soc. 89,* 1171
        (1967).
11. Szabo, L., Kalman, T. I., and Bardos, T. J. *J. Org. Chem. 35,*
        1434 (1970).

12. Bardos, T. J., Chakrabarti, P., Kalman, T. I., Mikulski, A. J., and Novak, L. 163rd National Meeting of the American Chemical Society, Boston (1972) MEDI 21.
13. Mikulski, A. J., Bardos, T. J., Chakrabarti, P., Kalman, T. I., and Zsindely A. *Biochim. Biophys. Acta 319,* 294 (1973).
14. Bardos, T. J., Novak, L., Chakrabarti, P., and Ho, Y. K. in *Synthetic Procedures Methods and Techniques in Nucleic Acid Chemistry* (L. Townsend ed.), in press.
15. Bruce, A. K. and Malchmen, W. H. *Radiation Res. 24,* 473 (1965).
16. Kotick, M. P., Kalman, T. I., and Bardos, T. J. *J. Med. Chem. 13,* 74 (1970).
17. Moffat, J. G. and Khorana, H. G. *J. Am. Chem. Soc. 83,* 649 (1960).
18. Ho, Y. K., Kung, M. P., and Bardos, T. J. 9th International Congress of Chemotherapy, London (1975), Abstracts, M-453.
19. Hardi, R., Hughes, R. G., Ho, Y. K., Chadha, K. C., and Bardos, T. J. *Antimicrobial Agents and Chemotherapy,10,* 682 (1976).
20. Munson, B. R. *Proc. Am. Assoc. Cancer Res. 17,* 157 (1976).
21. Kung, M. P., Ho, Y. K., and Bardos, T. J. *Cancer Res. 36,* 4537 (1976).
22. Chandra, P., Ebener, U., Bardos, T. J., Chakrabarti, P., Ho, Y. K., Mikulski, A. J., and Zsindely, A., *Ann. N. Y. Acad. Sci. 255,* 532 (1975).
23. Chandra, P., and Bardos, T. J. *Res. Commun. Chem. Pathol. Pharmacol. 4,* 615 (1972).
24. Chandra, P., Bardos, T. J., and Ebener, U. in *Progress in Chemotherapy* (G. K. Daikos, ed.) Vol. 3 Helenic Society for Chemotherapy, Athens, 1974, p. 182.
25. Srivastava, B. I. S. and Bardos, T. J. *Life Sci. 13,* 47 (1973).
26. Srivastava, B. I. S. *Biochim. Biophys. Acta 335,* 77 (1973).
27. Chandra, P., Ebener, U., and Gotz, A. *FEBS Lett. 53,* 10 (1975).
28. Chandra, P., Bardos, T. J., Ebener, U., Kornhuber, B., Gericke, D., and Gotz, A. in *Chemotherapy 11.* (K. Hellmann and T. A. Conners, ed.), Vol. 8, Plenum Press, New York, 1976, P. 191.
29. Bardos, T. J., Ho, Y. K., Aradi, J., and Zsindely, A. Xlth International Cancer Congress, Florence (1974) Abstracts, Vol. 3, p. 361.
30. Zsindely, A. and Bardos, T. J., unpublished data.
31. O'Malley, J., Ho, Y. K., Chakrabarti, P., DiBerardino, L., Chandra, P., Orinda, D. A. O., Byrd, D. M., Bardos, T. J., and Carter, W. A. *Mol. Pharmacol. 11,* 61 (1975).
32. Paffenholz, V., Le, H. V., Ho, Y. K., and Bardos, T. J. *Cancer Res. 36,* 1445, (1976).
33. Ho., Y. K., Le, H. V., Paffenholz, V., and Bardos, T. J. *Proc. Am. Assoc. Cancer Res. 17,* 322 (1976).
34. Chandra, P., Kornhuber, B., Gericke, D., Gotz, A., and Ebener, U. *Z. Krebsforsch. 83,* 239 (1975).

35. Chandra, P., Ebener, Y., Bardos, T. J. Gericke, D., Kornhuber,
    B., and Gotz, A. National Cancer Institute Monograph Ser-
    ies, 1976.
36. Ho, Y. K., Kung. M. P., and Bardos, T. J. *Fed. Proc. 35*, 1847
    (1976).

SIMPLE MODELS OF NUCLEIC ACID INTERACTIONS

J. ZEMLICKA, M. MURATA, AND J. OWENS

*Michigan Cancer Foundation and Department of Oncology,
Wayne State University School of Medicine,
Detroit, Michigan*

Progress in the chemistry of N^6-N^6 bridged adenine nucleosides is reviewed. The synthesis of 1,2-di(adenosin-N^6-yl)ethane (IIIb) and 1,4-di(adenosin-N^6-yl)butane (IIIc) was accomplished by reaction of 6-chloro-(9-β-D-ribofuranosyl)purine (I) with 1,2-diaminoethane (IIa) or 1,4-diaminobutane (IIb). N^6-(ω-alkyl) adenosines IVa and IVb were obtained as by-products. Compounds IIIb, IIIc, IVa, and IVb showed no *in vitro* activity in murine leukemia L 1210. *In vivo* tests with IIIb and IVa were also negative. However, derivative IVa inhibited the growth of human fibroblast cell culture at 50 µg/ml from 99.9%. Hypochromism and CD spectra of IIIb and IIIc were also studied. Thus, as expected, the hypochromism of IIIb at pH 7 was smaller than that of IIIc. However, CD spectra showed an opposite trend. In acid (pH 2) a considerable hypochromism was observed in IIIb, whereas that of IIIc virtually disappeared. Again, this trend was not reflected in CD spectra. The stability of nucleoside C_1'-N linkage of IIIb in acid is limited as compared with adenosine. This fact was also reflected in the reaction of IIIb with ethyl orthoformate catalyzed with HCl, which afforded 1,2-di(adenin-N^6-yl)ethane (VI). Conceivably, a proton transfer between N-3 and O-1'' in a stacked structure contributes to this effect. On the other hand, treatment of IIIb with ethyl orthoformate under catalysis with *p*-toluenesulfonic acid afforded the expected *bis*-ethoxymethylene derivative VIII. The latter was condensed with N^4,2',5'-O-triacetylcytidine 3'-phosphate (VII) to give, after deblocking and separation, oligonucleotides IXa and IXb. Both were degraded by pancreatic ribonuclease but were resistant toward snake venom

phosphodiesterase. A CPK model shows that in a stacked form both
3'-terminal vicinal glycol functions are very close, thereby preventing the approach of the enzyme. A possibility of differential
and selective functionalization of N^6-N^6 bridged ribonucleosides
was demonstrated by synthesis of compounds XVIIIa and XVIIIb, potential transition state analogs of protein synthesis. Thus,
coupling of intermediates XII and XIII or XIV and XV gave the
bridged nucleoside orthoesters XVIa and XVIb, which were hydrolyzed in acid to the corresponding acyl derivatives XVIIa and
XVIIb. The hydrogenolysis of the latter afforded the final products XVIIIa and XVIIIb. Reaction of N^6-(2-aminoethyl)adenosine
(IVa) with methyl orthoacetate in DMF catalyzed with HCl gave
the imidazoline derivative X. The latter afforded the parent
nucleoside XI on hydrolysis.

Interactions between the strands of nucleic acids are essential
for the biological role of both DNA and RNA. In principle, two
kinds of interactions are found in the nucleic acid molecules:
(a) hydrogen bonding between complementary bases (Watson-Crick
or "wobble" pairing) and (b) base-base interaction (stacking)
in DNA and RNA strands, where heterocyclic residues are held in
parallel planes in a sandwichlike arrangement. The intramolecular (intrastrand) base stacking in nucleic acids is extensively
documented by a number of studies. On the other hand, an inter-
molecular (interstrand) base stacking between different molecules
of nucleic acid or their portions has been investigated only to
a limited extent to date, despite the fact that such a situation
may occur in a number of biologically important processes.
    Let us turn our attention to one such possibility. It has
been established (1) that the mechanism of protein synthesis catalyzed by ribosomes involves the reaction of two distinct kinds
of transfer ribonucleic acids (tRNAs), one carrying a peptide
residue (donor or peptidyl tRNA) and another carrying an aminoacyl moiety (acceptor or aminoacyl tRNA). The transformation
can be described as a transfer of a peptidyl residue from peptidyl tRNA to an amino acid moiety of aminoacyl tRNA (Scheme 1).
In a more simplified fashion, the ribosome-catalyzed synthesis
of a peptide bond can be visualized as the reaction between the
two adenosine 3'-terminal units of peptidyl and aminoacyl tRNA.
The phenomenon is illustrated in formula A of Scheme 1 for a
hypothetical transition state of the peptide bond formation.
The peptide residue is represented by an acyl (acetyl) group,
whereas the glycine moiety is representative of the aminoacyl
function. It follows from an examination of space-filling CPK
(Corey-Pauling-Koltun) models that such a transition state is
perfectly feasible for a stacked structure of both adenine residues. Moreover, the CPK model also shows that the formation
of a "bridge" by inserting two or more methylene units between
exocyclic nitrogen atoms (amino groups) of both adenines is also

possible (Fig. 1). In addition, such a bridge could bring both adenosine portions closer and thereby facilitate the transfer of an acetyl group to the glycyl residue.

It was, therefore, of interest to study (a) the synthesis and base stacking of the corresponding "bridged" nucleoside models, (b) chemical and enzymic transformations which may be influenced by base stacking, and (c) to elaborate synthetic procedures for selective functionalization of such bridged nucleosides with an aim to prepare potentially useful models of biological situations, as described above.

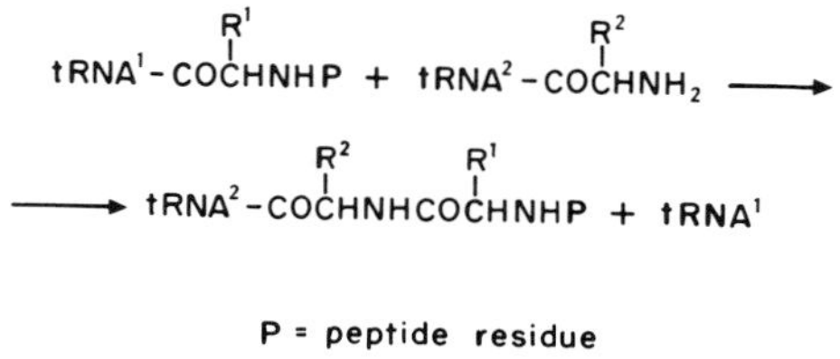

$$\text{tRNA}^1\text{-CO}\overset{\overset{\textstyle R^1}{|}}{\text{C}}\text{HNHP} + \text{tRNA}^2\text{-CO}\overset{\overset{\textstyle R^2}{|}}{\text{C}}\text{HNH}_2 \longrightarrow$$

$$\longrightarrow \text{tRNA}^2\text{-CO}\overset{\overset{\textstyle R^2}{|}}{\text{C}}\text{HNHCO}\overset{\overset{\textstyle R^1}{|}}{\text{C}}\text{HNHP} + \text{tRNA}^1$$

P = peptide residue

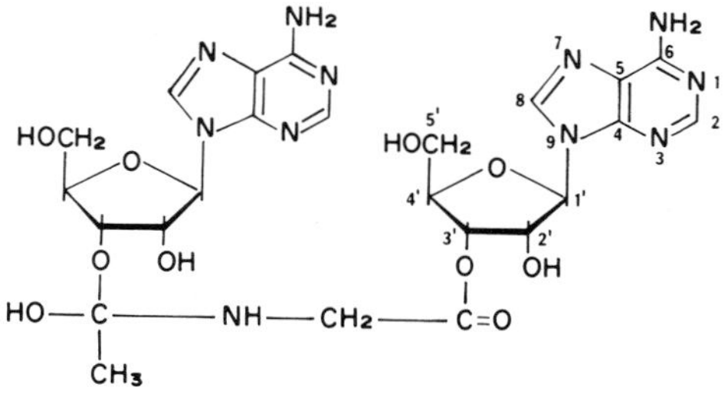

Scheme 1

One of the simplest models of bridged adenosine nucleosides--di∿(adenosin-N^6-yl)methane (IIIa)--is easily accessible by the reaction of adenosine with formaldehyde in buffered aqueous solution (2). Unfortunately, this compound has limited stability and solubility and therefore did not appear to be promising for our study. The homolog of IIIa, 1,2-di(adenosin-N^6-yl)ethane (IIIb) which was also reported some time ago (3) seemed more advantageous. The synthesis of IIIb was achieved by an improved method which essentially followed the general procedure for preparation of various 6-amino-substituted purine ribonucleosides (4). Thus, the readily available 6-chloropurine ribonucleoside I was treated with 1,2-diaminoethane (IIa) in the presence of triethylamine in DMF (dimethylformamide) at room

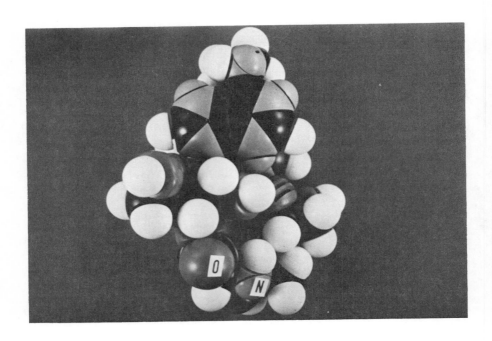

FIGURE 1. CPK model of a hypothetical transition state of the acetyl transfer from 3'-O-acetyladenosine to 3'-O-glycyladenosine as described by formula A. Both purine residues are stacked and in anti *conformation. Letters O and N denote carbonyl group (oxygen) and nitrogen atom of the original glycine moiety. Note the possibility of "bridge" formation (two methylene units) between N^6-N^6 atoms of adenine moiety. The bridge is not indicated in formula A.*

temperature to give products IIIb and IVa, easily separable by chro matography on Dowex 50 cation exchange resin in 26 and 21% yield, respectively (Scheme 2) (5). Compound IVa, which can be obtained in 80% yield from chloro nucleoside I, and excess 1,2-diaminoethan (IIa) by the above procedure (6) can also be used as a starting material for the preparation of the bridged nucleoside IIIb. In fact, the yield of IIIb is higher (40%) than in the case of the direct reaction of I with IIa. A similar procedure was employed for the synthesis of the tetramethylene bridged nucleoside IIIc with the replacement of 1,2-diaminoethane (IIa) by 1,4-diamino-butane (IIb). The yield of product IIIc was 36%.

Compounds IIIb, IIIc, IVa, and IVb were examined for growth inhibition in a murine leukemia L 1210 *in vitro* system and in a fibroblast culture. No activity was observed in the L 1210 system, whereas in human fibroblast culture a 48-hr incubation with IVa killed 99.9% of cells at a concentration of 50 µg/ml. The activity of the other three derivatives was much lower. A similar effect was observed in mouse fibroblast cells (6). On the other hand, murine leukemia L 1210 *in vivo* studies with IIIb and IVa were negative.

Ultraviolet and circular dichroism (CD) spectral studies provided a considerable body of information on base-base interactions (stacking) in bridged nucleosides IIIb and IIIc (5).

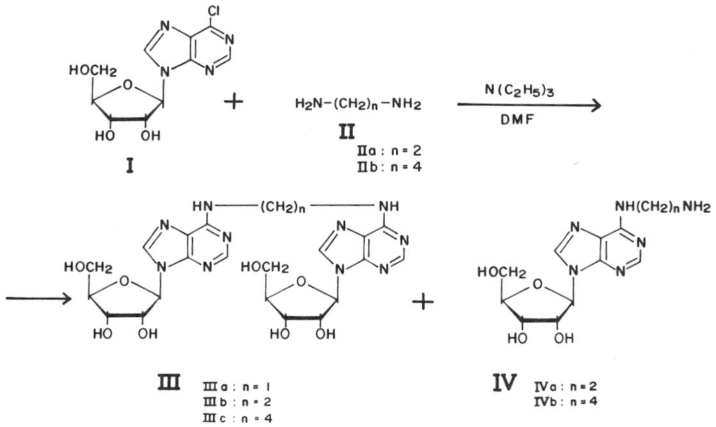

Scheme 2

The data on hypochromism *(H)*, hypochromicity *(h)*, and CD maxima are summarized in Table I. Complete spectra, including difference spectra, are available elsewhere (5). As expected, the hypochromism and hypochromicity of the bridged nucleoside with two methylene units (compound IIIb) at pH 7 are higher than those of the corresponding tetramethylene derivative IIIc. The data agree well with results obtained with 1,3-di(adenin-N^6-yl)propane (7), whose adenine residues are separated by three methylene units. The hypochromism of the latter is 15.5, which is approximately between the values for IIIb and IIIc. It is also of interest to note that hypochromism of IIIc is very close to that of ApA (adenylyl-3'→5'-adenosine) *(H=6.8)* (8), which indicates that the time-averaged separation of adenines in ApA and compound IIIc is similar.

The presence of optically active D-ribose moieties in IIIb and IIIc permitted the utilization of the CD spectral measurements for comparison with corresponding uv spectral data. At first glance it may seem surprising that the molar ellipticity of the tetra-

TABLE I, Uv and CD Spectral Characteristics, Oscillator Strengths (f), Hypochromism (H), and Hypochromicity (h) of the Bridged Nucleosides IIIb, IIIc and the Corresponding Reference Compounds Va and Vb[a]

Compound	uv maximum (nm)	ε_{max}	CD maximum (nm)	θ_{max}	f x 10^7	H	h	pH
IIIb	265[b]	24,800	278	-20,400	-0.62487504	19.2	30.6	7
IIIb	263[b]	26,300	275	-10,500	-0.68292171	14.8	29.3	2
IIIc	264	30,800	277	-26,100	-0.70341004	8.2	12.3	7
IIIc	264	38,100	270	- 9,700	--	None	None	2
Va	267	17,800	270	- 3,600	-0.38669886	--[c]	--[c]	7
Va	263	18,600	260	- 3,100	-0.40073898	--[c]	--[c]	2
Vb	268	17,600	270	- 3,900	-0.38292503	--[c]	--[c]	7
Vb	264	17,900	260	- 3,600	-0.38211177	--[c]	--[c]	2

[a] Summarized data from Ref. 5. For calculation of f, H, and h the following expressions were used: $f = 4.32 \times 10^{-9} \int \varepsilon_\lambda/\lambda^2 \, d\lambda$, $H \left[1- \frac{f^A}{2f^B}\right] \times 100$, where f^A is the oscillator strength of IIIb or IIIc and f^B the oscillator strengths of reference compounds Va and Vb, $h = (1-\frac{\varepsilon_{max}}{2\varepsilon_{max}}) \times 100$, where ε_{max} are the appropriate extinction coefficients at the uv maximum.

[b] Shoulder at ca. 277-278 nm.

[c] Compound Va served as a reference for IIIb, compound Vb for IIIc.

methylene derivative IIIc at pH 7 is appreciably higher than that
of IIIb, although the hypochromism and hypochromicity in both
compounds indicate an opposite trend (Table I). However, it
must be kept in mind that CD spectra may reflect more subtle
(complex) factors than simple time-averaged separation of bases,
e.·g., conformation of the base (anti-syn relationship) and of
the ribofuranose (puckering), mode of overlap of bases, their
orientation in the stack, etc.

A comparison of uv and CD spectra in acid (pH 2, 0.01 N HCl)
is also of interest. Generally, as observed, e.g., in the series
of oligonucleotides (8), hypochromism is suppressed on transition
from neutral to acid solution. However, considerable hypochromism and hypochromicity were observed for compound IIIb at pH 2,
whereas that of IIIc virtually disappeared (Table I). It is important to recognize that the addition of one proton to either
IIIb or IIIc need not necessarily lead to destacking. The addition of a second proton would probably destack the bases due to
electrostatic repulsive forces between both adenine rings (Scheme
3). For the same reasons it is also likely that the addition of
a second proton would be more difficult in compound IIIb, where
the bases are close together, relative to IIIc, whose time-averaged distance between the bases is appreciably greater. Thus,
this situation would be different from that found in some oligonucleotides (8) where the addition of second proton would be facilitated by the presence of a negatively charged phosphodiester
function.

$$V$$

V a : R = C_2H_5
V b : R = $CH_3(CH_2)_3$

The trend shown in the CD spectra of IIIb and IIIc at pH 2
is again somewhat different. Thus, the molar ellipticity is substantially decreased and no great difference is seen between the
two derivatives (Table I). The direct comparison of uv and CD
parameters is probably made difficult by subtle factors alluded
to in the discussion of corresponding spectra at pH 7.

It was also of interest to look for some chemical transformations which may result as a consequence of base stacking in
compounds IIIb and IIIc. One such relatively simple reaction is
the hydrolysis of the nucleoside bond. It has been found that

the *N*-glycosyl (nucleoside) bond in IIIb is rather unstable.
Thus, in 1 *N* HCl compound IIIb is hydrolyzed (9) to the corresponding base 1,2-di(adenin-N^6-yl)ethane (VI) with an approximate
half-life of 40 hr at room temperature. Even more surprising is
the instability of the nucleoside bond in IIIb which was observed
in the attempted reaction with ethyl orthoformate catalyzed by
anhydrous HCl in DMF at room temperature. Again, extensive cleavage of the nucleoside bond was observed and the corresponding
base, 1,2-di(adenin-N^6-yl)ethane (VI), was obtained in 30% yield
(Scheme 4). The latter was identical with an authentic sample
(10). Under similar conditions adenosine was perfectly stable,
giving 2',3'-O-ethoxymethyleneadenosine in 90% yield (11). Once
again, spectral studies of bridged nucleosides provided evidence
for the base stacking in IIIb at pH 2. It is therefore conceiv-
able that the hydrolysis of IIIb in acid may be facilitated by
an intramolecular proton transfer from adenine to the ribofuranose (1'') ring oxygen atom, followed by the opening of the ring.

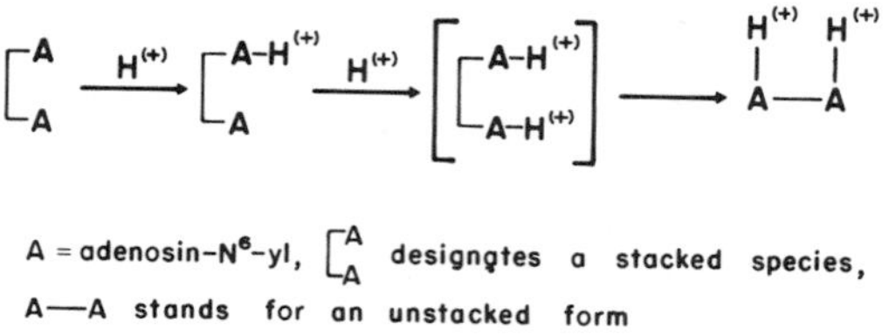

$$\left[\begin{matrix} A \\ A \end{matrix}\right] \xrightarrow{H^{(+)}} \left[\begin{matrix} A\text{-}H^{(+)} \\ A \end{matrix}\right] \xrightarrow{H^{(+)}} \left[\begin{matrix} A\text{-}H^{(+)} \\ A\text{-}H^{(+)} \end{matrix}\right] \longrightarrow \begin{matrix} H^{(+)} & H^{(+)} \\ | & | \\ A & \!\!-\!\! A \end{matrix}$$

A = adenosin-N^6-yl, $\left[\begin{matrix} A \\ A \end{matrix}\right]$ designates a stacked species,

A——A stands for an unstacked form

Scheme 3

CPK models indicate that a most likely species involved in such
a transfer would be a stacked structure with the proton attached
to the N-3 atom of the adenine ring (Fig. 2). It is recognized
that the prevalent site of protonation (12) of adenosine is N-1
and not N-3. However, the situation in the bridged nucleoside
IIIb may not exactly parallel that in adenosine. Moreover, if
an equilibrium among different protonated species of IIIb is considered, even the less abundant species can play an important
role provided that it is involved in a process capable of shift-
ing the original equilibrium. This may be the case in the acid-
catalyzed hydrolysis of the nucleoside bond in IIIb, where irreversible opening of the ribofuranose ring resulting after the
transfer of a proton from N-3 to O-1'' could affect the equilibrium by removing the N-3 protonated species. The interpretation
of the course of the reaction of IIIb with ethyl orthoformate in
DMF catalyzed by HCl can be countered with another objection,
namely that IIIb will be considerably destacked in DMF. However,
there is little data on how the stacking of the protonated form(s)
of the bases will be influenced by a denaturating solvent

(DMF). Moreover, even a relatively small equilibrium concentration of a protonated stacked species could, in principle, lead to an effective cleavage of the nucleoside bond following the reasoning outlined above.

The possible influence of the stacking of bases in IIIb or IIIc in certain enzymic reactions was next considered. This required the preparation of compounds IXa and IXb, the first example of oligonucleotides derived from a bridged nucleoside. The selection of cytidine 3'-phosphate as the nucleotide component followed our particular interest in oligonucleotide sequences related to the 3' terminal CpA (cytidylyl-3'→5'-adenosine) of tRNA. The oligonucleotide IXb would then represent two covalently joined CpA units of a tRNA.

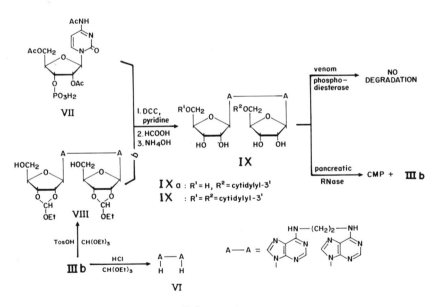

Scheme 4

N^4,2',5'-O-Triacetylcytidine 3'-phosphate (VII) obtained by the known procedure (13) was condensed with the *bis*-ethoxymethylene derivative (VIII) using DCC (dicyclohexylcarbodiimide) in pyridine as a reagent (Scheme 4). Starting protected nucleoside (VIII) was prepared from the bridged derivative IIIb, ethyl orthoformate, and *p*-toluenesulfonic acid as a catalyst in DMF at room temperature in 40% yield. As mentioned above, the reaction catalyzed by anhydrous HCl resulted in the formation of a substantial amount of the corresponding base 1,2-di(adenin-N^6-yl)-ethane (VI). Apparently, the cleavage of the nucleoside bond in

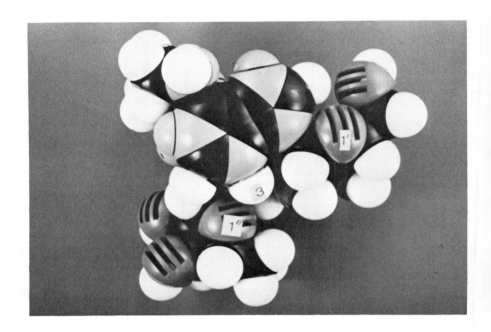

FIGURE 2.   CPK model of bridged nucleoside IIIb protonated
at the N-3 atom (for numbering see formula A).   Both purine residues are stacked and in anti conformation.   Note the closeness
of the protonated N-3 and the opposite ribofuranose ring oxygen
atom (O-1'').

IIIb is a function of the nucleophilicity of the acid anion.
The condensation mixture of VII and VIII was treated with 80%
formic acid (20 min at 0°C to hydrolyze the 2',3'-O-ethoxymethy-
lene groups) followed by a mild alkaline hydrolysis (NH_3 in methanol, 16 hr at room temperature) to remove the acyl groups. The
resultant complex mixture of products was chromatographed first
on a DEAE (diethylaminoethyl) cellulose column, and products IXa
and IXb were further purified by thin-layer chromatography on a
microcrystalline (Avicel) cellulose layer in 2-propanol - NH_4OH -
H_2O (7:1:2) mixture (compound IXa) and 1-butanol - acetic acid -
H_2O (5:2:3) system (compound IXb). Products IXa and IXb were
obtained in yields of 12 and 6%, respectively.
      Both oligonucleotides were virtually quantitatively (99%)
degraded by pancreatic ribonuclease, but were completely resistant to venom phosphodiesterase. The latter enzyme belongs to
the group of exonucleases which degrade the polynucleotide chain
from the 3'-terminal nucleotide to give a 3'-terminal nucleoside

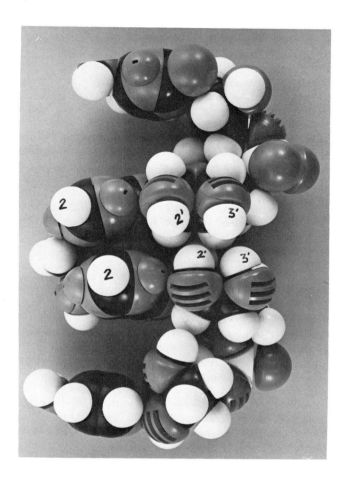

FIGURE 3. *CPK model of oligonucleotide IXb in a stacked form. All base residues are anti and the conformations of $C_{4'}$-$C_{5'}$ bonds are g,g. Note the closeness of both 3'-terminal ribofuranose residues (2',3' vicinal glycol functions).*

5'-phosphate and a neighboring oligonucleotide with a free 3'-hydroxy group (14). A free 3'-hydroxy group at the 3'-terminal nucleotide unit is an important structural requirement. In the present case, it is possible to invoke the base stacking of both adenine residues as an explanation for the observed resistance to degradation. The CPK model of IXb (cf. Fig. 3) shows that in a stacked structure both 2',3' vicinal groupings are very close together presumably preventing the requisite approach of the enzyme. It is obvious that such a situation would not depend on the number (one or two) of cytidine 3'-phosphate units in the molecule. Thus, both IXa and IXb are stable toward the phosphodiesterase. There remains the interesting question whether the extension of the bridge would lead to restoration of the capability of phosphodiesterase to cleave the corresponding oligonucleotide. Thus, the approximate dimension of the active site of the enzyme could be estimated. Work toward this objective is currently in progress.

We have next turned our attention to the selective functionalization of both ribose residues in bridged nucleosides IIIb and IIIc and accordingly undertook the synthesis of 2'(3')-O-acetyl-2''(3'')-O-glycyl derivatives of IIIb and IIIc (compounds XVIIIa and XVIIIb). It is obvious that derivatives XVIIIa and XVIIIb relate to a hypothetical transition state of an intramolecular transfer of acetyl group to the glycine moiety as given in formula A.

The introduction of acyl (aminoacyl) functions into bridged nucleosides IIIb or IIIc is possible, basically, by two different ways. In the first approach a selectively blocked bridged nucleoside is acylated and then deblocked to give the desired acyl (aminoacyl) derivative. An obvious drawback of this method is a necessity of blocking and deblocking of the bridged nucleosides by a multistep procedure. The choice of the proper protecting groups is extremely limited because of an inherent lability of 2'(3')-O-monoacyl functions in nucleosides, particularly in alkaline media. Despite these constraints it was possible to introduce in stepwise fashion 2'(3')-O-L-phenylalanyl and 2'(3')-O-L-leucyl residues into IIIb (15). More recently we achieved the selective substitution of IIIb with different aminoacyl functions (phenylalanyl and leucyl) on 2'- or 3'-hydroxy groups of both vicinal glycol groupings (16). The overall strategy resembles that employed in the chemical synthesis of oligonucleotides (17).

At the present time we would like to discuss an alternate approach, which consists of a substitution of 6-chloropurine riboside I and the corresponding N^6-(ω-aminoalkyl)adenosine, IVa or IVb, with a potential alkali-stable acyl function (orthoester) (18). Thus, (I) was reacted with methyl orthoacetate in the presence of anhydrous HCl in DMF to give 2',3'-cyclic orthoacetate X as a sirup in 80% yield. The reaction of N^6-(2-aminoethyl)adenosine (IVa) with methyl orthoacetate under the same conditions was more complex and afforded compound X in 45% yield. Product X obviously arose by an attack of the carbonium ion $CH_3(CH_3O)_2C^+$, derived from methyl orthoacetate, on the 2-aminoethyl function of IVa followed by cyclization. The structure of X followed from its NMR spectrum which showed three different methyl groups. Thus, the "imidazoline" methyl group appeared as a singlet, whereas two singlets (intensity ratio 2:3) were observed for each methyl group of the dioxolane grouping. It is therefore likely that compound X is a mixture of endo- and exodiastereoisomers in the ratio 2:3. This is in agreement with similar findings in other ribonucleoside 2',3'-cyclic orthoesters (19-21). Hydrolysis

of X in 80% acetic acid for 80 min at room temperature, followed
by treatment with ammonia in methanol (saturated at 0°C) for
approximately 2 hr at room temperature afforded the parent nucleoside XI in 80% yield (Scheme 5).

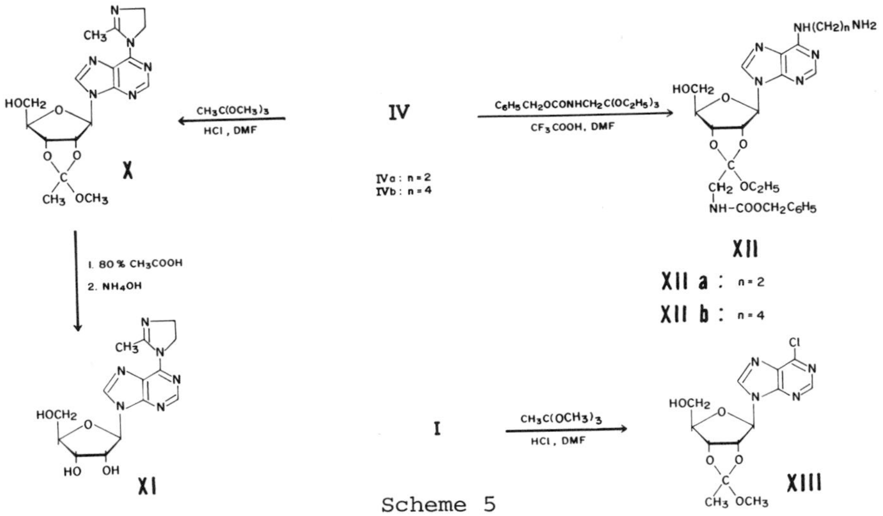

Scheme 5

By contrast, the acid-catalyzed orthoester exchange of
chloronucleoside I and compound IVa with ethyl *N*-benzyloxycarbony-
lorthoglycinate (22) in DMF proceeded as expected and afforded
the cyclic orthoesters XIIa and XIV in almost 40 and 85% yields,
respectively. Similarly, compound IVb afforded orthoester XIIb
in 40% yield (Scheme 5 and 6). An alternate approach was used
for the synthesis of cyclic orthoester XV. Thus, chloronucleoside orthoester XIII, on treatment with excess 1,4-diaminobutane
(IIb) with triethylamine in DMF for 16 hr at room temperature,
gave XV in 70% yield (Scheme 7). Thus, all four possible components for the synthesis of XVIb were obtained (Scheme 8, series b).

It was, therefore, of interest to compare the efficiency
of coupling with the pair XIIb and XIII *vis a vis XIV and XV*.
In both cases the usual conditions, triethylamine in DMF for 18
hr at room temperature, were used. The coupling of XIIb and
XIII afforded the bridged derivative XVIb in 40% yield, whereas
the interaction of XIV and XV gave XVIb in 60% yield. Hydrolysis
of XVIb in 80% formic acid at -15°C for 1 hr gave the diacyl derivative XVIIb in 80% yield. Hydrogenolysis of the latter over
Pd/BaSO$_4$ in 80% acetic acid for 4 hr gave the final product
XVIIIb in 65% yield (Scheme 8, series b).

Similarly, the coupling of XIIa and XIII, effected under
the conditions specified above, afforded the bridged nucleoside
XVIa in 16% yield. Hydrolysis of XVIa using 80% formic acid

(20 min at -15°C) gave the diacyl derivative XVIIa which, on
hydrogenolysis under the conditions described above, furnished
compound XVIIIa (Scheme 8, series a).

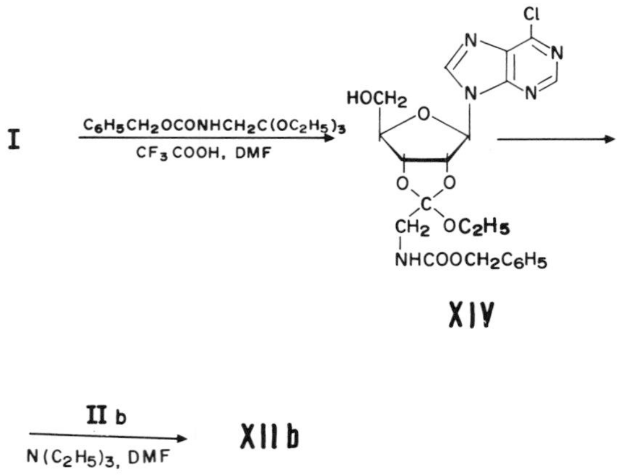

I $\xrightarrow[\text{CF}_3\text{COOH, DMF}]{\text{C}_6\text{H}_5\text{CH}_2\text{OCONHCH}_2\text{C}(\text{OC}_2\text{H}_5)_3}$

XIV

$\xrightarrow[\text{N}(\text{C}_2\text{H}_5)_3, \text{DMF}]{\textbf{II b}}$ **XII b**

Scheme 6

XIII $\xrightarrow[\text{N}(\text{C}_2\text{H}_5)_3, \text{ DMF}]{\textbf{II b}}$

XV

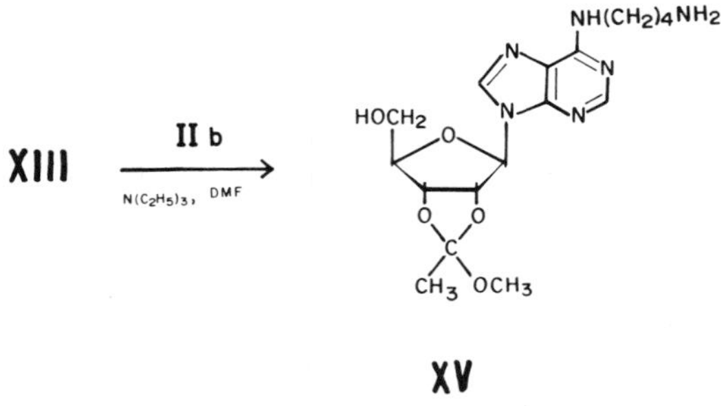

Scheme 7

It is obvious that the same approach, in principle, could
be used for the synthesis of other bridged nucleosides carrying
different acyl functions on both ribofuranose residues. Although
orthoesters derived from various carboxylic acids including D,
L-aminoacids (23,24) are available, the preparation of optically
pure (L-) amino acid orthoester derivatives, which would be of
particular interest, remains to be described.

ACKNOWLEDGMENTS

This study was supported in part by U.S. Public Health Research Grant No. GM-21093 from the National Institute of General Medical Sciences and in part by an institutional grant to the Michigan Cancer Foundation from the United Foundation of Greater Detroit. The authors are indebted to Dr. H. L. Chung and Mr. D. Marks for NMR spectroscopic measurements. Murine leukemia L 1210 *in vitro* tests were performed by Dr. D. Kessel; the corresponding *in vivo* examinations were done in the National Cancer Institute, Bethesda, Maryland.

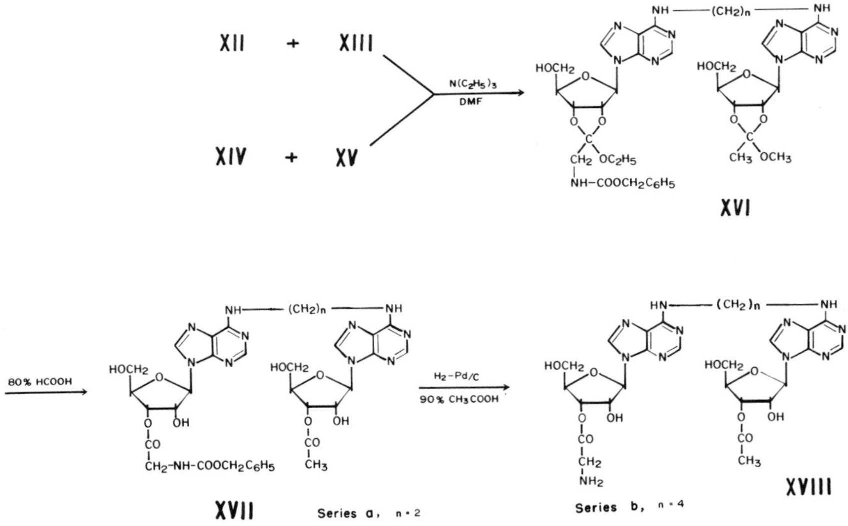

Scheme 8

For the sake of simplicity only 3' isomers are given here. Compounds XVII and XVIII can be mixtures of four positional isomers. The isomeric content of XVII and XVIII has not been established.

REFERENCES

1. Lengyel, P. and Söll, D., *Bacteriol. Rev. 33*, 264 (1969).
2. Feldman, M. Ya., *Biokhimiya 27*, 378 (1962).
3. Lettré, H. and Ballweg, H., *Justus Liebigs Ann. Chem. 656* 158 (1962).

4. Zemlicka, J. and Sorm, F., *Collect. Czech. Chem. Commun. 30,*
      1880 (1965).
5. Zemlicka, J. and Owens, J., *J. Org. Chem.42,* 517 (1977).
6. McCormick, J., Maher, V. M., and Zemlicka, J., *Biochem. Phar-*
      macol.   (1977).
7. Leonard, N. J. and Ito, K., *J. Am. Chem. Soc. 95,* 4010 (1973)
8. Warshaw, M.M. and Tinoco, I., Jr., *J. Mol. Biol. 20,* 29 (1966)
9. Zemlicka, J. and Owens, J., unpublished results.
10. Lister, J. H., *J. Chem. Soc.* 3394 (1960).
11. Zemlicka, J., in *Synthetic Procedures in Nucleic Acid Chem-*
      istry, Vol. 1 (W. W. Zorbach and R. S. Tipson, eds.),
      John Wiley & Sons, New York, 1968, p. 202.
12 Kochetkov, N. K. and Budovskii, E. I. (Eds.), *Organic Chem-*
      istry of Nucleic Acids. Part A, Plenum Press, New York,
      1971, p. 149.
13. Lohrmann, R. and Khorana, H. G., *J. Am. Chem. Soc. 86,* 4188
      (1964).
14. Davidson, J. N., in *The Biochemistry of Nucleic Acids,* Aca-
      demic Press, New York, 1972, p. 191.
15. Li, C. and Zemlicka, J., *J. Org. Chem.42,* 706 (1977).
16. Li, C. and Zemlicka, J., unpublished results.
17. Kössel, H. and Seliger, H., in *Progress in the Chemistry of*
      Organic Natural Products,(W. Herz, H. Grisebach, and G. W.
      Kirby, eds.), Springer, New York, 1975, p. 297.
18. Murata, M., Owens, J., and Zemlicka, J., unpublished results.
19. Fromageot, H.P.M., Griffin, B. E., Reese, C. B., and Sulston,
      J. E., *Tetrahedron Lett. 23,* 2315 (1967).
20. Zemlicka, J., *J. Am. Chem. Soc. 97,* 5896 (1975).
21. Zemlicka, J., Chladek, S., Ringer, D., and Quiggle, K., *Bio-*
      chemistry 14, 5239 (1975).
22. Zemlicka, J. and Chladek, S., *Collect. Czech. Chem. Commun.*
      31, 3775 (1966).
23. Graham, W. H., *Tetrahedron Lett. 25,* 2223 (1969).
24. Zemlicka, J. and Murata, M., *J. Org. Chem. 41,* 3317 (1976).

SYNTHESIS OF NUCLEOTIDES WHICH INACTIVATE
ENZYMES UTILIZING ADENINE NUCLEOTIDES

ALEXANDER HAMPTON

*The Institute for Cancer Research, The Fox
Chase Cancer Center, Philadelphia, Pennsylvania*

ABSTRACT

Replacement of the nucleotide 5'-methylene group by a carbonyl
group gave the carboxylic-phosphoric anhydride isoteres of AMP
and ATP, which behaved as adenine nucleotide site-directed inactivators of adenylosuccinate lyase, AMP amino-hydrolase, and
pyruvate kinase. The N^6-*o*- and *p*-fluorobenzoyl derivatives of
ATP were prepared because of their potential to react "inside"
ATP sites (via their amide carbonyl) or "outside" (with displacement of fluorine) and were found to act as ATP-site-directed
reagents for rabbit, pig, and carp muscle AMP kinases. 2',3'-
O-isopropylidene adenosine was converted to 5'-*C*-acylaminomethyl
and 5'-*C*-aroylaminomethyl derivatives of AMP. The enzyme affinities of these compounds disclosed a structural difference between pig and rabbit AMP kinases near their AMP sites. Some N^6-
$(CH_2)_n NHCOOH_2 I$ derivatives of ATP were studied with rabbit, pig,
and carp AMP kinases; when *n* was 6, selective ATP-site-directed
inactivation of the rabbit enzyme occurred.

Adenine nucleotides, in particular adenosine 5'-triphosphate (ATP),
are substrates or sometimes effectors of a relatively large proportion of enzymes. Recent work in this laboratory has been directed toward the synthesis of adenine nucleotide analogs and derivatives designed to function as adenine nucleotide site-directed
reagents for enzymic amino acid residues situated either at the
adenine nucleotide binding sites or at various distances from
those sites. Nucleotide derivatives that react at or near enzymic
adenine nucleotide sites are potentially useful as probes of the
structure and mechanism of catalysis of the enzymes, whereas

85

those nucleotides designed to act as exosite reagents have, as
pointed out by B. R. Baker in 1959 (1), the additional potential
of acting as species- or tissue-selective inactivators of substrate-identical enzymes because of structural differences in
areas outside the active site of such enzymes. An account is
given here of (a) the synthesis of two classes of adenine nucleotide derivatives that behave as reagents for the adenine nucleotide sites of several enzymes, (b) the synthesis of a series of
5'-*C*-substituted AMP derivatives the enzyme affinities of which
point to a structural difference between two enzymes from similar
species exo to their AMP sites, and (c) the synthesis of an ATP
derivative that behaves as a species-specific ATP-site-directed
exosite enzyme reagent.

The carboxylic-phosphoric mixed anhydride isoteres of AMP
(1) and ATP (II) (Scheme 1) were prepared as potential adenine
nucleotide site reagents in view of their structural similarity
to the normal substrates and because of their potential to either
phosphorylate or acylate nucleophilic amino acid residues (2,3).

In addition, it seemed likely that adsorption of I or II to the
substrate sites might frequently be accompanied by partial neutralization of the phosphoryl charge due to the presence of basic
amino acids in the phosphate binding site, and that this would
enhance the reactivity of the anhydride in the enzyme-bound analog, thereby tending to increase the selectivity of labeling of
the substrate site. The synthesis of I was based on Michelson's
anion-exchange method for the conversion of 5'-nucleotides to
phosphoanhydride derivatives (4,5). Treatment of the sodium salt
of 9-(β-D-ribofuranosyluronic acid)adenine in DMF with one molecular equivalent of diphenylphosphorochloridate with careful ex-
clusion of moisture produced the diphenyl ester of I (Scheme 2)
and this, upon treatment in pyridine solution with two molecular
equivalents of tri-*n*-butylammonium phosphate, yielded I in 90% over
all yield. The compound was too moisture sensitive to character-

ize by the usual methods, and its identity and homogeneity were established by treatment with dry ammonia in DMF when the only uv-absorbing material produced was the authentic carboxamide of the starting material. The anhydride (II) was similarly prepared by reaction of the diphenyl ester of I with tri-*n*-butylammonium tripolyphosphate. Compounds I and II were more stable as 1% solutions in dry DMF at -25°C than as solids. After two days the DMF solution of I upon dilution into water showed uv spectral changes indicative of N^6 acylation, and concomitantly the ability (described below) of the solution of I to inactivate AMP aminohydrolase decreased sharply. At pH 7.7, I hydrolyzes several

NaOC–O–Ad + (PhO)$_2$POCl $\xrightarrow[5\,\text{hr, 22°}]{\text{DMF}}$ (PhO)$_2$POC–O–Ad (HO OH)

Tri-*n*-butylammonium tripolyphosphate in pyridine, 5hr, 22°

Tri-*n*-butylammonium phosphate in pyridine, 5hr, 22°

(HO)$_2$POPOPOC–O–Ad $\xrightarrow[\text{DMF}]{\text{NH}_3}$ H$_2$NC–O–Ad $\xleftarrow[\text{DMF}]{\text{NH}_3}$ (HO)$_2$POC–O–Ad

orders of magnitude faster than acetyl phosphate, and it has been suggested (2) that the enhanced reactivity may be associated with intramolecular assistance from N-3. Attempts to prepare the analogous carboxylic-phosphoric mixed anhydride analogs of UMP and TMP in similar fashion gave none of the desired compounds, and preliminary studies indicated that the intermediate diphenyl ester did not form or was unstable.

Freshly prepared analog I inactivated two AMP-utilizing enzymes tested. A third AMP-utilizing enzyme (rabbit muscle AMP kinase) was not inactivated by a 1 mM nominal initial concentration of I, but a two-day old sample was employed and the result is inconclusive. Exposure of *E. coli* adenylosuccinate AMP-lyase to 80 µm of I caused a 95-99% loss of enzyme activity (2). The effect was prevented by a 30-sec hydrolysis of I in the assay buffer (pH 7.7) or by prior addition of 120 µm adenylosuccinate (K_m = 20 µM). Rabbit muscle AMP aminohydrolase was ∿50% inactivated by a total of 540 µm I added in three portions (3). Inactivation was abolished by 15-sec hydrolysis of I prior to the addition of enzyme or by the presence of 50 µM AMP (K_m = 400 µM). These data indicate that both enzymes are subject to rapid acylation or phosphorylation at their AMP sites.

The freshly prepared ATP analog II was tested with two ATP-utilizing enzymes (3). AMP kinase of rabbit muscle was unaffected by a 1 mM nominal initial level of II, whereas rabbit muscle pyruvate kinase was 50% inactivated by 100 μM II added in three increments. The enzyme was protected by a 15-sec hydrolysis of II or by the presence of any of the substrates ATP (100 μM), ADP (2.5 mM), or phosphoenolpyruvate (1.5 mM), and the effect therefore appears to be associated with acylation or phosphorylation at the ATP site.

N^6-o-Fluorobenzoyladenosine 5'-triphosphate (Scheme 3) and its p-fluoro analog were synthesized as potential reagents for amino acid residues at or near enzymic ATP sites (6). The fluoro

N^6-o-F-Bz-ATP

benzamide residue was considered of interest because it is relatively stable in physiological buffers and because fluorine attached to aromatic systems is more reactive than the other halogens. Furthermore, the two ATP derivatives have the potential of undergoing an attack by enzymic nucleophiles not only at their ortho and para carbons, respectively, but also at their amide carbons, thus significantly increasing the probability that they could function at ATP-site-directed reagents for any given ATP-utilizing enzyme. The N^6-o- and p-fluorobenzoyl derivatives of AMP were obtained in 80–85% yield by the treatment of pyridinium AMP in pyridine solution with a twentyfold excess of the respective fluorobenzoyl chlorides followed by deblocking of the ribose hydroxyl and phosphoryl groups and partial deblocking of N^6 with pyridine-aqueous sodium hydroxide under carefully controlled conditions. That the aroyl substituent is at N^6 and not at N-1 is indicated from the finding of Anzai and Matsui (7) that both benzoyls of N-N-dibenzoyl-9-methyl adenine are at N^6. The N^6-fluorobenzoyl derivatives of AMP were converted in 50–60% yield to the N^6-o- and p-fluorobenzoyl derivatives of ATP by Michelson's anion-exchange reaction (4, 5).

N^6-Benzoyl-ATP and its o- and p-fluoro derivatives were excellent substrates of hexokinase (Table I) and the reaction pro-

TABLE I Substrate and Inhibitor Properties of Some N^6-Aroyl
Derivatives of AMP, ADP, and ATP

Enzyme	Compound	V_{max} (rel. %)	Inhibition type[a]	K_i[b] (μM)
Hexokinase	N^6-Bz-ATP	56		--
(yeast)	N^6-o-F-Bz-ATP	53		--
	N^6-p-F-Bz-ATP	70		--
	ATP	100		--
Pyruvate	N^6-Bz-ADP	0.5	C	630
kinase	N^6-o-F-Bz-ADP	0.5		--
(rabbit	N^6-p-F-Bz-ADP	0	C	53
muscle)	ADP	100		800
Adenylate	N^6-Aroyl-AMP	0	NC	--
kinase	N^6-Bz-ATP	28	--	990
(rabbit	N^6-o-F-Bz-ATP	58	--	500
muscle)	N^6-p-F-Bz-ATP	0	C	--
	AMP	100		500
	ATP	100		330

[a]C: competitive with respect to the corresponding normal
substrate, NC: noncompetitive.

[b]K_i = enzyme-inhibitor dissociation constant.

vided a convenient method for obtaining the corresponding ADP
derivatives. No inactivation of the hexokinase was detected under these conditions and more searching tests for inactivation
could not be made because the enzyme was unstable in the absence
of glucose. The three N^6-aroyl derivatives of ADP were able to
absorb effectively to the ADP site of pyruvate kinase (reaction
catalyzed: ADP + phosphoenolpyruvate → ATP + pyruvate), as shown
by their substrate and competitive inhibitor properties (Table I).
However, the enzyme was not inactivated at 0°C for 5 hrs by
N^6-p-fluorobenzoyl-ATP (100 μM), N^6-p-fluorobenzoyl-ADP (260 μM),
or N^6-o-fluorobenzoyl-ADP (3 mM) (6). N^6-Benzoyl-ATP and N^6-o-
and p-fluorobenzoyl-ATP possessed affinity for the ATP site of
rabbit muscle AMP kinase (reaction catalyzed: AMP + ATP → 2 ADP)
as evidenced by substrate or competitive inhibitor properties
(Table I); parallel findings were obtained with AMP kinase of pig
or carp muscle (6). All three kinases were inactivated by the
N^6-o- and p-fluorobenzoyl ATP derivatives. The loss in enzyme
activity followed pseudo-first-order kinetics and the rates were

TABLE II   *Inactivation of Rabbit Muscle AMP Kinase by N^6-Aroyl Derivatives of ATP*

Aroyl group	Concentration of ATP analog	Concentration of ATP (mM)	$L_{\frac{1}{2}}$ of kinase at 0^0 (min)
o-F-C_6H_4	2.2 mM	0	200
o-F-C_6H_4	2.2 mM	0.55	1000
p-F-C_6H_4	50 μM	0	100
p-F-C_6H_4	50 μM	2.64	210

slower in the presence of ATP in the case of both inactivators, as illustrated in Table II for the rabbit enzyme. N^6-Benzoyl-ATP did not inactivate this enzyme at a 100-fold higher concentration and 400-fold longer time than required by its p-fluoro derivative to produce measureable inactivation, and the effect thus appears to involve covalent bond formation between the more strongly electrophilic fluorobenzoyl ATP derivatives and a nucleophilic group of the enzyme. The identity of the leaving group (whether F^-, ADP, or ATP) was not investigated. That the fluorobenzoyl ATP derivatives were acting as ATP-site-directed reagents was further supported by the finding that the corresponding AMP derivatives did not inactivate the AMP kinases. The kinetics of inactivation were studied in the most detail with N^6-o-fluorobenzoyl-ATP, for which the inverses of the apparent first-order rate constants at four different inactivator levels were plotted against the inverse of the inactivator levels (Fig. 1). The inactivation followed saturation kinetics, indicating that the formation of a bond between the inactivator and the enzyme is not a random bimolecular process but is preceded by reversible binding of the inactivator to the enzyme. The pseudo-first-order rate constants varied over only a seven fold range for the pig, rabbit, and carp AMP kinases, indicating that N^6-o-fluorobenzoyl-ATP might react with the same amino acid residue near the ATP site of each enzyme.

In order to design substrate-site-directed adenine nucleotide derivatives that have the ability to form bonds with enzymic amino acids situated some distance from the substrate site, it is necessary first to determine at which nucleotide atoms a substituent can be attached without preventing formation of the enzyme-substrate complex. In the course of carrying out such

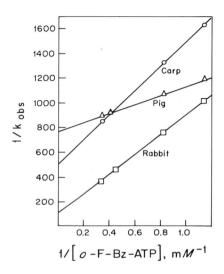

FIGURE 1. *Rate saturation effect in the inactivation of muscle AMP kinases by N^6-o-fluorobenzoyl-ATP; k_{obs} is the apparent first-order rate constant for inactivation by a given concentration of the inactivator.*

studies with AMP-utilizing enzymes, we developed a route for the synthesis of a series of 5'-*C*-acylaminomethyl derivatives of AMP (8). This involved a catalytic reduction of the *allo* epimer of 5'-*C*-nitromethyl-2',3'-*O*-isopropylideneadenosine (compound I, Scheme 4), treatment of the resulting aminomethyl nucleoside II with one equivalent each of phenyl chloroformate and triethylamine to give the phenylurethane III, phosphorylation of the latter with 2-cyanoethyl phosphate-DCC, and removal of the 2-cyanoethyl and phenoxycarbonyl groups with aqueous sodium hydroxide followed by removal of the isopropylidene group with aqueous acid. This gave the allo epimer of 5'-*C*-aminomethyl-AMP (IV) (in homogeneous form in 10% overall yield). The triethylammonium salt of IV was treated in methanolic solution with carboxylic acids that had been converted *in situ* to mixed carbonic anhydrides by the action of *N*-ethoxycarbonyl-2-ethoxy-1,2-dihydroquinoline, to afford the required 5'-*C*-acylaminomethyl AMP derivatives V in 70-90% yield.

The derivatives of structure V were found to be linear competitive inhibitors of pig and rabbit muscle AMP kinases with respect to AMP (9). The enzyme-inhibitor dissociation constants (K_i values) (Table III) are hence taken to be a measure of the affinity of each compound for the AMP sites. The affinity for the rabbit AMP site increased when the lipophilicity was increased by replacing a methyl group by ethyl, phenyl, or tolyl groups; conversely, the affinity decreased when the lipophilicity was

decreased by replacing ethyl by chloromethyl or phenyl by *p*-
fluorophenyl, thus indicating interaction of the 5'-*C*-substituents
with a lipophilic region of the rabbit enzyme. The data for the
pig enzyme permit the same general conclusion; however, the 5'-
benzamidomethyl-AMP binds to the pig enzyme only one-third as
strongly as the 5'-acetamidomethyl-AMP, whereas it binds to the
rabbit enzyme approximately twice as well as the 5'-acetamido-
methyl-AMP. Further evidence that the enzymes are structurally

dissimilar at these phenyl-interacting regions is that attachment of fluorine to the phenyl group enhances affinity for the
pig enzyme but decreases affinity for the rabbit enzyme. These
findings illustrate that structural differences between substrate-
identical enzymes from closely related species can be detected
by varying the lipophilic-hydrophilic balance and steric properties of substituents attached to the substrate. Such information
could be used as a basis for the design of species-selective enzyme inactivators in accord with the concepts elaborated by
Baker (1).

We recently synthesized, as potential ATP-site-directed
exosite enzyme reagents, a series of ATP derivatives bearing
the iodoacetylamino-*n*-alkyl substituents $(CH_2)_nNHCOCH_2I$ on N^6
(10). This route involved the reaction of 6-chloropurine ribonucleoside 5'-phosphate with an excess of an aliphatic diamine,
iodoacetylation of the resulting N^6-aminoalkladenosine 5'-phos-
phate with *N*-iodoacetoxysuccinimide, and subsequent conversion
to the required ATP derivatives by the Hoard-Ott procedure (11).
The compounds were tested as inactivators of AMP kinases of rabbit, pig, and carp muscle. When *n* was 5 no enzyme was inactivated. (Table IV), when *n* was 6 or 7 a selective inhibition of

TABLE III *Reversible Inhibition of Muscle AMP Kinases*

$$CH_2NHCOR$$
$$H_2O_3PO-C-H$$

(structure: ribose ring with O, Ad, HO, OH)

R	K_i value (μM)	
	Rabbit	Pig
CH_3	1900	1380
C_2H_5	930	560
CH_2Cl	5300	4100
C_6H_5	1120	3950
$C_6H_4-o-CH_3$	620	1150
$C_6H_4-m-CH_3$	850	1560
$C_6H_4-p-CH_3$	560	1020
C_6H_4-p-F	1660	3100

the rabbit enzyme was observed, and when n was 8 all three enzymes were weakly inactivated. When n was 6 the selective effect
was very pronounced (Table IV) and was therefore characterized
in detail. No inactivation occurred when iodine was replaced
by hydrogen (Table IV), indicating that alkylation of the enzyme is probably required. This selective inhibition was also
observed in mixtures of the rabbit and pig enzymes, and no evidence could be found from such experiments for activation of the
hexamethylene derivative by the rabbit enzyme or for its deactivation by the pig enzyme. The inactivation of the rabbit enzyme
was concluded to be ATP-site-directed because (a) the inhibitor
was itself a substrate, (b) ATP retarded the inactivation, (c)
iodoacetamide did not inactivate at similar levels (Table IV),
(d) removal of the β and γ phosphate residues abolished ATP-pro-
tected inactivation, and (e) the activation exhibited saturation
kinetics as judged by the type of study illustrated in Fig. 1,
thus indicating that the inhibitor binds reversibly to the en-

TABLE IV Inactivation of AMP Kinases by N^6-$(CH_2)_nNHCOCH_2R$ Derivatives of ATP

Enzyme source	% loss of activity[a]					
	n=5,R=I	n=6,R=I	n=6,R=H	n=7,R=I	n=8,R=I	Iodoacetamide
Rabbit muscle	0(1.0)[b]	76(0.79)	0(0.89)	14(0.97)	12(1.0)	0(3.0)
Pig muscle	0(1.35)	0(2.76)	--	0(0.97)	15(1.0)	0(3.0)
Carp muscle	0(1.0)	0(2.76)	--	0(0.97)	11(1.0)	0(3.0)

[a] Measured after 6 hr at 0°C and corrected for % loss of activity in the control.
[b] Figures in parentheses indicate the concentration (mM) of each ATP derivative.

zyme prior to alkylating it. It was concluded that the species-
selective inactivation is due to alkylation of an amino acid
residue of the rabbit enzyme that is absent, inaccessible, or
less reactive in the pig and carp enzymes.

These results afford additional evidence that substrate-
site-directed agents capable of reacting with an amino acid residue outside the substrate site can be designed so as to
inactivate a selected enzyme but not to inactivate a substrate-
identical enzyme from another biological source. Baker and his
co-workers have previously reported instances of such selective
enzyme inactivation involving rabbit muscle and beef heart lac-
tate dehydrogenases (12), mouse liver and mouse L1210 leukemia
cell dihydrofolate reductases (13), and rat liver and rabbit
liver guanine deaminase (14). In addition, Mertes and co-workers
described the selective inactivation of Ehrlich ascites thymidylate synthetase with respect to the calf thymus enzyme (15).

REFERENCES

1. Baker, B. R., *Design of Active-Site-Directed Irreversible
       Enzyme Inhibitors*, Wiley, New York, 1967.
2. Hampton, A. and Harper, P. J., *Arch. Biochem. Biophys. 143*,
       340 (1971)
3. Hampton, A., Harper, P. J., and Sasaki, T., *Biochem. Bio-
       phys. Res. Commun. 65*, 945 (1975).
4. Letters, R. and Michelson, A. M., *Bull. Soc. Chim. Biol. 45*,
       1353 (1963).
5. Michelson, A. M., *Biochim. Biophys. Acta 91*, 1 (1964).
6. Hampton, A. and Slotin, L. A., *Biochemistry 14*, 5438 (1975).
7. Anzai, K. and Matsui, M., *Bull. Chem. Soc. Jpn. 46*, 3228
       (1973).
8. Kappler, F. and Hampton, A., *J. Org. Chem. 40*, 1378 (1975).
9. Hampton, A., Slotin, L. A., Kappler, F., Sasaki, T., and
       Perini, F., *J. Med.* Chem. *19*, 1371 (1976).
10. Hampton, A., Slotin, L. A., and Chawla, R. C., *J. Med. Chem.
       19*, 1279 (1976).
11. Hoard, D. E. and Ott, D. G. *J. Am. Chem. Soc. 87*, 1785 (1965).
12. Baker, B. R. and Patel, R. P., *J. Pharmacol. Sci. 53*, 714
       (1964).
13. Baker, B. R. and Vermeulen, *J. Med. Chem. 13*, 1154 (1970).
14. Baker, B. R. and Wood, W. F., *J. Med. Chem. 12*, 216 (1969).
15. Barfknecht, R. L., Huet-Rose, R. A., Kampf, A., and Mertes,
       M. P., *J. Am. Chem. Soc. 98*, 5041 (1976).

SYNTHESIS, ANTIVIRAL ACTIVITY, AND MECHANISM OF ACTION OF A NOVEL SERIES OF PYRIMIDINE NUCLEOSIDE ANALOGS*

WILLIAM H. PRUSOFF, TAI-SHUN LIN, MING-SHEN CHEN, GEORGE T. SHIAU

Department of Pharmacology

DAVID C. WARD

Departments of Human Genetics and Molecular Biophysics and Biochemistry, Yale University School of Medicine, New Haven, Connecticut

ABSTRACT

The synthesis and biological properties of a series of 5'-amino and 5'-azido pyrimidine nucleosides is described. Several of these compounds, e.g., 5-iodo-5'-amino-2,5'-dideoxyuridine (AIU), 5-iodo-5'-amino-2',5'-dideoxycytidine (AIC), and 5'-amino-5'-deoxythymidine, inhibit herpes simplex virus (HSV) replication in the absence of detectable cytotoxicity. Studies on the mechanism of AIU inhibition have shown that AIU is phosphorylated only in HSV-infected cells. The major acid-soluble metabolite, the 5'-triphosphate of AIU (AIdUTP), is subsequently utilized for the incorporation of AIU into both viral and host cell DNA. AIdUTP is a potent allosteric inhibitor of *E. coli* thymidine kinase and undergoes degradation to free AIU via a novel first-order acid-catalyzed phosphorolysis reaction. While the molecular mechanism of AIU inhibition has yet to be established, it appears that both its antiviral activity and lack of cell toxicity may be a consequence of the fact that AIU phosphorylation is a specific herpes virus induced event.

*This work was supported by Public Health Service Grants CA-05262 and CA-16038 from the National Cancer Institute and the US Energy Research and Development Administration Contract E (11-1)-2468.

97

The search for effective nontoxic antiviral agents has been vig-
orously pursued during the past few decades. Since many of the
nucleosides with antiviral activity evolved from programs directed toward the development of anticancer drugs, it is not surprising that most of these agents exhibited some cellular toxicity when administered systemically. Antiviral nucleosides that
were incorporated into DNA also posed an additional problem, a
potential to be mutagenic or carcinogenic. Therefore, the design of compounds that would inhibit the biosynthesis of DNA yet
not be incorporated into the DNA polymer appeared to be a desirable objective.

The finding by Langen and his co-workers (1) that 5'-halo-
genated analogs of thymidine were not incorporated into DNA yet
markedly inhibited thymidylate kinase, was of considerable sig-
nificance (Fig. 1). The substitution of a halogen for the hydroxyl group in the 5'-position of the nucleoside presumably precluded phosphorylation to the triphosphate derivative which is
required for biosynthesis of DNA. Such nucleoside inhibitors of
thymidylate kinase would also produce an inhibition of DNA biosynthesis, since the formation of dTTP from dTMP is blocked.
Both pathways leading to the formation of dTTP via dTMP would be
affected, that is, the *de novo* pathway via dUMP as well as the
"salvage" pathway via thymidine (Fig. 2). Of additional interest is the recent report by Galegov *et al.* (2) which indicated
that 5'-fluorothymidine inhibited the replication of vaccinia
virus.

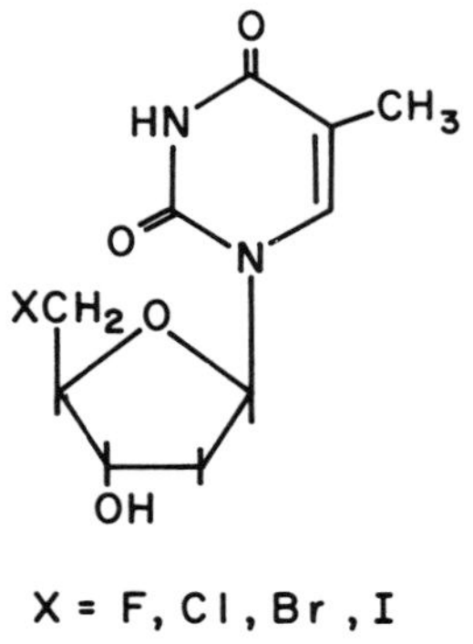

X = F, Cl, Br, I

Figure 1.

$dUMP \diagdown \quad de\ Novo$

$dTMP \rightarrow dTDP \rightarrow dTTP \rightarrow DNA$

$\diagup salvage$

Thymidine

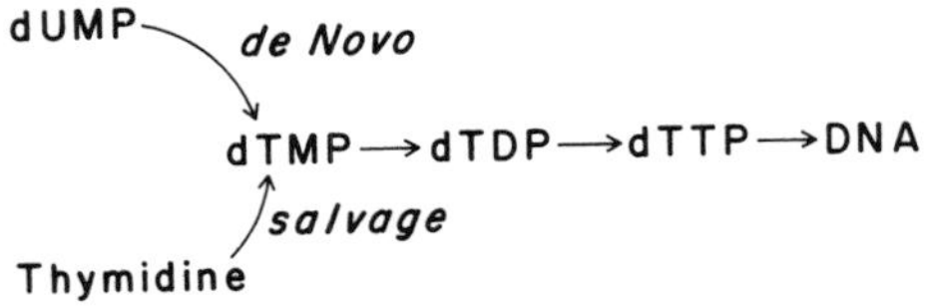

Figure 2.

Our decision to study the biological properties of 5'-amino
sugar nucleosides was stimulated not only by the original synthesis of 5'-amino-5'-deoxythymidine (Fig. 3) by Horwitz and his
co-workers (3), but also by the finding by Neenan and Rhode (4),

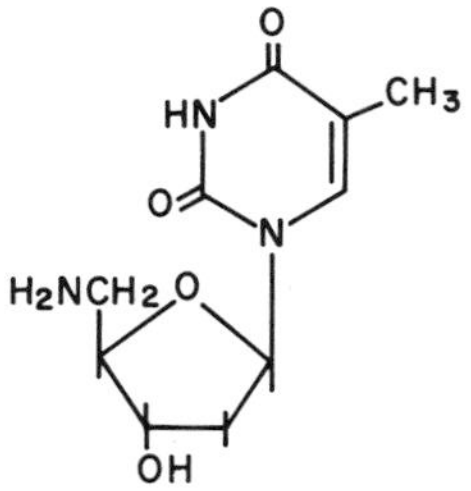

Figure 3.

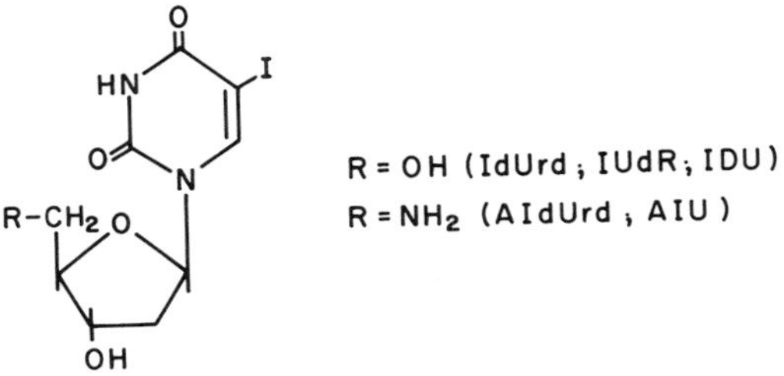

R = OH (IdUrd; IUdR; IDU)
R = NH₂ (AIdUrd ; AIU)

Figure 4.

as well as ourselves (5), that this compound is a potent inhibitor of mammalian thymidine kinase (K_i = 3 µM), an enzyme that
shows enhanced activity in many tumor and virus-infected systems.
The compound also exhibited a modest inhibition of dTMP kinase
(K_i = 130 µM) (4,5). Further encouragement came from our observation that 5'-amino-5'-deoxythymidine produced a significant
inhibition of the replication of herpes simplex virus in culture (6).

Since 5-iodo-2'-deoxyuridine (IdUrd) is a potent antiviral
agent, but poorly soluble and highly toxic when given systemically, our first synthetic effort was the preparation of the 5'-
amino analog of IdUrd (AIU, AIdUrd) (Fig. 4). Our objective was
to retain the antiviral activity of IdUrd while decreasing its
toxicity and solubility problems.

The method of Horwitz and co-workers (3), which had successfully afforded the 5'-amino analog of thymidine from thymidine,
was initially applied to the synthesis of AIU, utilizing IdUrd
as the starting material (Fig. 5). Although the 5'-azido analog
of IdUrd formed in good yield, subsequent catalytic reduction
of the ·azido moiety to the amino group resulted in replacement
of the 5-iodo group of the pyrimidine ring by a hydrogen atom,
thus giving 5'-amino-2',5'-dideoxyuridine as the major product.
Similar results were obtained using either palladium on charcoal
or PtO₂ as catalysts.

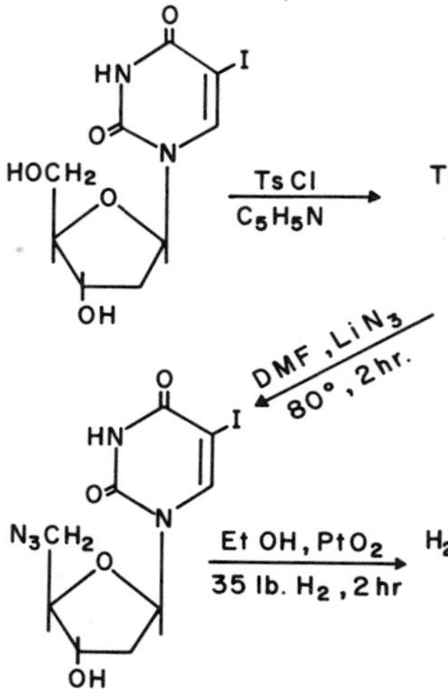

Figure 5.

Preliminary attempts to iodinate 5'-amino-2',5'-dideoxy-
uridine by various procedures were not successful, even though
these methods worked well when applied to deoxyuridine.   However,
the method of halogenation developed by Dale *et al.* (7) was found
to be successful when applied to 5'-amino-2',5'-dideoxyuridine
(Fig. 6).   Although this procedure affords the desired product,
only a limited amount of material can be conveniently processed
at one time.   While this procedure is excellent for the preparation of radioactive (^{125}I)-labeled AIdUrd, it is of limited
utility for large-scale preparations.

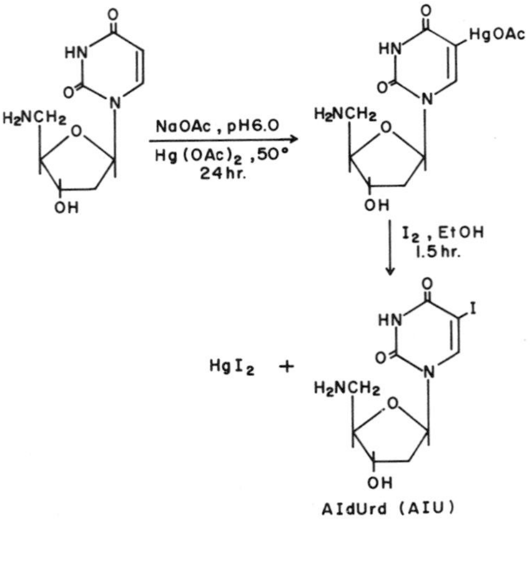

Figure 6.

Our present procedure for the preparation of bulk quantities of AIdUrd utilizes triphenylphosphine for the reduction of the 5'-azido analog of IdUrd to the corresponding 5'-amino analog (AIdUrd). This procedure not only affords large-scale synthesis, but also does not result in replacement of the iodine from the pyrimidine ring. Thus, the preparation of 5-iodo-5'-azido-2', 5'-dideoxyuridine follows the procedure of Horwitz *et al.* (3), while the subsequent conversion of the azido to the amino moiety involves a reaction with triphenylphosphine, as described by Mungall *et al.* (8) for the conversion of 5'-azido-5'-deoxythymidine to 5'-amino-5'-deoxythymidine. This reaction sequence is depicted in Fig. 7.

Recently Hata *et al.* (9) described a one-step procedure for preparing the 5'-azido derivatives of a number of pyrimidine and purine nucleosides (thymidine, uridine, N^4-benzoylcytidine) in a very high yield (Fig. 8). This could be of value since it eliminates the tosylation reaction that is required as an intermediate step in converting the 5'-hydroxyl of a nucleoside to the 5'-azido moiety. However, the requirement for purification by column chromatography imposes a limitation on the amount of product than can be readily obtained.

The 5'-amino analogs of 5-iodo-2'-deoxyuridine (AIU, AIdUrd), 5-bromo-2'-deoxyuridine, 5-chloro-2'-deoxyuridine, 5-fluoro-2'-deoxyuridine, 5-iodo-2'-deoxycytidine (AIC), 5-trifluoromethyl-2'-deoxyuridine, and arabinosylcytosine were synthesized (Fig. 9) and some of their biological activities were determined (10,11). While the 5'-amino analogs were less potent inhibitors on a molar basis, they had a more favorable therapeutic index than the corresponding 5'-hydroxylated parent analog.

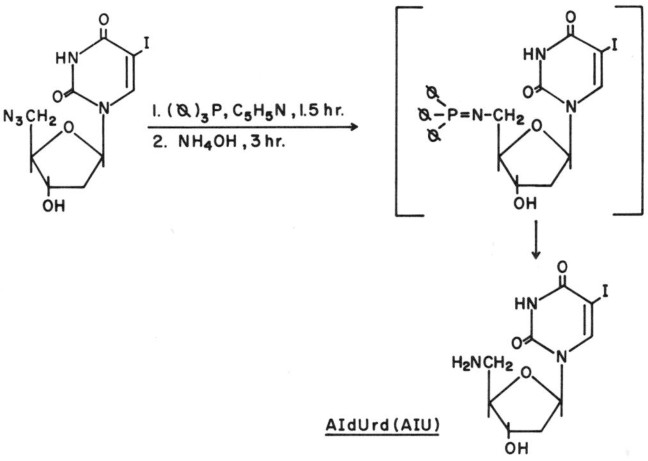

Figure 7.

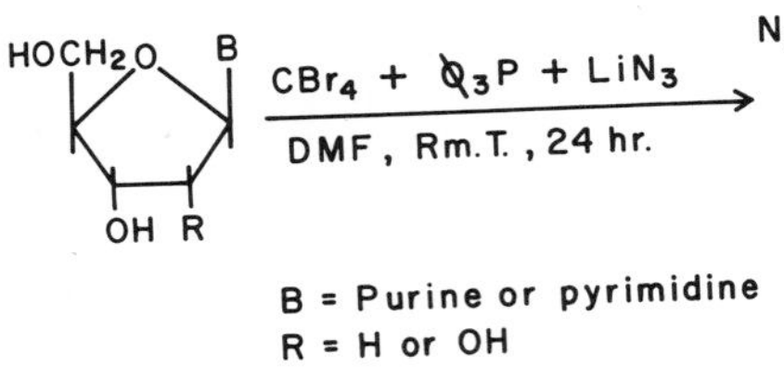
HOCH2 O B
CBr4 + ϕ3P + LiN3
DMF, Rm.T., 24 hr.
N3-CH2 O B
OH R
OH R
B = Purine or pyrimidine
R = H or OH

Figure 8.

A serious problem encountered with some of these 5'-amino nucleoside analogs was the presence of trace amounts of the starting compound, which could not be detected by elemental analysis or NMR. This problem, which became apparent upon chromatography of radioactive ^{125}I-5-iodo-5'-amino-2',5'-dideoxycytidine, was particularly pronounced in nucleosides derived from a parent compound that had a markedly greater potency. However, repeated recrystallization of the 5'-azido intermediate rather than of the 5'-amino product effectively removes the 5'-hydroxyl starting compound. Recrystallization of the 5'-azido intermediate is repeated until the 5'-hydroxy- or 5'-tosyl-precursors cannot be detected by TLC or paper electrophoresis at several pH values. Because of the possibility of trace contamination,

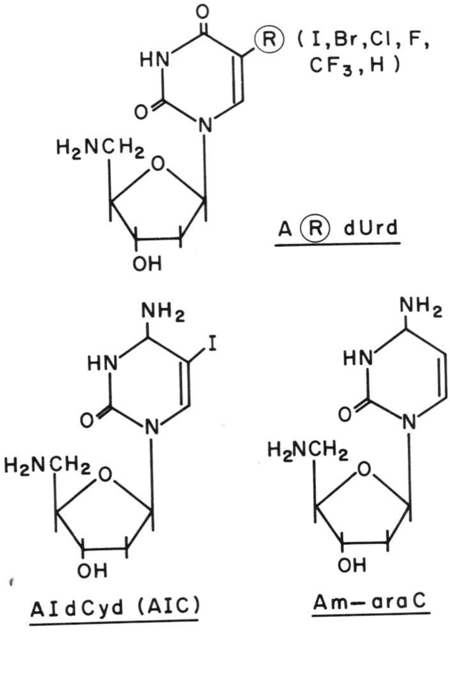
(R) (I,Br,Cl,F,
CF3,H)
A(R)dUrd
AIdCyd (AIC)
Am–araC

Figure 9.

evaluation of the antiviral potential of such nucleosides must
be done with highly purified compounds. Both AIU and AIC, after
extensive purification, still exhibit significant antiviral activity against herpes simplex virus.

The 5'-amino analog of IdUrd has been the most extensively
studied of the 5'-amino nucleosides synthesized thus far. AIU
has a highly restricted spectrum of antiviral activity. Of all
the DNA virus examined to date, only herpes simplex virus type
I and guinea pig herpeslike virus are significantly inhibited.
Viruses such as vaccinia (SV40), adeno, and minute virus of mice
(MVM) are not affected. A high degree of selectivity is also
seen with RNA viruses. While murine leukemia viruses are sensitive, Rous sarcoma virus, polio, sendai, influenza A, and measles
virus are resistant to inhibition by AIU.

A very important finding is the total absence of detectable
toxicity toward a variety of cells in culture, including those
of murine, avian, simian, and human origin. Whereas IdUrd is
mutagenic to L5178 cells in culture, AIU was found to be nonmut-
agenic (12). Studies in newborn and eight-day-old mice revealed
no evidence of gross or histological toxicity.

The apparent lack of toxicity both *in vitro* and *in vivo* en--
couraged studies of the therapeutic effect of AIU on experimental
herpetic keratitis in rabbits. Successful therapy of herpetic
keratitis in these animals was obtained with AIU, but at an eight-
fold greater concentration than that required with IdUrd (13).
Because of the absence of any toxicity when given systemically
to mice, studies are in progress on the evaluation of its efficacy in experimental herpes encephalitis and cutaneous herpes
infections. If successful we hope to extend such studies to man.

Studies of the mode of AIU inhibition produced some rather
unexpected findings. AIU inhibits neither the uptake of labeled
uridine into RNA, nor of labeled amino acids into proteins, but
it does inhibit the uptake of labeled thymidine into DNA of HSV-1
infected cells. This was accompanied by a marked increase in the
pool size of dATP, dGTP, and dCTP, but a decrease in the otherwise
large increase in the pool size of dTTP normally found in herpes
virus infected cells.

Since our working hypothesis attributed the lack of host cell
toxicity to a probable inability of the 5'-amino nucleoside analog to be phosphorylated, we had envisioned a direct interaction
of AIU with either HSV-induced DNA polymerase or the HSV template.
However, studies with ^{125}I-labeled AIU showed that AIU is indeed
phosphorylated but *only* in HSV-infected cells (14). The failure
of uninfected cells to phosphorylate AIU is not due to a HSV-
induced alteration of the cellular membrane, since lysates of
only the HSV-infected cells will phosphorylate AIU. Infections
of the host cells (Vero or A9) with vaccinia, SV40, or MVM virus
also does not lead to phosphorylation of AIU. Thus, AIU uptake
is an HSV-1 induced phenomenon.

More than 98% of the (^{125}I)-AIU present in HSV-1 infected
Vero cells comigrated with a synthetic preparation of 5'-*N*-tri-
phosphate of AIU (15), prepared by a modification of the procedure of Letsinger (Fig. 10) (16). Our data (14) show that the
5'-amino moiety of AIU is phosphorylated in the HSV-infected
cell to the corresponding triphosphate analog (AIdUTP), which
subsequently is incorporated into both viral and host cell DNA,
as determined by isopycnic density centrifugation in CsCl. Analysis of the DNA by both enzymic digestion with pancreatic DNase
I, spleen DNase II, micrococcal nuclease, spleen and venom phosphodiesterase, as well as acid hydrolysis, demonstrate that AIU
is being incorporated internally into the DNA polymer. The incorporation of AIU into DNA, therefore, does not involve merely
end addition nor does it result in chain termination.

The biochemical consequence of the presence in DNA of a
phosphoramidate (P-N) bond, which is known to be extremely acid
labile, is under active investigation. The CsCl profiles of the
(^{125}I)-AIU-DNA exhibited a very high background of ^{125}I throughout the gradient, which may be caused by the extensive fragmentation of the AIU-DNA during its isolation and purification.
This high background was not found with (^{125}I)-IdUrd-DNA. What
effect the presence of the phosphoramidate bond-linked AIU in
DNA has on other physical chemical properties of DNA as well as
on the process of transcription and translation of such substituted DNA is under active investigation.

AIdUTP

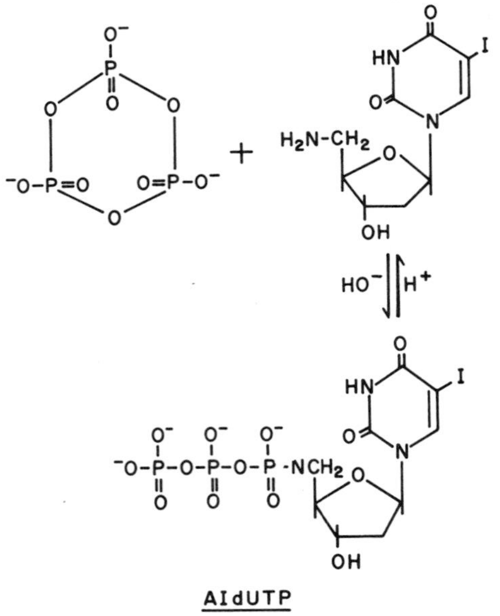

Figure 10.

The biochemical and chemical properties of AIdUTP are also being studied and compared with those of dTTP and IdUTP (Fig. 11). Letsinger and co-workers (16,17) have shown that the 5'-N'-triphosphate of 5'-amino-5'-deoxythymidine is a substrate for *E. coli* DNA polymerase in an *in vitro* enzyme system. dTTP, in addition to being a precursor of DNA, exerts an allosteric inhibition of thymidine kinase, dCMP deaminase, and ribonucleoside diphosphoreductase. A comparison of the inhibitory effect of AIdUTP and dTTP on *E. coli* thymidine kinase (15) has shown that AIdUTP is sixty-fold more potent than dTTP at pH 7.8. Both dTTP and AIdUTP appear to exert their inhibition effect by converting the active monomer form of thymidine kinase (3.6S) into an inactive dimer (5.8S). AIU is a poor inhibitor of *E. coli* thymidine kinase with a K_i of 240 μM; hence AIdUTP (as an allosteric effector) is about 360 times more effective than AIU (as a competitive substrate) in inhibition of *E. coli* thymidine kinase.

Although AIdUTP is stable in alkaline solutions, below pH 8 it undergoes degradation by a novel first-order acid-catalyzed phosphorolysis. Presumably the secondary ionized oxygen on the phosphate is protonated prior to phosphorolysis of AIdUTP to AIU and trimetaphosphate. Since neither dTTP nor IdUTP has a phosphoramidate linkage, they both exhibit a much greater pH stability.

In conclusion, we have synthesized the 5'-amino analog of IdUrd, which, although less potent an antiviral agent than IdUrd, has the unique feature of being nontoxic not only to cells in culture but also to animals. Both the antiviral effect and lack of toxicity may be explained by the fact that the AIU phosphorylation required for drug activation, is a herpes-virus-induced event (Fig. 12). The phosphorylated derivative, being highly charged, is not readily transported from the infected cell into normal tissues. Hence the virus is responsible for its own demise as well as for protection of the normal cell from the toxicity that would otherwise result if the phosphorylated form of AIU were transported into normal uninfected tissues. Although AIU is incorporated into both viral and host cell DNA, this can be manifested only in the herpes simplex virus-infected cell. Whether other nucleosides with similar specificity of action for other viruses can be synthesized remains to be seen.

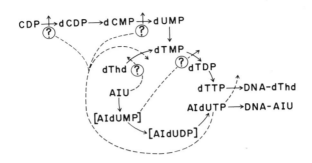

Figure 11. *Potential sites of inhibition by 5-iodo-5'-amino-2',5'-dideoxyuridine (AIU) and its phosphorylated derivatives.*

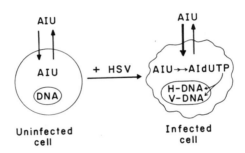

Figure 12. *Phosphorylation of 5-iodo-5'-amino-2',5'-dideoxyuridine (AIU) and subsequent incorporation into DNA occurs only in the herpes simplex virus infected cell.*

REFERENCES

1. Langen, P., G. Etzold, and G. Kowollik, *Acta Biol. Med. Germ.* 23, 19 (1969).
2. Galegov, G. A., T. A. Bekemirov, T. V. Veselovskaya, and S. V. Lavrov, *Problems of Virology* No. 1, 47 (1975).
3. Horwitz, J. P., A. J. Tomson, J. A. Urbanski, and J. Chua, *J. Org. Chem.* 27, 3045 (1962).
4. Neenan, J. P. and W. Rhode, *J. Med. Chem.* 16, 580 (1973).

5. Cheng, Y-C. and W. H. Prusoff, *Biochemistry 12,* 2612 (1973).
6. Cheng, Y-C., B. Goz, J. P. Neenan, D. C. Ward, and W. H. Prusoff, *J. Virol. 15,* 1284 (1975).
7. Dale, R. M. K., D. C. Ward, D. C. Livingston, and E. Martin, *Nucleic Acid Res. 2,* 915 (1975).
8. Mungall, W. S., G. L. Greene, G. A. Heavner, and R. L. Letsinger, *J. Org. Chem. 40,* 1659 (1975).
9. Hata, T., I. Yammato, and M. Sekine, *Chem. Lett.* p. 997 (1975).
10. Lin, T-S, J. P. Neenan, Y-C. Cheng. W. H. Prusoff, and D. C. Ward, *J. Med. Chem. 19,* 495 (1976).
11. Lin, T-S., C. Chai, and W. H. Prusoff, *J. Med Chem. 19,* 915 (1976).
12. Capizzi, R. L., unpublished results.
13. Albert, D. M., M. Lahav, P. N. Bhatt, T. W. Reid, R. E. Ward, R. C. Cykiert, T. S. Lin, D. C. Ward, and W. H. Prusoff, *J. Invest. Ophthal. 15,* 470 (1976).
14. Chen, M. S., D. C. Ward and W. H. Prusoff, *J. Biol. Chem., 251,* 4833 (1976).
15. Chen, M. S., D. C. Ward and W. H. Prusoff, *J. Biol. Chem., 251,* 4839 (1976).
16. Letsinger, R. L., J. S. Wilkes and L. B. Dumas, *J. Am. Chem. Soc. 94,* 292 (1972).
17. Letsinger, R. L., J. S. Wilkes, and L. B. Dumas, *Biochemistry 15,* 2810 (1976).

THE CHEMISTRY AND BIOLOGICAL ACTIVITY OF C-NUCLEOSIDES RELATED TO PSEUDOURIDINE*

DEAN S. WISE, ROBERT A. EARL, AND LEROY B. TOWNSEND

*Division of Medicinal Chemistry,
Department of Biopharmaceutical Sciences and
Department of Chemistry,
University of Utah, Salt Lake City*

ABSTRACT

The chemistry of pseudouridine, a pyrimidine nucleoside of interest because it contains a carbon to carbon linkage between the algycon and the carbohydrate moiety, has been explored in an effort to produce derivatives that may find use as antitumor or antiviral agents. Effort was particularly concentrated on the modification of the algycon moiety. Pseudouridine has been acetylated and chlorinated to furnish, 2,4-dichloro-5-(2,3,5-tri-O-acetyl-β-D-ribofuranosyl)pyrimidine. From this, key intermediate analogs of the known modified minor pyrimidine nucleosides found in various RNAs in which the carbohydrate moiety is attached to the N-1 nitrogen atom have been prepared and will be reported. A good yield of the 5'-O-tosyl derivative of 2,3-O-isopropylidinepseudouridine has been prepared and used as a starting material for the preparation of various 5'-substituted pseudouridine derivatives, which will be described. Other modifications of the carbohydrate moiety will be discussed. The activity of several of these derivatives against Leukemia L-1210 will also be discussed.

*
This investigation was supported by Grant No. CA-11147, awarded by the National Cancer Institute, DHEW, and by American Cancer Society Grant No. CI89.

The existence of pseudouridine (ψ) was first observed (1) in
1951.   In 1957, ψ was rediscovered (2,3) and the structure tentatively reported (4) as 5-D-ribosyluracil (1) in 1960.   The
structural features of anomeric configuration and ring size of
the carbohydrate moiety were subsequently established (5) as *beta*
and furanose (1) in 1962.   This furnished the first example of
a new class of nucleosides, which has now been termed C-nucleosides.   Since that time, the chemical modification of pseudouridine has been confined primarily to a few of the standard procedures, e.g., acetylation, methylation, cyanoethylation, and thiation (5,7).   The recent synthesis of C-ribosyl derivative, ethyl
2- (D-ribofuranosyl)-2-formylacetate, has been reported (8) and
this derivative was ring annulated with guanidine to affort the
interesting 5-(β-D-ribofuranosyl)-isocytosine.

Structure 1

     We have initiated a program of extensive research in this
specific area and would now like to report our progress in the
modification of pseudouridine, *per se*.   The aims of this research
may be categorized into the following three areas: (1) reactions
involving the carbohydrate moiety, (2) reactions involving the
heterocyclic moiety, and (3) reactions involving both the carbohydrate moiety and the heterocyclic moiety.
     When we first began this study we found that the axiom about
the pass play in football has a correlated problem in the chemical modification of pseudouridine: In pseurouridine chemistry,
during the course of the reaction, four things can happen, three
of which are bad.   The bond between the ribosyl moiety and the
C-5 position of uracil imparts an allylic property to the anomeric carbon, which in essence gives the anomeric carbon a carbonium ion character as postulated by Cohn (4) and Shapiro and
Chambers (9).   This ring opening and recyclization during a reaction at this active site of the molecule may give any of four
products only one of which has the desired β configuration in
conjunction with the ribofuranoxyl ring.
     In order to investigate reactions designed to effect speci-
fic transformations of the uracil moiety, which would also mod-

ify the carbohydrate moiety of pseudouridine, it was necessary
for us initially to repeat most of the published procedures for
selectively blocking the carbohydrate moiety. These procedures
were not found to be very satisfactory in our hands for large-

Structures 2-5

scale reactions due to incomplete reactions and/or indications
that anomerization and/or isomerization were occurring, Acetona-
tion of pseudouridine using acetone and a catalytic amount of perchloric acid produced an approximate 1:1 anomeric mixture of the
isopropylidene derivative. However, a very good preparation
(essentially quantitative) of 5-(2,3,-O-isopropylidene-β-D-ribo-
furanosyl)uracil (6) was developed using a mixture of DMF, acetone, 2,2-dimethoxypropane, and a catalytic amount of concentra-

ted hydrochloric acid. Subsequent tosylation with *p*-toluene
solfonyl chloride in pyridine gave a good yield of the 5'-*O*-
tosyl derivative (7). We used the 5'-*O*tosyl derivative as our
starting material for the preparation of various 5'-substituted
pseudouridine derivatives. Treatment of 7 with sodium iodide
in acetone at reflux afforded a 55% yield of the 5-(5-deoxy-5-
iodo-2,3-*O*-isopropylidene-β-D-ribofuranosyl)uracil (8a). In a
similar manner the 5'-bromo (8b) and 5'-chloro (8c) derivatives
were prepared by treatment of the 5'-*O*-tosyl with lithium bromide and lithium chloride, respectively. However, preparation
of the 5'-fluoro derivative (11) proved to be somewhat more difficult since treatment of the 5'-*O*-toysl derivative using most
of the standard reactions, e.g., treatment of 7 with tetraethy-
lammonium fluoride in acetonitrile, resulted in extensive decomposition. However, we found that treatment of 5-(2,3-*O*-isopropy-
lidene-β-D-ribofuranosyl)uracil with the reagent, 1,1,2-trifluoro
2-chlror-1-diethylaminoethane (10) at room temperature provided
at 68% yield of the blocked 5'-deoxy-5'-flurorpseudouridine (10).
Deblocking of these nucleosides, without anomerization, was accom
plished with trifluoroacetic acid in methanol.

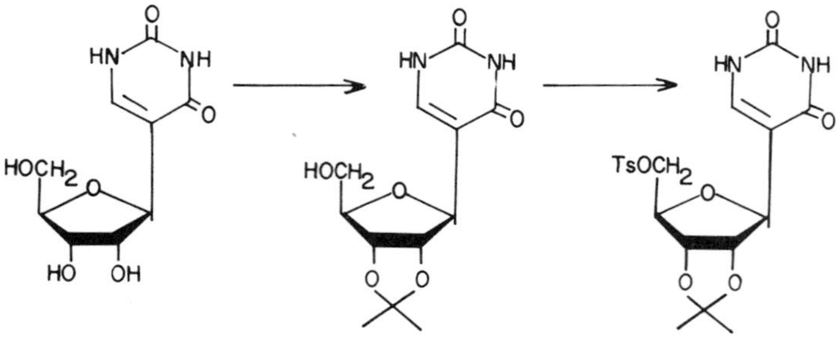

Structure 6 & 7

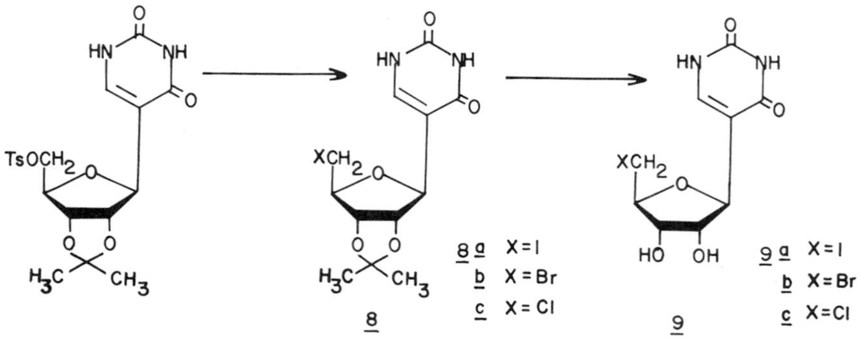

Structure 8 & 9

Treatment of **8a** with hydrogen in the presence of palladium on carbon and subsequent deblocking with methanolic ammonia gave a good yield of the 5'-deoxypseudoouridine (**13**).

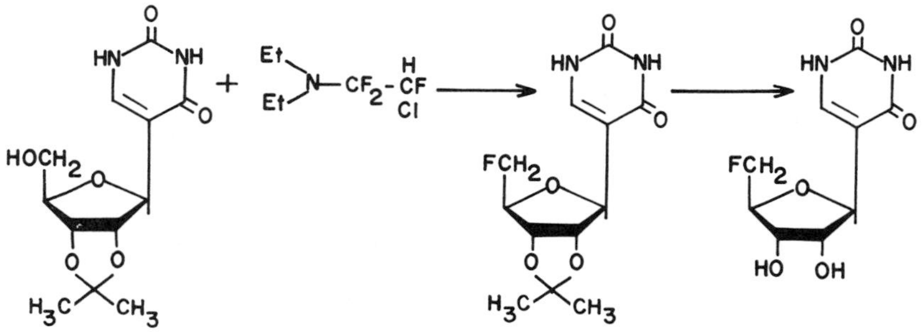

Structure 10 & 11

We first began modifications involving the heterocyclic portion of pseudouridine by reinvestigating alkylations of the uracil moiety at the N-1 position. Although the products shown here have been previously reported (4,11), the procedure used has not been reported and may be of interest.

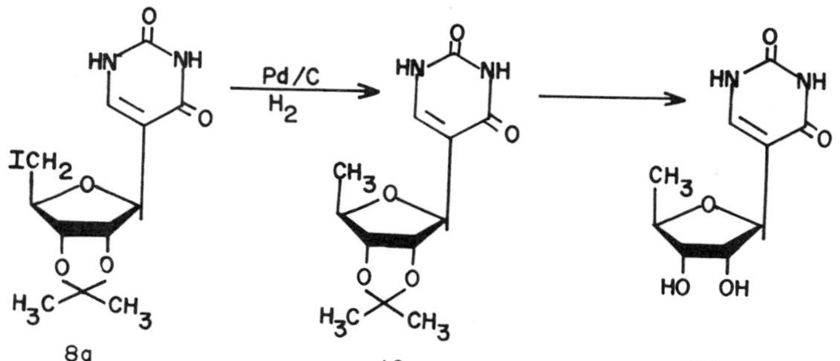

Structure 12 & 13

In each case 2,3,5-tri-*O*-acetylpseudouridine (**14**) was silylated with *bis*-trimethylsilyl acetamide in dichloromethane at room temperature. Although the reaction appeared to be immediate, generally a half-hour was allowed to assure complete silylation. The alkylating agent, i.e., methyl iodide, 2,3,5-tri-*O*-acetyl-β-D-ribofuranosyl chloride, or ethyl iodide, was then added to this mixture without removing the by-products of the silylation. Reactions were generally complete within 18 hr and the yields were good. We then initiated research designed to furnish a ψ derivative with groups at C2 and C4 that would be amenable toward facile nucleophilic displacement.

We have now prepared what we consider the ultimate starting
material for functional group transformations on the heterocyclic moiety of pseudouridine.   2,4-dichloro-5-(2,3,5-tri-O-acetyl-
β-D-ribofuranosyl)pyrimidine (19) was prepared by treatment of

Structure 14-18

2',3',5-tri-O-acetylpseudouridine with phosphoryl chloride in
the presence of diethylaniline hydrochloride in 70 - 75% yields.
This derivative (19) lends itself easily to various modifications
Of the commonly used nucleophiles, the alkoxide and the alkyl
mercaptides were investigated first.   Treatment of 15 with 1.1

equivalents of sodium ethylate at room temperature for 1 hr gave
a good yield of 2-chlror-4-ethoxy-5-(β-D-ribofuranosyl)pyrimi-
dine (20).

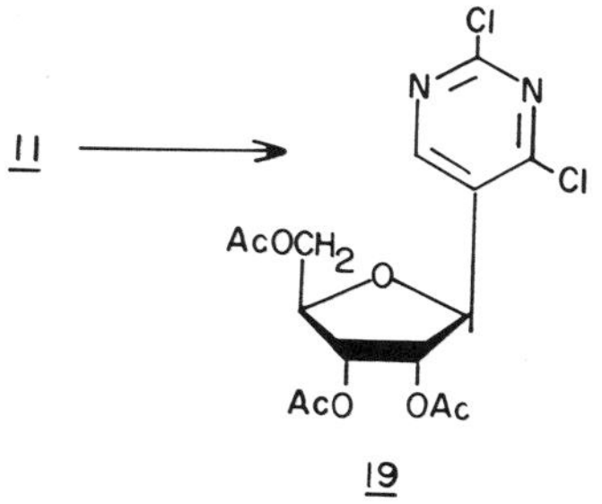

Structure 19

A good yield (essentially quantitative) of 2,4-dimethoxy-5-
(β-D-ribofuranosyl)pyrimidine (21) was obtained by simply increasing the concentration of the alkoxide and the reaction temperature. The reaction of 19 with sodium methyl mecaptide was
also found to be concentration dependent. Treatment of 19 with
5 equivalents of sodium methyl mercaptide in methanol at reflux
and continuous saturation with methanethiol produced a near quantitative yield of the dimethylthio derivative 23. However, if
methanethiol was not added throughout the reaction, side products
were produced that were identified as either the 4-ethoxy-2-meth-

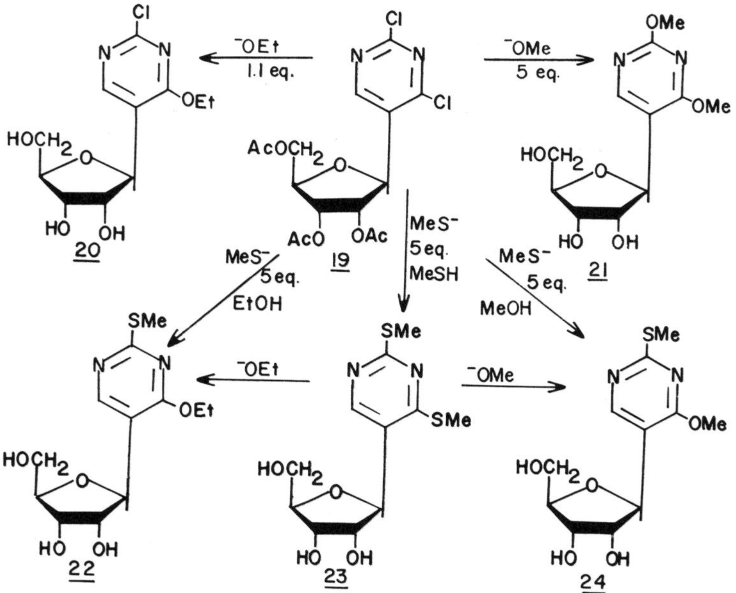

Structures 20-24

ylthio derivative 22, when ethanol was used as the reaction solvent, or the 4-methoxy-2-methylthio derivative 24, when methanol
was used as the reaction solvent. These products, 22 and 24,
could also be prepared by treatment of the dimethylthio derivative 23 with either ethoxide or methoxide ion. It is of interest to note that both 21 and 23 can be selectively hydrolyzed
with 1 N sodium hydroxide to their respective 4-one derivatives
25 or 26 without any observable anomerization. This substanti-
ates the site at which the ethoxide or methoxy moiety resides in
22 or 24.

Structures 25 & 26

Compound 19 was treated with hydrogen in the presence of 5%
palladium on charcoal to afford the intermidiate 27. The nucleoside (27) was deblocked with methanolic sodium methoxide to provide the interesting compound 5-β-D-ribofuranosylpyrimidine (28).

Possibly the most important investigations in this area were
those involving treatment of 19 with ammonia. When 19 was reacted with liquid ammonia at room temperature for 24 hr a mixture of 4-amino-2-chloro-5-(β-D-ribofuranosyl)pyrimidine (29)
and 2-amino-4-chloro-5-(β-D-ribofuranosyl)pyrimidine (29) was
produced in 78% yield. The isomer ratio (29:30) was approximately 1:1. When the conditions of the reaction were modified by
raising the temperature to 100°C, the sole product was 2,4-di-

thione derivative 34 in good yield.   This intermediate was deblocked with sodium methoxide in methanol to yield 5-(ß-D-ribofuranosyl)pyrimidin-2,4-dithione (35) (2,4-dithiopseudouridine).
Treatment of 35 with ammonia at 100°C for 24 hr gave a good
yield of 32b which was identical with a sample prepared from 29.

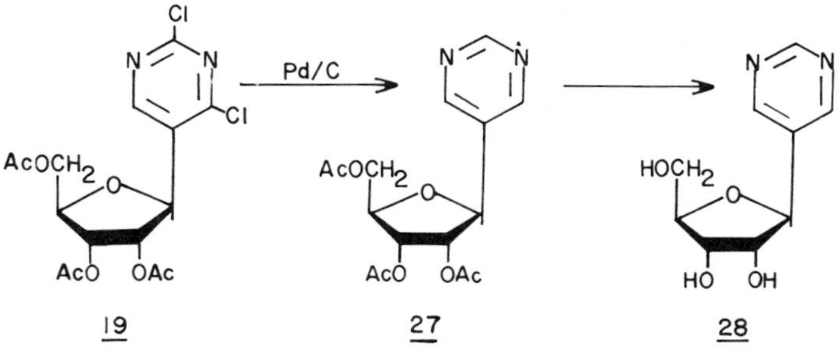

Structure 27 & 28

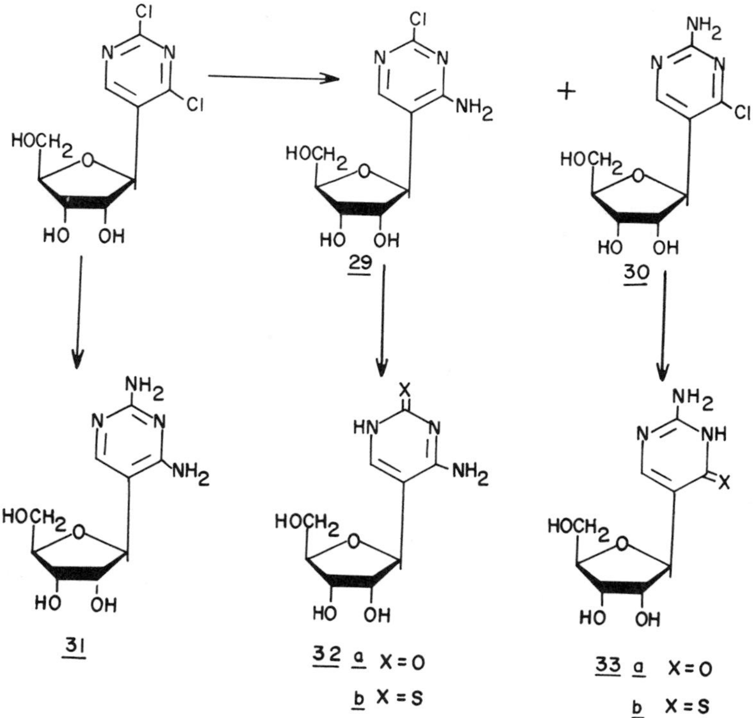

Structure 29-33

We have also turned our attention to reactions that involve
both the heterocyclic and the carbohydrate portion of pseudouridine. We investigated the reaction of 1 with 2-acetoxyisobuty-
ryl chloride in a variety of solvents. In every instance, we
obtained a mixture mainly composed of two products. However,
these two compounds proved to be very difficult to separate.
The reaction of 1 with 2-acetoxyisobutyryl chloride in the absence of any solvent proceeded rapidly at 130°C to produce only
one product, which was shown by spectral evidence to be 3'-O-

Structures 34 & 35

acetyl-2'-chloro-2'-deoxypseudouridine (36). Treatment of 36
with methanolic sodium methoxide gave crystalline 2'-chloro-2'
deoxypseudouridine (38) in good yield, while treatment of 36
with hot methanolic sodium methoxide gave a good yield of 4, $O^{2'}$-
cyclopseudouridine (37). Compound 37 could be opened easily with
dilute acid to produce an α/β mixture of pseudouridine arabinoside 39 or with an aqueous base to produce only the β-isomer 39.
      These studies have only slightly opened the door to the pos-
sible modifications of pseudouridine, and research designed to
explore the above areas in more detail as well as research in
several new areas is under active investigation. No significant
biological activity has been found for any of the compounds thus
far tested in this study, however, 32a has been reported as active against various mouse leukemias including (ara-C)-resistant
mouse currently undergoing preclinical toxicological evaluation.

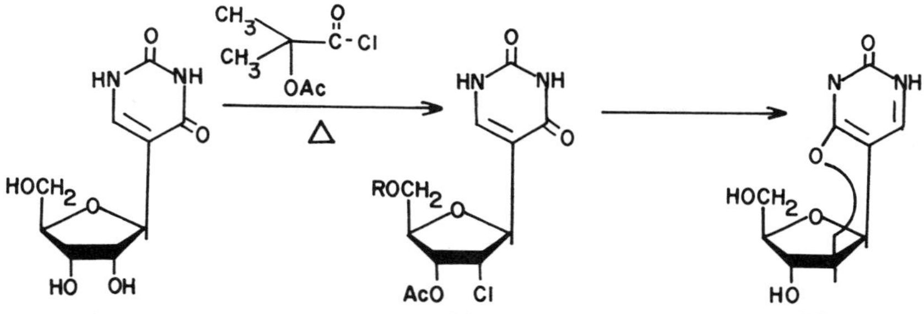

Structure 36 & 37

Certain other derivatives have demonstrated some marginal activity or cytotoxicity (10^{-4}M) against a L-1210 cell culture systems in our laboratory but we have observed essentially no activity (anticancer) *in vivo* up to the present time.

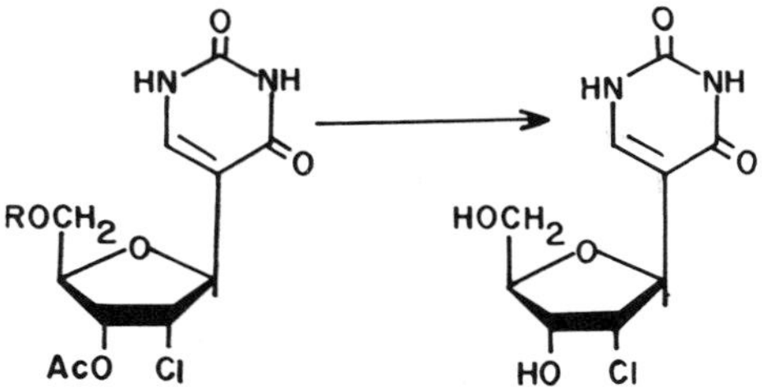

Structure 38

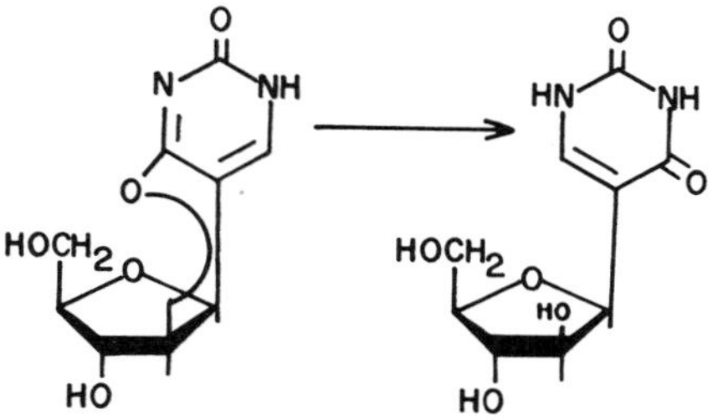

Structure 39

REFERENCES

1. Cohn, W. E. and Volkin, E., *Nature 167*, 483 (1951).
2. Davis, F. F. and Allen, F. W., *J. Biol. Chem. 227*, 907 (1957).
3. Cohn, W. E., *Fed. Proc., Fed. Am. Soc. Exp. Biol. 16*, 166 (1957).
4. Cohn, W. E., *J. Biol. Chem. 235*, 1488 (1960).
5. Michelson, A. M., and Cohn, W. E., *Biochemistry 1*, 490 (1962).
6. Chambers, R. W., *Prog. Nucl. Acid Res. Mol. Biol., 5*, 349 (1966); Ref. 27 and references cited therein.
7. Wigler, P. W., Bindslen, B., and Breitmen, R. T., *J. Carbohydr. Nucleosides Nucleotides, 1*, 307 (1974).
8. Chu, C. K., Watanabe, K. A., and Fox, J. J., *J. Heterocycl. Chem. 12*, 817 (1975).
9. Shapiro, R. and Chambers, R. W., *J. Am. Chem. Soc. 83*, 3920 (1961).
10. Yarovenko, N. N. and Raksha, M. A., *J. Gen. Chem. 29*, 2125 (1959).
11. Dlugajazyk, A., and Eiler, J. J. *Biochim. Biophys. Acta 119* 11 (1966).
12. Burchenal, J. F., Ciovacco, K., O'Toole, T., Kiefner, R., Dowling, M. D., Chu, C. K., Watanabe, K. A., Wempen, I., and Fox, J. J. *Cancer Res. 36*, 1520 (1976).

THE CHEMISTRY AND BIOLOGICAL ACTIVITY OF
CERTAIN 4-SUBSTITUTED AND 3,4-DISUBSTITUTED
PYRAZOLO(3,4-d)PYRIMIDINE NUCLEOSIDES*

RAYMOND P. PANZICA,[+] GANAPATI A. BHAT, ROBERT A. EARL,
JEAN-LOUIS G. MONTERO, LINDA W. ROTI ROTI,
AND LEROY B. TOWNSEND

Division of Medicinal Chemistry, Department of
Biopharmaceutical Sciences and Department of
Chemistry, University of Utah
Salt Lake City, Utah

ABSTRACT

In our laboratories we have several on-going research programs involving the synthesis of 4-substituted and 3,4-disubstituted pyrazolo(3,4-d)pyrimidine nucleosides. One synthetic program that deals with the preparation of 4-substituted pyrazolo(3,4-d)pyrimidine nucleosides utilizes 4-methylthio- and 4-chloro-1-(2,3,5-tri-o-acetyl-β-D-ribofuranosyl)pyrazolo(3,4-d)pyrimidines as key intermediates. From these two nucleosides we have synthesized a number of 4-substituted derivatives, e.g., 4-benzylamino-, 4-dimethylamino-, 4-hydrazino-, 4-(hydroxyamino)-, 4-(methylamino)-, 4-mercapto-, 4-methoxy-, 4-(p-nitrobenzythio)-1-(2,3,5-tri-o-acetyl-β-D-ribofuranosyl) pyrazolo(3,4-d)pyrimidines, etc. Another program involving syntheses of 3,4-disubstituted pyrazolo(3,4-d)pyrimidines uses as one of its precursors methyl 4-amino-1-(β-D-ribofuranosyl)pyrazolo(3,4-d)pyrimidine-3-carboximidate. Chemical modification of the carboximidate group provided a new series of nucleosides having a variety of substituents on the 3 position of the aglycon. The biological activities of these compounds were studied against experimental tumors in vitro and in vivo.

*
This research was supported by Research Contracts N01-CM-23710 and N01-CM-43806 from the Division of Cancer Treatment, National Cancer Institute, National Institutes of Health, Department of Health Education and Welfare.

[+]
Present Address: Department of Medicinal Chemistry, University of Rhode Island, Kingston, Rhode Island.

121

CHEMISTRY

4-SUBSTITUTED PYRAZOLO(3,4-d)PYRIMIDINE NUCLEOSIDES

4-Amino-1-(β-D-ribofuranosyl)pyrazolo(3,4-d)pyrimidine (4-APP riboside, 1) is a nucleoside that resembles the naturally occurring nucleoside adenosine (2) and has been synthesized by three different glycosylation procedures. The first synthetic route (1) used the chloromercuri procedure and furnished two nucleosides. The major component was assigned structure 1 although this assignment was made without any unequivocal proof for either the site of ribosylation or the anomeric configuration. A subsequent investigation (2), using the acid-catalyzed fusion procedure, also provided two nucleosides but in this investigation the sites of ribosylation were unequivocally established as N1 and N2 on the basis of uv spectral comparisons with model methyl compounds.

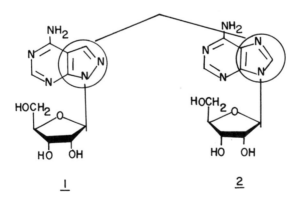

A more recent synthesis, which was reported (3) from our laboratories, corroborated the site of ribosylation of 1 as N1 and established the anomeric configuration as β.

Interest in 4-APP riboside (1) and related analogs has been further stimulated by several reports that 4-APP riboside will function as a substrate for anabolic (adenosine kinase (4,5) and adenylate kinase (5,6)) and catabolic (adenosine deaminase (5,7, 8)) enzymes. In addition, experimental observations indicate that 4-APP riboside is a more effective inhibitor of both cell growth and the purine *de novo* biosynthetic pathway than 4-aminopyrazolo

(3,4-*d*)pyrimidine (4-APP) per se. Such findings would suggest
that intracellular 4-APP ribonucleotide, i.e., the actual inhibitor, accumulates significantly faster with 4-APP riboside as a
precursor that with 4-APP as a precursor (7). This might be expected since it has been reported that 4-APP riboside is an excellent substrate for adenosine kinase (4,5) while 4-APP is a poor substrate for adenine phosphoribosyltransferase (9).

In view of this reported biological data and the fact that
very few synthetic 4-substituted pyrazolo(3,4-*d*)pyrimidine nucleosides had been prepared, we initiated a synthetic program that
would provide pertinent nucleosides in this area. Using allopurinol riboside (10-13) (3, 1-(β-D-ribofuranosyl)pyrazolo(3,4-
d)pyrimidin-4-one) as our starting material, we have now synthesized a number of selected 4-substituted pyrazolo(3,4-*d*)pyrimi-
dine nucleosides. The first step in our synthetic sequence involved the acetylation (pyridine/acetic anhydride) of allopurinol riboside (Scheme 1). A near-quantitative yield of 1-(2,3,
5-tri-*O*-acetyl-β-D-ribofuranosyl)pyrazolo(3,4-*d*)pyrimidin-4-one
(4) was obtained. This nucleoside was thiated with P$_2$S$_5$-dioxane
(14) and the desired thionucleoside 5 was obtained in excellent
yield. Removal of the *O*-acetyl blocking groups was accomplished with methanolic ammonia and furnished 1-(β-D-ribofuranosyl)
pyrazolo(3,4-*d*) pyrimidin-4-thione (6); an isomer of 6-mercapto-
purine riboside. Treatment of 6 with either methyl iodide, *p*-
nitrobenzyl bromide, allyl bromide, or 5-chloro-4-nitro-1-methyl-
imidazole under basic reaction conditions afforded 7, 8, 9, or 10 ,
respectively.

The 4-methylthio group of 7 was found to be amenable to
nucleophilic displacements (steam-bath temperature) and provided
good yields of the 4-methylamino (11), 4-dimethylamino (12), 4-
hydrazino (13), and 4-hydroxylamino (14) derivatives (Scheme 2).
In addition, treatment of 7 with Raney nickel furnished 1-(β-D-
ribofuranosyl)pyrazolo(3,4-*d*)pyrimidine (15), an analog of the
nucleoside antibiotic nebularine.

Another pyrazolo(3,4-*d*)pyrimidine nucleoside, 4-chloro-
(2,3,5-tri-*O*-acetyl-β-D-ribofuranosyl)pyrazolo(3,4-*d*)pyrimidine
(16), proved to be an equally important precursor of certain derivatives. The preparation of 16 via the silyl fusion method
was reported (3) earlier from our laboratories. However, this
procedure was found to be unsuitable for the large-scale production of 16. We have now prepared 16 from 4 using dimethylchloromethyleneammonium chloride (15) a mild chlorinating agent.

The 4-chloro substituent of 16 was very susceptible to nucleophilic displacements and provided another pathway (Scheme 3)
requiring less steps and time to the nucleosides 11 through 15.
In addition, treatment of 16 with sodium methylthiolate (in
methanol) provided an alternate route to 7. Also, treatment
of 16 with methoxide furnished the interesting 4-methoxy-1-(β-
D-ribofuranosyl)pyrazolo(3,4-*d*)pyrimidine (17).

The extreme reactivity of the 4-chloro group was demonstrated with the following amines: methylamine, benzylamine, furfurylamine, and aniline. Treatment of 16 with a 25% aqueous
methylamine solution (steam bath) effected a rapid conversion
(≈10 min) of 16 to 11. Using benzylamine or furfurylamine,
the 4-chloro group was still rapidly displaced (t.l.c.), however,
deacetylation of the carbohydrate moiety occurred at a much
slower rate. In each case, two nucleosides were isolated from
the reaction; the 5'-O-acetylated derivatives (18 and 20) and
the completely deblocked analogs (19 and 21). With aniline, an

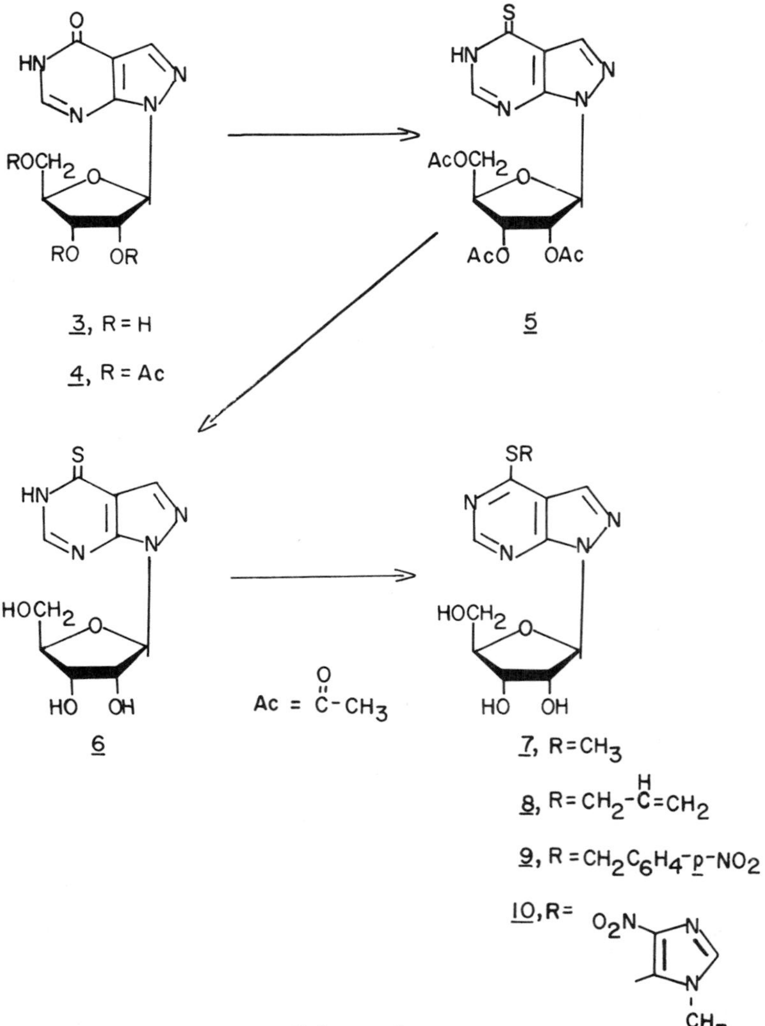

Scheme 1

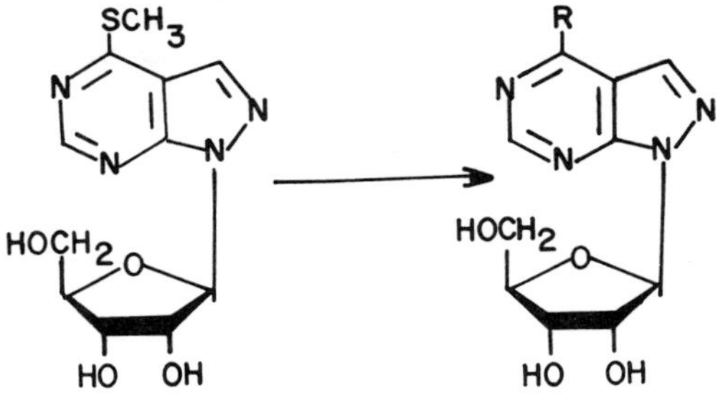

Scheme 2

11, R = NHCH₃ ... $R = \overset{H}{N}CH_3$

14, R = NOH ... $R = \overset{H}{N}OH$

12, R = N(CH₃)₂

15, R = H

13, R = NNH₂ ... $R = \overset{H}{N}NH_2$

Ac = $\overset{O}{\overset{||}{C}}CH_3$

1, 7, 11-15

16

17

18, R = Ac

19, R = H

20, R = Ac

21, R = H

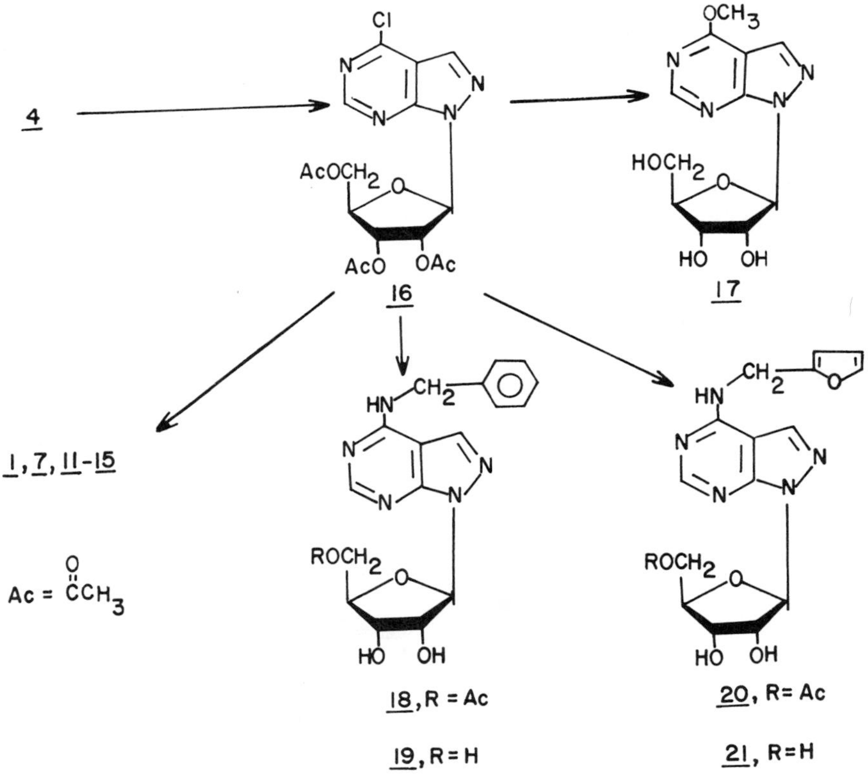

Scheme 3

amine that is less basic than the aforementioned group (several magnitudes), only displacement of the chloro group occurs and the acetyl groups residing on the ribose moiety are left intact. The sole product from this reaction is 4-anilino-1-(2,3,5-tri-O-acetyl-β-D-ribofuranosyl)pyrazolo(3,4-d)pyrimidine (22). Thus, as one decreases the basicity of the nucleophile, deacetylation does not occur, whereas replacement the chloro group still takes place.

22   23

Conversion of 22 to 23 was achieved by treatment with methanolic ammonia.

Several other 4-alkylamino derivatives of possible therapeutic interest were prepared from 15 in moderate yields. Reaction with 3-methyl-2-butenylamine or ethylenimine provided 24 and 25, respectively. It is worthwhile mentioning that the site of substitution and β-configuration of all nucleosides synthesized in this study was confirmed by the conversion of 16 to 4-APP riboside (1): a nucleoside whose structure has been rigorously established (5).

$$24, R = \overset{H}{N} - CH_2CH=C(CH_3)_2$$

$$25, R = N\text{-aziridinyl}$$

Treatment of 4-APP riboside with dimethylformamide dimethyl
acetal (16) furnished a good yield of 4-dimethylaminomethylenea-
mino-1-(β-D-ribofuranosyl)pyrazolo(3,4-d)pyrimidine (26) a derivative of 1 that is remarkably soluble in aqueous solutions.
Another interesting nucleoside in this area is benzaldehyde-*p*-
(*bis*(2-chloroethyl)amino)-1-(β-D-ribofuranosyl)pyrazolo(3,4-d)
pyrimidin-4-yl-hydrazone (27) which was synthesized by a condensation of 12 with *p-N,N-bis*(2-chloroethyl)aminobenzaldehyde (17).

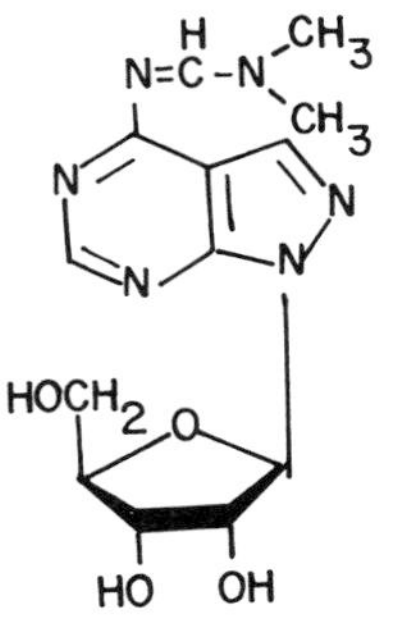

26

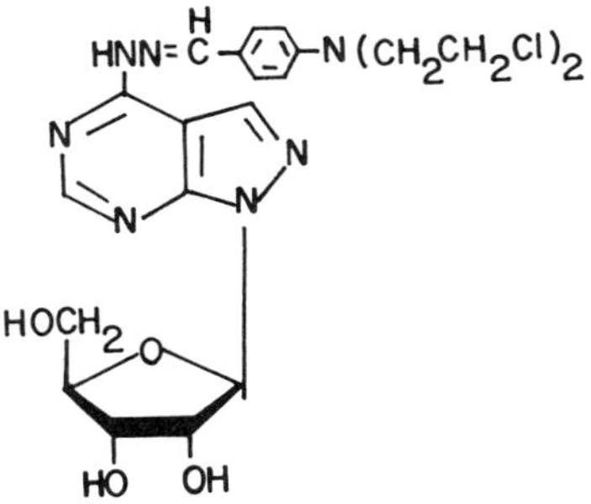

27

3,4-DISUBSTITUTED PYRAZOLO(3,4)PYRIMIDINE NUCLEOSIDES

The pyrrolopyrimidine nucleoside antibiotics tubercidin (28),
toyocamycin (29), and sangivamycin (30) have been reported (18,19,
20) to possess significant biological and chemotherapeutic activity. A visual examination of structure 31 revealed that the
C3 position of the pyrazolo(3,4-d)pyrimisine moiety could be
functionalized in a manner similar to the C5 position of toyocamycin (29) and sengivamycin (30). Therefore, we initiated a syn-

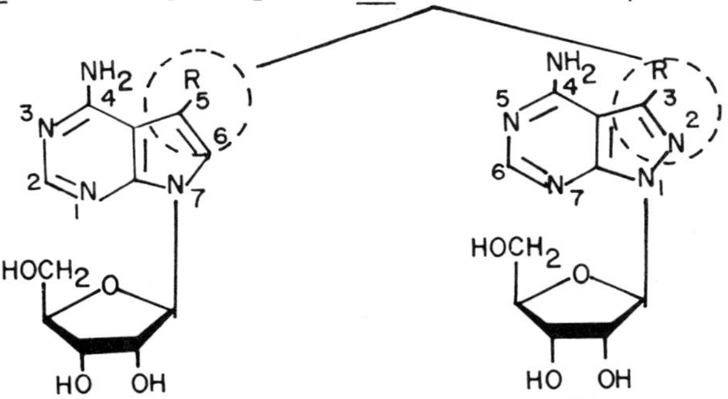

28, R = H

29, R = CN

30, R = $\overset{\overset{\text{O}}{\|}}{\text{C}}NH_2$

31

thetic program to provide "6-aza" derivatives of the above pyr-
rolo(2,3-*d*)pyrimidine nucleoside antibiotics (29 and 30).

 The glycosylation of 4-acetamido-3-cyanopyrazolo(3,4-*d*) pyrimidine (32) (21) with crystalline 2,3,5-tri-*O*-acetyl-β-D-ribo-
furanosyl chloride (33) (22) furnished a good yield (56%) of 4-
acetamido-3-cyano-1-(2,3,5-tri-*O*-acetyl-β-D-ribofuranosyl)pyra-

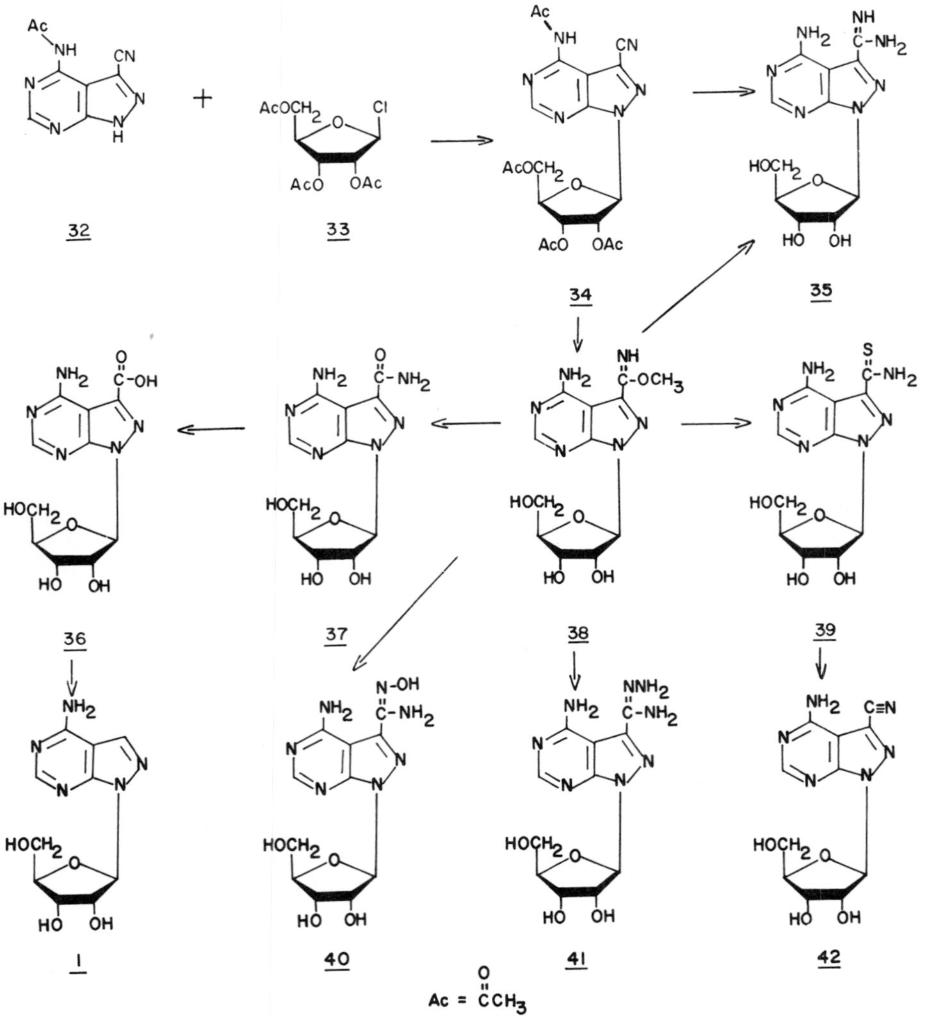

Scheme 4

zolo(3,4-d)pyrimidine (34) (Scheme 4). The 3-cyano group was
found to be very reactive and prevented a direct conversion of
34 into 6-azatoyocamycin (42) by a simple removal of the O- and
the N-acetyl protecting groups. For example, treatment of 34
with sodium methoxide in methanol afforded the imidate 38 instead of 42. Similarly, treatment of 34 with liquid ammonia or
methanolic ammonia provided the amidine (35), rather than the
desired cyano nucleoside (42).

However, methyl 4-amino-1-(β-D-ribofuranosyl)pyrazolo(3,4-
d)pyrimidine-3-formimidate (38) proved to be a versatile precursor for the preparation of the desired analogs of toyocamycin
(42) and sangivamycin (37) as well as several other key derivatives (40 and 41) in this series, Stirring in dry methanol
containing an equivalent amount of sodium hydrosulfide (room
temperature) effected a rapid conversion of 38 to the thiocarboxamide derviative (39). Treatment of 39 with mercuric chloride
and triethylamine in dimethylformamide furnished the desired
4-amino-3cyano-1-(β-D-ribofuranosyl)pyrazolo(3,4-d)pyrimidine
(6-azatoyocamycin).

A catalytic amount of sodium hydroxide in water (room temperature) effected a facile conversion of 38 to 4-amino-1-(β-
D-ribofuranosyl)pyrazolo(3,4-d)pyrimidine-3-carboxamide (37, 6-
azasangivamycin). This target nucleoside played a significant
role in establishing the site of ribosylation and anomeric configuration of this series of 3,4-disubstituted nucleosides.
Treatment of 37 with an excess of hot aqueous sodium hydroxide
afforded the carboxylic acid derivative 36, which was subsequently decarboxylated in sulfolane to furnish 4-APP riboside (1);
thus establishing the site of glycosylation as N1 and the configuration as β for this entire series of 3,4-disubstituted py-
razolo(3,4-d)pyrimidine nucleosides.

BIOLOGICAL AND CHEMOTHERAPEUTIC EVALUATION

All 3,4-disubstituted pyrazolo(3,4-d)pyrimidine nucleosides
described herein have shown some antitumor activity with 35 and
37-40 significantly increasing the life span of L1210 bearing
mice (Table I). The results, particularly for 35 and 37, represent a substantial improvement over the observed antitumor activity of 4-APP riboside (1) (Table II).

Compounds 37 and 42 were selected for an *in vitro* study in
our laboratory, in an effort to discover the cellular and biochemical basis for this significant increase in activity. Concentrations of these compounds required for 50% reduction of
growth rate (ID$_{50}$'s) were somewhat lower than that observed for
1 (Tables I and II), but this difference did not seem to be large
enough to explain the striking difference in antitumor activity

TABLE I,   Activity of Certain 3,4-Disubstituted Pyrazolo(3,4-d)Pyrimidine Nucleosides Against L1210 Mouse Leukemia, *In Vitro* and *in Vivo*

	in vitro[a]	*in Vivo*[b]			
Compound	ID_{50} (M)	Dose (mg/kg)	Schedule	T/C	Cures
35		50	1-5	--	
		25	1-5	196	2/6
37	1×10^{-7}	100	1-5	258	2/6
		100	1-9	---	
38		400	1-5	164	
39		12.5	1-5	--	
		3.12	1-9	163	
40		100	1-5	--	
		25	1-5	185	
41		166	1-5	131	
42	8×10^{-8}	100	1-9	127	

[a] ID_{50} values are the minimum concentrations required to reduce the growth rate of L1210 to 50% of the control rate.   A dash (--) indicates lack of significant growth inhibition with 10^{-1}M concentration of the compound, and a blank space is left for compounds that were not evaluated.   Data presented in this table were obtained in our laboratory.

[b] The dose indicated was administered on the days, after inoculation of the mice with tumor cells, listed under "Schedule. Results are expressed as the ratio (%) of survival time of the treated animals to untreated, control animals under "T/C".   A dash (--) in this column indicates a test in which host toxicity precluded the possibility of antitumor activity.   "Cures" lists the number of animals surviving for 30 days, the period of evaluation, over the total number of animals in the group.   Data were selected from testing results obtained under the auspices of Drug Research and Development Branch of the Division of Cancer Treatment, NCI, NIH, PHS, DHEW, according to the protocols described in Cancer Chemother. Rept. 25:3 (1972).   Selection of data includes the most effective dose, and the lowest dose found to be toxic to the host, if host toxicity was found.   A T/C value $\geq$125 is required for evaluation of a compound as active.

TABLE II, Activity of Certain 4-Substituted Pyrazolo(3,4-d)pyrimidine Nucleosides Against L1210 Mouse Leukemia, *in Vitro* and *in Vivo*

Compound	in Vitro[a] ID_{50} (M)	Dose (mg/kg)	in Vivo[b] Schedule	T/C	Cures
1	4×10^{-7}	133	1	148	1/6
		66	1-9	--	
3	--	100	1-5	108	
4		25	1-5	104	
5		50	1-5	104	
6	--	12.5	1-5	103	
7	9×10^{-6}	100	1-5	122	
8	--	200	1-5	--	
		100	1-5	102	
9	1×10^{-4}	200	1-5	101	
10	2×10^{-5}	Not available			
11	$>10^{-4}$	200	1-5	--	
		50	1-5	104	
12	--	50	1-5	107	
13	6×10^{-5}	400	1-9	--	
		200	1-9	112	
14	1×10^{-5}	25	1-5	--	
16		100	1-5	112	
17	--	200	1-5	107	
18	--	50	1-5	102	
19	--	25	1-5	103	
20	2×10^{-5}	25	1-9	109	
21	$>10^{-4}$	400	1-9	106	
22	--	400	1-9	100	
23	--	25	1-5	100	
26	6×10^{-7}	100	1-5	--	
		50	1-5	113	
27		100	1-9	102	

[a] See footnote a, Table I.

[b] See footnote b, Table I.

in the case of 37. Therefore, we further investigated the cellular effects of these compounds by determining the viability of
cells after treatment with the nucleoside analogs, for various
time intervals. Cells were incubated with the analogs at equi-
toxic concentrations, i.e., concentrations that would ensure complete inhibition of growth. After this treatment, the cells were
returned to a normal medium and the cell number was followed with
time to ascertain if any fraction of the population remained vi-
able. After treatment with 4-APP riboside for up to 72 hr, cell
growth resumed. This result indicated that growth inhibition by
this compound is cytostatic in nature, at least for a fraction
of the population (23). In contrast, incubation of the cells
with the 3-carboxamide derivative 37 for 36 hr or longer revealed
that 37 was cytocidal, i.e., no increase of cell number was detected during the 8-day period of observation after the cells
were returned to a normal medium.

Biochemical studies with human erythrocytes (5) showed that
4-APP riboside (1) was deaminated by adenosine deaminase, and
could be converted to the mono-, di-, and triphosphate derivatives. The 3-carboxamide derivative (37) of 4-APP riboside, on
the other hand, appeared to be converted only to the monophosphate derivative. The di- and triphosphate derivatives of 37
were not observed nor could deamination of 37, per se, be detected. These results suggest that the cytocidal nature of
growth inhibition by 37 might be due to an accumulation of the
5'-monophosphate derivative. However, more detailed studies
using L1210 cells will be required before definite conclusions
can be made.

Seven of the 4-substituted pyrazolo(3,4-d)pyrimidine nucleosides, discussed in this presentation, have been found to inhibit growth of L1210 in vitro, with an ID_{50} of less than 10^{-4} M
(Table II). 4-APP riboside (1) and the 4-dimethylaminomethylen-
eamino derivative (26) had ID_{50} values of less than 10^{-6} M. The
similarity of the results from in vitro growth inhibition studies for 1 and 26 suggested that the 4-substituent of 26 might
be readily hydrolyzed to release the free $-NH_2$ compound (1). An
aqueous solution of 26 was allowed to stand at room temperature
for 2 hr and nucleoside 1 could then be detected by thin-layer
chromatography as the major product.

Replacement of the exocyclic amino group of adenosine and
certain of its analogs with a hydroxylamino substituent furnished
a series of compounds that showed toxicity in experimental animals
and/or man. For example, hydroxylaminoformycin (NSC 124165) and
4-hydroxylamino-1-(β-D-ribofuranosyl)pyrazolo(3,4-d)pyrimidine
(14) had significant host toxicity in L1210-bearing mice (data
from the Division of Cancer Treatment, NCI). On daily doses as
low as 50 mg/kg, none of the animals survived for 5 days, the

designated time for host toxicity evaluation, and treatment on days 1-9 with nontoxic doses did not result in any detectable antitumor activity. On the other hand, 4-hydroxylaminopurine riboside (HAPR) had antileukemic activity in mice at doses of 120-270 mg/kg, administered on days 1-10 after inoculation with tumor (24), but caused methemoglobin formation in dogs and was a potent hemolytic agent in man (25). In view of the species variation in response to HAPR and other hemolytic agents (25), further evaluation of the potential antitumor activity of NSC 124165 and <u>14</u> in animal species other than mouse would be of interest.

ACKNOWLEDGMENTS

The authors would like to thank Mr. Steven J. Manning and staff for the large-scale prepartion of certain intermediates, and Mrs. Suzanne Mason and Miss LaRee Jones for technical assistance.

REFERENCES

1. Davoll, J. A., and K. A. Kerridge, *J. Chem. Soc.* 2589 (1961).
2. Montgomery, J. A., S. J. Clayton, and W. E. Fitzgibbon, Jr., *J. Heterocycl. Chem. 1,* 215 (1964).
3. Revankar, G. R., and L. B. Townsend, *J. Chem. Soc. C,* 2440 (1971).
4. Schnebli, H. P., D. L. Hill, and L. L. Bennett, Jr., *J. Biol. Chem. 242,* 1997 (1967).
5. Agarwal, R. P., G. W. Crabtree, K. C. Agarwal, R. E. Parks, Jr., and L. B. Townsend, *Proc. Am. Assoc. Cancer Res. 17,* 214 (1976).
6. Henderson, J. F., and I. G. Junga, *Cancer Res. 21,* 118 (1961).
7. Bennett, L. L., Jr., P. W. Allan, D. Smithers, and M. H. Vail, *Biochem. Pharmacol. 18,* 725 (1969).
8. Hecht, S. M., R. B. Frye, D. Werner, T. Fukui, and S. D. Hawrelak, *Biochemistry 15,* 1005 (1976).
9. Montgomery, J. A., *Prog. Med. Chem. 7,* 69 (1970).
10. Earl, R. A., R. P. Panzica, and L. B. Townsend, *J. Chem. Soc. Perkin Trans. I,* 2672 (1972).
11. Lichtenthaler, F. W., P. Voss, and A. Heerd, *Tetrahedron Lett.* 2141 (1974).
12. Cuny, E. and F. W. Lichtenthaler, *Nucleic Acid Res.,* Special Publication No. 1, S25 (1975).
13. Schmidt, R. R., J. Karg, and W. Guilliard, *Angew. Chem. Int. Ed. 14,* 64 (1975).

14. Falco, E. A., B. A. Otter, and J. J. Fox, *J. Org. Chem. 35,* 2326 (1970).

15. Zemlicka, J., and F. Sorm, *Collect. Czech. Chem. Commun. 30,* 3052 (1965).

16. Zemlicka, J., and A. Holy, *Collect. Czech. Chem. Commun. 32,* 3159 (1967).

17. Elderfield, R. C., I. S. Covey, J. B. Geiduschek, W. L. Meyer, A. B. Ross, and J. H. Ross, *J. Org. Chem. 23,* 1749 (1958).

18. Townsend, L. B., B. C. Hinshaw, R. L. Tolman, R. K. Robins, and J. F. Gerster, 156th ACS Meeting, Atlantic City, New Jersey, Sept. 1968, MEDI 29.

19. Schram, K. H., Ph. D. dissertation. University of Utah, Salt Lake City, 1973.

20. Suhadolnik, R. J. *Nucleoside Antibiotics.* Wiley, New York, (1970).

21. Earl, R. A., and L. B. Townsend, *J. Heterocycl. Chem. 11,* 1033 (1974).

22. Earl, R. A., and L. B. Townsend, *J. Carbohyd. Nucleosides Nucleotides, 1,* 77 (1974).

23. Roti Roti, L. W., L. B. Townsend, and R. E. Parks, Jr., *Proc. Am. Assoc. Cancer Res. 17,* 205 (1976).

24. Burchenal, J. H., M. Dollinger, J. Butterbaugh, D. Stoll, and A. Giner-Sorolla, *Biochem. Pharmacol. 16,* 423. (1967).

25. Dollinger, M. R., and I. H. Krakoff, *Clin. Pharmacol. Therap. 17,* 57 (1975).

THE CHEMISTRY AND BIOLOGY OF SOME NEW NUCLEOSIDE ANALOGS ACTIVE AGAINST TUMOR CELLS

MIROSLAV BOBEK AND ALEXANDER BLOCH

Department of Experimental Therapeutics, Grace Cancer Drug Center, Roswell Park Memorial Institute, Buffalo, New York

Various structural analogs of the natural purines and pyrimidines, including nucleoside analogs, have proven their value in the therapy of cancer or have shown promising activity against experimental tumors (1-4). Thus, there is ample justification to continue exploring this group of compounds with the aim of discovering new analogs with useful antitumor activity.

Since little definitive information is available concerning the biochemical parameters that distinguish the cancer cell from normal cells, the design of new, hopefully selective inhibitors is based essentially on empiricism tempered by the recognition of some structure-activity relationships derived from the activity displayed by previously prepared compounds.

One approach that has been used rather successfully for the preparation of antitumor agents involves the isosteric replacement of one or more atoms or functional groups of a natural metabolite. The design of two newly prepared compounds was based on this approach.

The first compound, 1,2-dihydro-2-oxo-5-methyl pyrazine 4-oxide, was synthesized by the sequence of reactions outlined in Fig. 1. To protect the nitrogen-1 of the previously prepared 1,2-dihydro-2-oxo-5-methylpyrazine(5) from oxidation, the compound was first benzoylated using benzoylchloride in pyridine. Oxidation of the resulting intermediate with m-chloroperoxybenzoic acid, followed by hydrolysis with sodium methoxide, furnished the pyrazine analog of thymine(6).

Synthesis of the corresponding 2'-deoxy-ribonucleoside* was

*The preparation of this nucleoside was carried out in collaboration with Drs. Bardos and Berkowitz of SUNY/Buffalo(10).

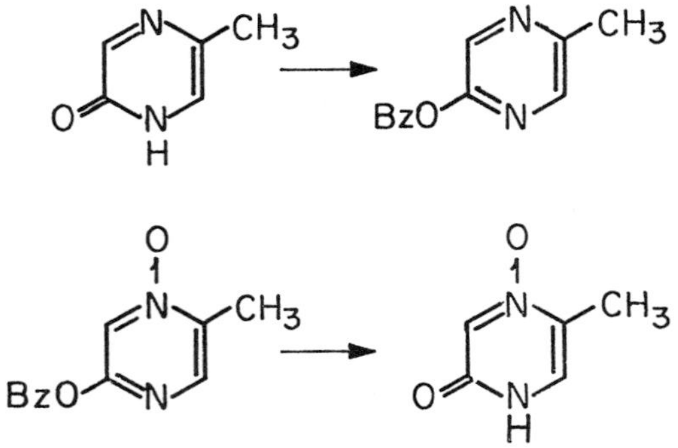

Figure 1.

carried out by the procedure summarized in Fig. 2. Two reaction
conditions were used. In one, the trimethylsilyl derivative of
2-oxo-5-methyl-pyrazine 4-oxide was condensed with blocked
2-deoxy-D-*erythro*-pentofuranosyl chloride in the presence of
molecular sieves, since it had been shown in the case of 6-meth-
ylcytosine(7) that molecular sieves enhance formation of the β-
anomer upon condensation of the silylated base with the blocked
sugar chloride. Employing this technique, the crude anomeric
mixture of blocked nucleosides was obtained in 54% yield, from
which the β-anomer was isolated in 10% and the α-anomer in 25%
yield. Thus, under our experimental conditions, molecular sieves
did not promote extensive formation of the β-anomer.

    Since the rapid removal of trimethylsilyl chloride, a by-
product of the condensation reaction, had been suggested to favor formation of the β-anomer (8), a mixture of $HgO-HgBr_2$ (1:1)
was subsequently used in the condensation reaction in place of
the molecular sieves. HgO reacts rapidly with trimethylsilyl
chloride to give hexamethyldisiloxane and mercuric chloride (9).
By this method the yield of the blocked anomeric mixture of the
nucleosides was 71%, from which the blocked β-anomer was isolated
in 44% and the α-anomer in 24% overall yield.

    Removal of the blocking groups was effected by sodium meth-
oxide in methanol. When *p*-chlorobenzoyl groups were used for
blocking, the yield of free nucleoside was 66%; with *p*-toluoyl
groups it was 50%. This difference is ascribable to the relative instability of the substituted pyrazine *N*-oxide in the
presence of sodium methoxide. A blocking group such as *p*-chloro-
benzoyl, which is more readily hydrolyzed by this agent than
the more stable *p*-toluoyl group, provides for an improved product yield.

    In Table I, the cell growth inhibitory activity of the new-
ly prepared deoxyribonucleoside is compared with that of some
previously synthesized derivatives of the parent compound, 1,2-
dihydro-2-oxopyrazine 4-oxide. Whereas the ribofuranosyl and,
more extensively, the 2'-deoxyribofuranosyl derivatives of this

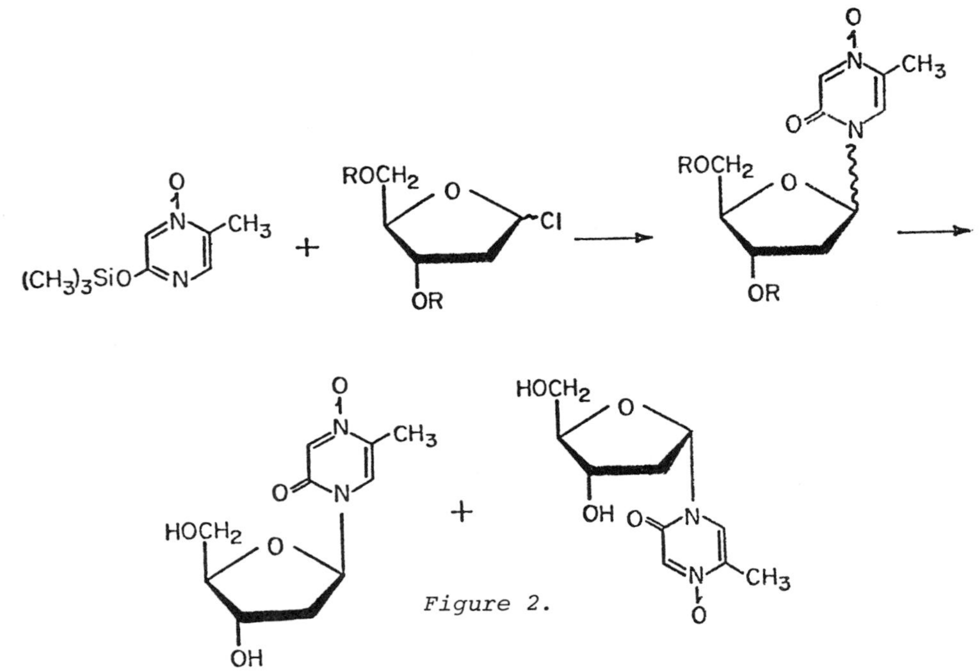

Figure 2.

base, were markedly active in the bacterial system, they were
essentially inactive against the leukemic cells.   In contrast,
the pyrazine analog of thymidine was effective against the tumor,
but was only marginally active against the bacterial cells.   This
reverse selectivity, that of the deoxyuridine analog for bacteria
and of the thymidine analog for tumor cells, serves to document
the difficulty of predicting structure-activity relationships.

Because of its *in vitro* effectiveness, the pyrazine analog
of thymidine was also evaluated for its antitumor activity *in
vivo* and, as shown in Table II, it significantly increases the
life span of mice bearing leukemia L-1210.   It should be noted
that even at the highest dose used no overt toxicity was encountered, as is also shown by the fact that the treated mice did not
suffer appreciable weight loss.

Based upon the isosteric replacement approach, a second
group of compounds encompassing the 5-ethynyl derivatives of uracil and its nucleosides was prepared(11).   Because the acetylene
group has a dimension (1.9 Å) that closely approaches that of
the methyl group (2 Å), these compounds were envisioned to function as acetylenic analogs of thymine and its nucleosides.   The
synthesis of 5-ethynyluracil was carried out as outlined in Fig. 3.
5-Formyluracil was treated with (dibromomethylene)triphenylphos-
phorane to give dibromovinyluracil.   To achieve the elimination
of HBr, this intermediate was treated with phenyllithium, which
provided 5-ethynyluracil in 52% yield.

TABLE I,   Effect of various Pyrazine Analogs of Natural Pyrimi-
           dines on Leukemic and Bacterial Cell Growth *in Vitro*

Derivative of 1,2-dihydro-2-oxo-pyrazine 4-oxide	Molar concentration for 50% growth inhibition of	
	Leukemia L-1210	S. faecium
Reference compound	$> 10^{-4}$	8×10^{-6}
1-(β-D-ribofuranosyl)[a]	$> 10^{-4}$	8×10^{-6}
1-(2-Deoxy-β-D-erythro-pentofuranosy)[b]	$> 10^{-4}$	5×10^{-11}
5-Methyl[a]	$> 10^{-4}$	8×10^{-4}
5-Methyl 1-(2-deoxy-β-D-erythro-pentofuranosyl)[c]	9×10^{-7}	8×10^{-5}

[a] M. Bobek and A. Bloch, J. Med. Chem. 15, *164*, 1972.

[b] P. T. Berkowitz, T. J. Bardos, and A. Bloch, *J. Med. Chem.* 16, 183, 1973.

[c] M. Bobek, P. T. Berkowitz, T. J. Bardos, and A. Bloch J. Med. Chem., 20, *458 (1977)*.

TABLE II,   Effect[a] of 1,2-Dihydro-1-(2-*deoxy-β-D-erythro*-pento-
            furanosyl)-2-oxo-5-methylpyrazine 4-*oxide* on Leukemia
            L-1210 *in Vivo*

Dose (mg/kg/day x 6)	ILS (% above control)	Δ Weight change (gm-day 6)
Control (tumored)	--	+2.2
100	18	+1.9
150	26	+1.5
200	30	+0.9
400	55	+0.2

[a] Female DBA/2J CR mice (5-7 weeks old, 17-19 gm), were inoculated i.p. with 1×10^6 L-1210 cells. I.P. treatment was begun 24-hr later.

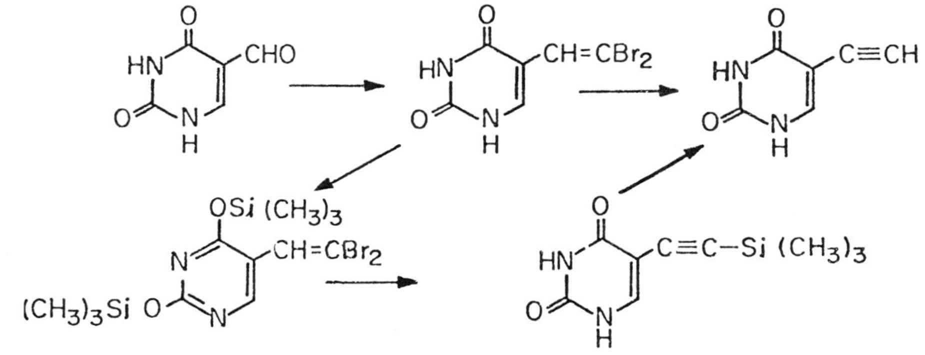

Figure 3.

In an effort to improve this yield, dibromovinyluracil was silylated before treatment with phenyllithium. This reaction, which gave the acetylenic base in 80% yield, was accompanied by an interesting rearrangement that resulted in the silylation of the acetylenic group. The facile rearrangement of the trimethylsilyl group has previously been shown to occur upon metalation of trimethylsilyloxybenzylalcohols (12), and upon metalation of (bis)trimethylsilyloxy-5-bromouracil(13). In the former instance, the rearrangement was postulated to occur via a 1,2-shift of silicone(12), whereas a 1,3-ionic shift was suggested in the latter case(13). It appears that the 1,5-rearrangement that we have observed is intramolecular, since it does not occur when the acetylenic function is present at the 6-position of silylated uracil.

The deoxyribonucleoside, 5-ethynyl-2'-deoxyuridine, was prepared in two ways (Fig. 4). In one, the (bis) trimethylsilyl derivative of dibromovinyl uracil was condensed with the p-toluoyl protected sugar chloride to give the anomeric mixture of the blocked nucleosides at an α:β ratio of 1:3. The β-anomer was deblocked with sodium methoxide in methanol, and after silylation with hexamethyldisilazane and trimethylchlorosilane in toluene, it was treated with phenyllithium and hydrolyzed to give 5-ethynyl-2'-deoxyuridine(11). Since the overall yield of the reaction procuct was only 19%, an alternative synthesis, involving the condensation of the fully silylated 5-ethynyluracil with blocked sugar chloride in the presence of $HgBr_2$-HgO, was carried out. After removal of the protecting groups with sodium methoxide in methanol, the nucleoside was obtained in an α:β ratio of 1:3, from which the β-anomer was isolated in 40% yield.

The ribonucleoside derivative of 5-ethynyluracil was prepared by essentially the same procedure. Condensation of silylated dibromovinyluracil with tetraacetylribofuranose gave pro-

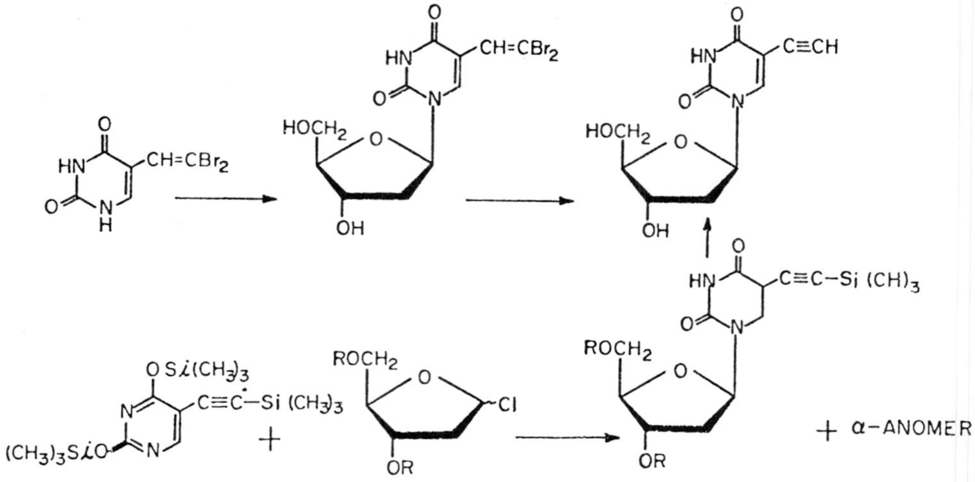

Figure 4.

tected dibromovinyluridine, which was converted to 5-ethynyl-
uridine by treatment with phenyllithium.

Because modification of the 6-position of uracil as, for
example, in 6-azauracil and 6-azauridine, led to marked antitumor
activity(14), the synthesis of 6-ethynyluracil was undertaken
(Fig. 5) by converting 6-formyluracil via the 6-dibromovinyl intermediate to the 6-ethynyl derivative. As noted above, treatment of (bis)trimethylsilyl-6-dibromovinyl uracil with phenyllithium did not result in the 1,5 migration seen in the case of
the 5-substituted analog.

The effects that these newly synthesized agents exert on
the *in vitro* growth of tumor and bacterial cells is shown in
Table III. The vinyl and dibromovinyl groups at the 5-position
of uracil or 2'-deoxyuridine did not give rise to inhibitory
activity against leukemia L-1210 or *S. faecium* cells. In contrast, the 5-ethynyl group produced inhibitory action when substituted in uridine, and 2'-deoxyuridine. When placed at the
6-position of uracil, both the dibromovinyl and the ethynyl
group gave rise to significantly active inhibitors.

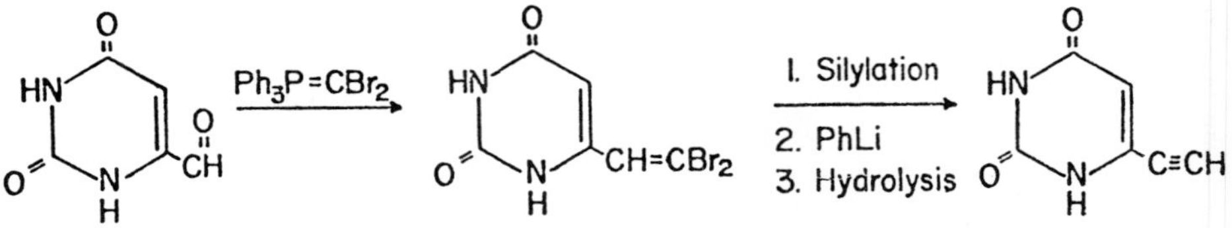

Figure 5.

TABLE III, Effect on Various 5- and 6-Substituted Pyrimidines on the *in Vitro* Growth of Bacterial and Tumor Cells

	Molar concentration for 50% growth inhibition of	
Compound	L-1210	S. faecium
5-Vinyluracil	$> 10^{-4}$	$> 10^{-3}$
5-Vinyl-2'-deoxyuridine	$> 10^{-4}$	$> 10^{-3}$
5-Dibromovinyluracil	$> 10^{-4}$	$> 10^{-3}$
5-Dibromovinyl-2'-deoxyuridine	$> 10^{-4}$	$> 10^{-3}$
5-Ethynyluracil	$> 10^{-4}$	4×10^{-4}
5-Ethynyluridine	6×10^{-5}	2×10^{-5}
5-Ethynyl-2'-deoxyuridine	2×10^{-8}	2×10^{-6}
5-Fluoro-2'-deoxyuridine	2×10^{-9}	1×10^{-11}
5-Bromo-2'-deoxyuridine	6×10^{-5}	--
6-Dibromovinyluracil	2×10^{-6}	4×10^{-5}
6-Ethynyluracil	4×10^{-6}	3×10^{-4}

It is of interest to consider some selected physical-chemical characteristics of the newly synthesized compounds in relation to the extent of their biological effect. As shown in Table IV, the inhibitory activity of the 5-substituted derivatives appears to correlate rather well with their pK_as, which are lower than those of the corresponding natural metabolites. The approximate size of the 5-substituent may also be of importance in determining the relative extent of inhibition, greater activity being encountered when the size is close to that of the substituent found in the natural metabolite (in this case hydrogen or the methyl group). Obviously, in view of the very limited sample used, this generalization must be accepted with some caution.

To gain some idea concerning the metabolic site(s) at which the new analogs may act, various natural pyrimidines were added to the L-1210 cell culture medium containing the drug. As shown in Table V, the inhibitory action of 5-ethynyl-2'-deoxyuridine, as well as the nucleoside derivative of 5-methylpyrazine 4-oxide, is reversed by the pyrimidines. Thymidine, followed by 2'-deoxyuridine, were the most effective reversing agents, suggesting that the analogs exert their inhibitory effects along the metabolic path leading to DNA, or possibly act by virtue of being incorporated into this macromolecule. Initial data obtained with 5-ethynyl-2'-deoxyuridine show (Fig. 6) that the compound interferes with the activity of a cytoplasmic thymidine kinase, partially purified from leukemia L-1210 cells(15). Upon phosphorylation by this enzyme, the resulting nucleotide interferes with thymidylate synthetase activity (Fig. 7)(15).

TABLE IV   Some Physical-Chemical Parameters of the 5-Substituted Derivatives

Compound	pK_a	Approximate size of 5-substituent $\overset{\circ}{A}$
Uracil	9.45	1.2
5-Fluorouracil	8.15	1.35
5-Ethynyluracil	8.20	1.9
5-Methyluracil (thymine)	9.85	2.0
5-Trifluoromethyluracil	7.35	2.4

TABLE V   Reversal by Natural Pyrimidines of the Inhibition of Leukemia L-1210 Cell Growth Caused by Some Structural Analogs

Inhibitor (I)	Metabolite (S)	$(I)/(S)_{50}$ [a]
5-Ethynyl-2'-deoxyuridine	dTR	0.8
	dUR	0.01
	dCR	0.005
	UR	0.002
	CR	0.001
1,2-Dihydro-1-	dTR	50
(2-deoxy-β-D-erythro-	dUR	2
pentofuranosyl)-2-oxo-5-	dCR	0.2
methylpyrazine 4-oxide	UR	0.04
	CR	0.02

[a] *Ratio of inhibitor concentration to substrate concentration providing 50% growth, as determined at 1 x 10^{-5} M substrate.*

In view of the effectiveness of the acetylenic function at the 5-position of 2'-deoxyuridine, other nucleoside derivatives bearing the ethynyl group at various positions in the carbohydrate moiety of pyrimidines and purines, as well as at the purine heterocycle, have been or are being prepared. The approach used for the synthesis of 5-ethynyl nucleosides lends itself equally well to modification of the carbohydrate moiety, as exemplified by Figs. 8 and 9, where the synthesis of the 5'-acetylenic derivatives of uridine and thymidine is outlined.

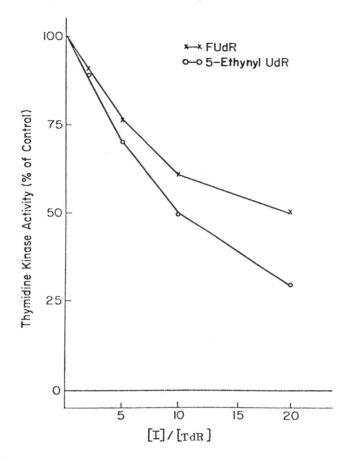

Figure 6.

    Inhibition of thymidine kinase partially purified from leu-
kemia L-1210 cells[15], by 5-ethynyl 2'-deoxyuridine as compared
to 5-fluoro-2'-deoxyuridine. The incubation mixture contained
100μM thymidine. The assay procedures were as described.[16]

    The pronounced inhibitory potency of a structural analog is
not in itself a measure of its potential utility. For example,
as shown in Fig. 10, we have prepared the 4'-thio analog of guanosine by condensation of the silylated base with the protected
sugar acetate in the presence of stannic chloride, followed by
removal of the blocking groups with methanolic ammonia. Whereas

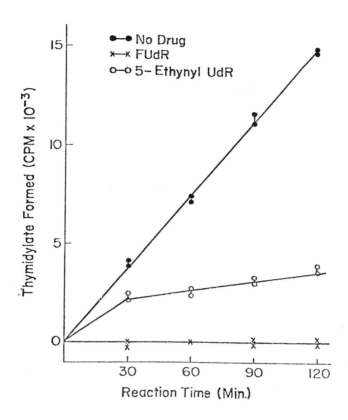

Figure 7.
    Inhibition of thymidylate synthetase partially purified from
leukemia L-1210 cells[15] by the 5'-phosphates of 5-ethynyl 2'-
deoxyuridine and 5-fluoro-2'-deoxyuridine. The analogs were
first incubated for 20 hrs in the presence of thymidine kinase
and 4 mM ATM-Mg[2+]. Portions of the reaction products were
then added to the thymidylate synthetase assay mixture[16,17]. At
the nucleoside level the analogs did not inhibit the reaction.

this compound, at concentrations up to 10^{-4} M, showed no inhibitory activity in either the L-1210 or the *S. faecium* systems, it
has the capacity to strongly synergize 6-thionosine, as shown by
an isobologram (Fig. 11) based upon data obtained with *S. faecium.*
The pronounced deviation from the dotted line, which represents

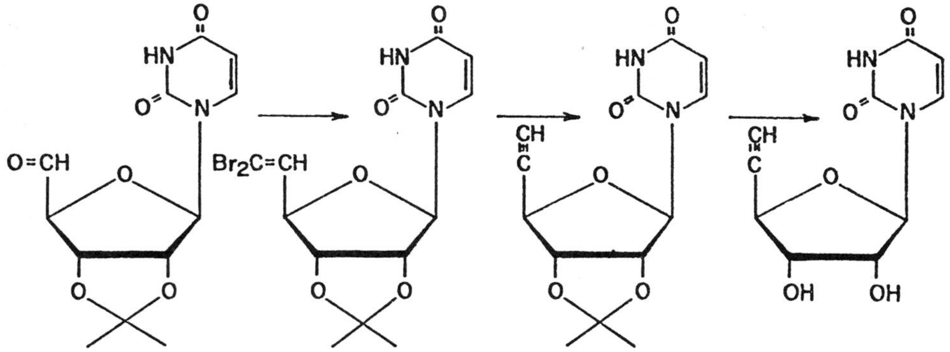

Figure 8.

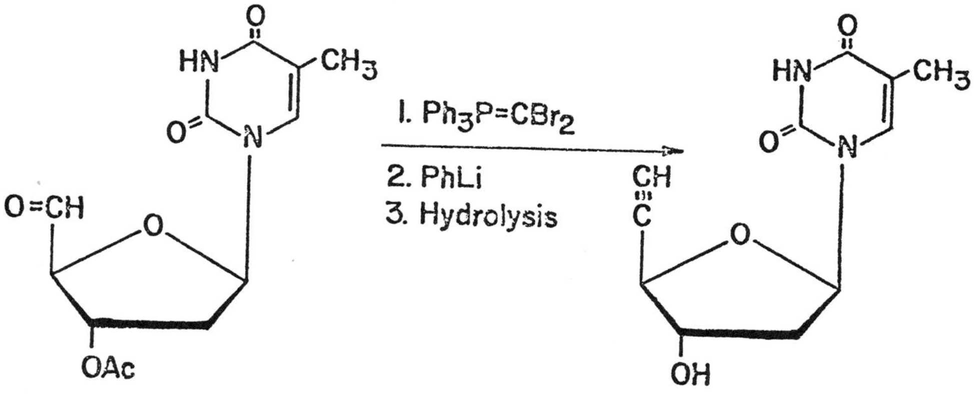

Figure 9.

theoretical additiveness, is a reflection of the strong potenti-
ating action of the compound in this assay system.

The design of purine or pyrimidine analogs should not, however, be limited to the modification of only their basic structure. Their specificity of action might be expected to be enhanced by conjugation with carriers that have affinity for specific tissues. Based upon this rationale, FUDR was coupled to
the 3-position of estradiol as outlined in Fig. 12. Estradiol
was treated with chloroacetic acid in the presence of NaOH.
Formylation, followed by treatment with $SOCl_2$ gave estradiol

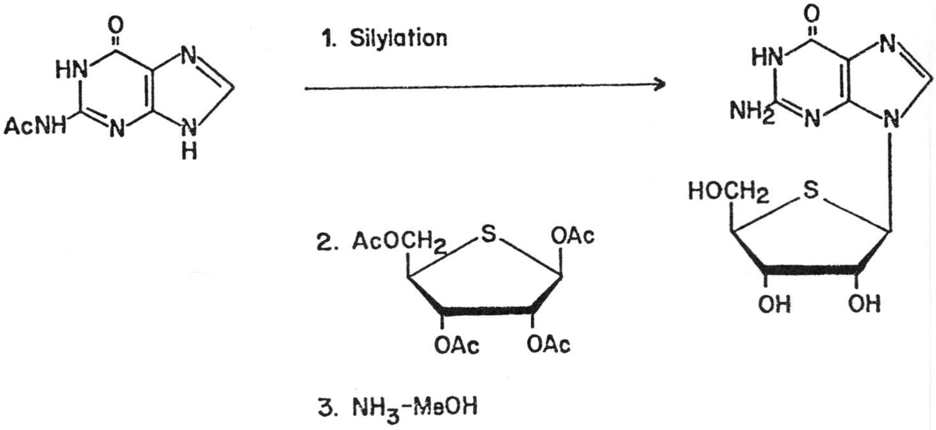

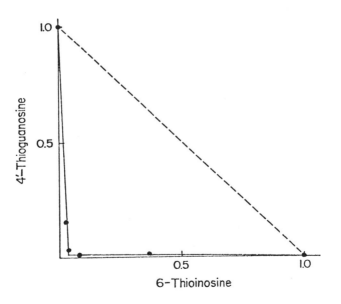

Figure 10.

Potentiation of the Inhibitory Activity of 6-Thioinosine
by 4'-Thioguanosine

Figure 11.

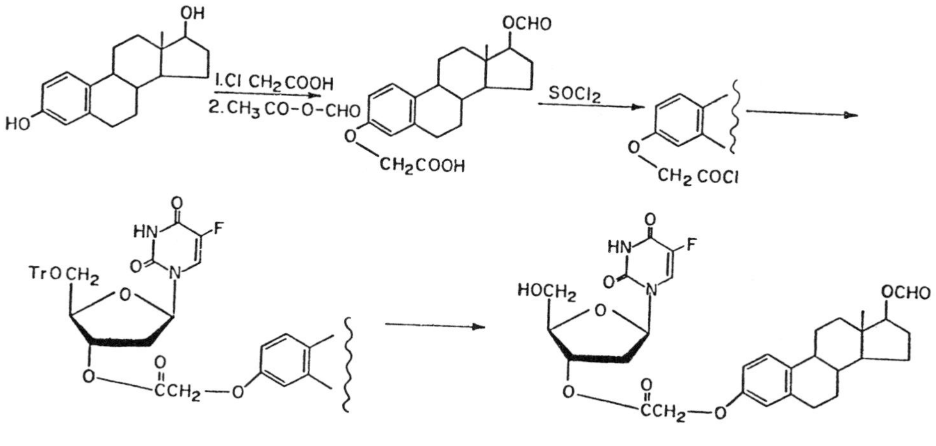

Figure 12.

acetyl chloride, which was then condensed with 5'-trityl FUDR. Deblocking of the trityl group was effected with silica gel.

To permit evaluation of the importance of the site of attachment upon biological activity, conjugation by means of a maleoyl bridge was also carried out at the 2-position of estradiol. 2-nitroestradiol was protected with tetrahydropyranyl groups, reduced and condensed with 5'-trityl-3'-maleoyl-FUDR followed by removal of the protecting groups under mild acid conditions.

The evaluation of these agents against tumors that contain estrogen receptors is currently in progress. In summary, nucleoside analogs continue to show encouraging promise of utility in the treatment of neoplastic diseases.

ACKNOWLEDGMENTS

The contribution by Drs. T. Y. Cheng, Y. L. Fu, J. Perman, and R. A. Sharma to various phases of this work is gratefully acknowledged. The studies were aided by Grants CI-125 from the American Cancer Society and CA-12585 and CA-13038 from the National Cancer Institute.

REFERENCES

1. R. B. Livingston and S. K. Carter, *Single Agents in Cancer Chemotherapy*, Plenum, New York, 1970.
2. G. Brule, S. J. Eckhardt, T. C. Hall, and A Winkler, *Drug Therapy of Cancer*, WHO, Geneva, 1973.
3. A. Bloch, The design of biologically active nucleosides, in *Drug Design*, Vol. IV (A. J. Ariens, ed), Academic Press, New York, 1974.
4. A. Bloch (ed.), Chemistry, biology and clinical uses of nucleoside analogs, *Ann. NY Acad. Sci. 255*, 1-610 (1975).
5. G. Karmas and P. E. Spoerri, *J. Am. Chem. Soc. 74*, 1580 (1952).
6. M. Bobek and A. Bloch, *J. Med. Chem. 15*, 164 (1972).
7. M. W. Winkley and R. K. Robins, *J. Org. Chem. 33*, 2822 (1968).
8. M. P. Kotick, C. Szantay, and T. J. Bardos, *J. Org. Chem. 34*, 3806 (1969).
9. E. Wittenburg, *Chem. Ber. 101*, 1095 (1968).
10. M. Bobek, P. Berkowitz, T. J. Bardos, and A. Bloch, *J. Med. Chem., 20*, 458 (1977).
11. J. Perman, R. A. Sharma, and M. Bobek, *Tetrahedron Lett.* 2427 (1976).
12. A. Wright and R. West, *J. Am. Chem. Soc. 96*, 3214 (1974).
13. I. Arai and G. D. Daves, Jr., *J. Chem. Soc. Chem. Comm.* 69 (1976).
14. R. F. Handschumacher, P. Calabresi, A. D. Welch, V. Bono, H. Fallon, And Frei, *Cancer Chemother. Rep. 21*, 1 (1962).
15. M. Bobek, J. Perman, R. A. Sharma, Yung-chi Cheng, and A. Bloch, manuscript in preparation.
16. L.-S Lee and Y.-C. Cheng. *J. Biol. Chem. 251*, 2600 (1976).
17. D. Roberts, *Biochemistry 5*, 3546 (1976).

POTENTIATION OF THE CHEMOTHERAPEUTIC ACTION
OF ANTINEOPLASTIC AGENTS BY NUCLEOSIDES

H. OSSWALD

Institute for Toxicology and Chemotherapy, German
Cancer Research Center, Heidelberg, F.R.G.

ABSTRACT

Certain nucleosides enhance the chemotherapeutic action of antineoplastic agents without increasing the toxicity when these
combination partners are applied at different intervals. The
prevention of toxic effects of methotrexate with thymidine
during the rescue phase indirectly increases the curative action
of this combination. Thymidine or inosine potentiate the chemotherapeutic action of cyclophosphamide. Cytidine or guanosine
acts synergistically in combination with 5-fluorouracil. Bleomycin exerts the best chemotherapeutic synergism except when
administered simultaneously. No distinct correlation exists
between the optimal interval of the combination partners and
the generation time of the used murine tumors.

INTRODUCTION

Recent experiments have revealed that when thymidine and
cyclophosphamide (2,4,5), or vinblastine and thymidine (2,5)
are administered in combination (at different times), their
chemotherapeutic effect on several types of transplantable murine tumors was enhanced without increasing the toxicity (2,3,4,
5). The clinical application of these combinations (1,8) has
furnished interesting results in patients with advanced mammary
carcinoma, ovarian carcinoma, and cervix carcinoma, after failure with the usual treatment. This first clinical trial was
characterized by long-lasting remissions (predominantly more
than a year) following the course of treatment with one of these

combinations. Two patients who suffered from ovarian carcinoma (stage IV) are alive without any signs of relapse after four years. Consequently, these results have encouraged us to study other combinations of nucleosides and antineoplastic agents.

MATERIAL AND METHODS

SPF random bred female Swiss mice with body weights of 30 ± 1 gm (Broekman Instituut, Helmond, Netherlands) were housed in Makrolon cages (type II), there were 15 animals per cage and they received an Altromin-M pellet diet and water *ad libitum*. All experiments were performed in animal laboratories with controlled temperature ($26^{o}C$), humidity (60%), and a ninefold change of the air per hour. In regard to the chemotherapeutic sensibility of the antineoplastic agents used different transplantable murine tumors (Ehrlich-carcinoma, SPA-reticulum-cell-sarcoma, and HRS-reticulum-cell-sarcoma) were chosen for the experiments. The tumor-bearing mice were autopsied at the end of the experiments and the tumors evaluated by weight. "The term cured animals" in the following tables is related to a complete tumor regression, which lasted three months after the end of the experiment without a relapse. The results were controlled by the four-table-test with the degree of 2α equal to 0.05. All experiments were reproduced to prevent casual errors. The nucleosides were prepared as pyrogen-free substances by Pharma Waldhof, Mannheim.

RESULTS

The combinations of thymidine and cyclophosphamide were tested on the SPA-reticulum-cell-sarcoma, a particularly resistant tumor strain. Therefore, high doses of cyclophosphamide were necessary in these experiments, with an average tumor weight of 3.8 gm at the onset of treatment (5).

The results in Table I show remarkable differences between the monotherapy with cyclophosphamide and the cyclophosphamide/thymidine combinations. The monotherapy with cyclophosphamide gave no curative action when the same dosage was applied as in the combination (group 3). The doubled dose of cyclophosphamide affected a curative action in 10 out of 15 animals (group 4). However, the combination of thymidine and cyclophosphamide induced a curative effect in 12 out of 15 mice (group 1), and the thymidine/cyclophosphamide/thymidine combination (group 2) caused a complete curative action.

A comparison of the efficiency of the cyclophosphamide/in-
osine and the thymidine/cyclophosphamide/thymidine combination (8)
on intramuscularly implanted Ehrlich-carcinoma (average weight
of tumor 6.0 gm at onset of treatment) is shown in Table II.   In
this experiment, in which the tumor weight amounted to nearly
one-fifth of the animals, the curative results were not impressive.   Only the cyclophosphamide/inosine combination (group 2)
showed significant differences in a comparison with the cyclophosphamide monotherapy (group 4), but in this experiment the treatment was started at an advanced stage of tumor-growth.

The combinations of 5-fluorouracil (5-FU) with various nucleosides, (Table III) in various sequences of the combinations
revealed a distinct difference in curative action in some groups
of treatment.   Thymidine, applied at an interval of 18 hr after
5-FU administration, induced an increased curative action only
when it was given at the two following days.   The addition of
cytidine or guanosine gave significant curative actions in comparison to 5-FU alone, if these nucleosides were used 16 or 18
hr after 5-FU administration.

Of particular interest was the methotrexate/thymidine combination in regard to the decrease of methotrexate toxicity.
This was observed when thymidine was applied during the rescue
phase combined with tetrahydrofolic acid (citrovorum factor).
The increase of the curative action of this combination (group
2) seemed to be correlated, mainly with the decrease of toxicity
(Table V).   The average of body weight between onset and end of
experiment showed a remarkable difference from methotrexate monotherapy.   In these experiments a low dosage of tetrahydrofolic
acid was used to test the rescue ability of thymidine.   Mean-
while, clinical experiences are corroborating these experiments.
The painful methotrexate stomatitis was prevented when thymidine
was administered during rescue phase.

The most interesting curative results could be those obtained by the combination of bleomycin and inosine (7).   Several
other nucleosides in combination with bleomycin also revealed
synergistic effects, but the extent was not so impressive.   Surprisingly, the bleomycin/inosine combination, in which no interval between the combination partners was used, showed the best
curative action.   Contrary to the other nucleosides used, in all
the experiments inosine induced an unspecific tumor inhibition
in the dosage used (Table V).   Also the dosage of bleomycin monotherapy gave unspecific tumor inhibition.   On the contrary, curative action and tumor inhibition of the bleomycin/inosine combinations are significant.   The clinical application of a metho-
trexate/tetrahydrofolic acid/thymidine bleomycin/inosine combination yielded interesting results in squamous cell carcinomas of
the oral cavity and the tongue.   Not only was the usual stomatitis prevented, but also most of the other usual side effects of

TABLE I. Differences in the Chemotherapeutic Action of Different Combinations of Thymidine and Cyclophosphamide on SPA-Tumor Implanted Intramuscularly in Swiss-Mice. (Treatment 8 Days after Implantation. Average Weight of Tumor: 3.8 Grams. Duration of Treatment: 2 Weeks).

Group Nr.	Scheme of Treatment Single Doses per Week					Ratio of Cured Animals to Total Number	Average Weight of Tumor in Grams	Average of Body-Weight Change between Onset and End of Experiment in Grams
	Nucleoside	Time-Interval	Alkylating Agent	Time Interval	Nucleoside			
1	Thymidine 3x100 mg/kg sc, 3 Hours Interval	14 H	Cyclophosphamide 120 mg/kg sc.			12/15	0.8 ± 0.3	- 1.7
2	Thymidine 3x100 mg/kg sc, 3 Hours Interval	14 H	Cyclophosphamide 120 mg/kg sc	24 H	Thymidine 3x100 mg/kg sc. 3 Hours Interval	15/15	0	- 1.9
3			Cyclophosphamide 120 mg/kg sc.			0/15	2.7 ± 0.7	- 1.7
4			Cyclophosphamide 240 mg/kg sc.			10/15	0.8 ± 0.9	- 3.1
5	Untreated Control					0/15	8.2 ± 0.8	+ 6.6

From reference 5, by permission of Thieme Verlag Stuttgart.

TABLE II. Chemotherapeutic Action of Combinations of Thymidine or Inosine with Cyclophosphamide on Ehrlich-Carcinoma Implanted Intramuscularly in Swiss-Mice (Treatment: 14 Days after Implantation. Average Weight of Tumor 6.0 Grams. Duration of Treatment: 5 Weeks).

Group Nr.	Scheme of Treatment (Single Doses per Week)					Ratio of Cured Animals to Total Number	Average Weight of Tumor in Grams	Average of Body-Weight between Onset and End of Experiment (%)
	Nucleoside	Time-Interval	Alkylating Agent	Time Interval	Nucleoside			
1			Cyclophosphamide 140 mg/kg sc.		Inosine 300 mg/kg orally	8/15	1.2 ± 0.5	− 3.3
2	Inosine 300 mg/kg orally	18^h	Cyclophosphamide 140 mg/kg sc.	2^h	Inosine 300 mg/kg orally	9/15	2.0 ± 0.8	− 3.1
3	Thymidine 2 x 250 mg/kg sc.	13^h	Cyclophosphamide 140 mg/kg sc.	24^h	Thymidine 2 x 250 mg/kg sc.	6/15	3.2 ± 0.9	− 2.6
4			Cyclophosphamide 140 mg/kg sc.			1/15	3.1 ± 1.1	− 5.7
5			Cyclophosphamide 210 mg/kg sc.			3/15	1.9 ± 0.6	− 11
6	Untreated Control					0/15	22.4 ± 5.7	− 18

From reference 8 by permission of Thieme Verlag Stuttgart.

TABLE III. Overadditive Chemotherapeutic Action of Nucleoside-5-Fluorouracil-Combinations on RS-Reticulum-Cell-Sarcoma Implanted Intramuscularly in Swiss-Mice. (Onset of Treatment: 5 Days after Implantation Average of Tumor-Weight 3.8 Grams). Duration of Treatment: 5 Weeks.

Group Nr.	Scheme of Treatment (Single Doses per Week)					Ratio of Cured Animals to Total Number	Average Weight of Tumor in Grams	Average of Body-Weight Change between Onset and End of Experiment (%)
	Nucleoside	Time Interval	Antimetabolite	Time Interval	Nucleoside			
1	Cytidine 3 x 100 mg/kg s.c.	16^h	5-FU 50 mg/kg s.c.	18^h	Thymidine 2 x 4 x 100 mg/kg orally	8/15	2.9 ± 2.0	− 4.8
2	Cytidine 3 x 100 mg/kg s.c.	16^h	5-FU 50 mg/kg s.c.			8/15	2.1 ± 1.6	− 1.4
3			5-FU 50 mg/kg s.c.	16^h	Cytidine 3 x 100 mg/kg s.c.	9/15	3.1 ± 1.8	− 1.8
4	Guanosine 3 x 80 mg/kg orally	18^h	5-FU 50 mg/kg s.c.			8/15	3.6 ± 1.1	− 1.7
5			5-FU 50 mg/kg s.c.	18^h	Guanosine 3 x 80 mg/kg orally	9/15	3.3 ± 1.4	− 0.9
6			5-FU 50 mg/kg s.c.			2/15	5.8 ± 4.1	− 2.5
7	Untreated Control					0/15	13.5 ± 3.4	− 6.2

TABLE IV. Influence of Nucleoside-Timing on the Chemotherapeutic Action of Methotrexate (MTX) on Sarcoma 180 Implanted Intramuscularly in Female Swiss-Mice. (Treatment: 7 Days after Implantation. Average Weight of Tumor at Onset of Experiment: 2.5 Grams) Duration of Treatment: 3 Weeks.

Group Nr.	Scheme of Treatment (Single Doses per Week)				Ratio of Cured Animals to Total Number	Average of Body-Weight between Onset and End of Experiment (%)
	Antimetabolite	Timing of Citrovorum-Factor after MTX-Treatment	Timing of Nucleosides after MTX-Treatment	Nucleoside		
1	MTX 2 x 200 mg/kg i.v. 3^h Interval	CF 2 x 10 mg/kg s.c. 18^h, 21^h			3/20	− 15.4
2	MTX 2 x 200 mg/kg i.v. 3^h Interval	CF 2 x 10 mg/kg s.c. 18^h, 21^h	18^h 21^h	Thymidine 450 mg/kg s.c. Thymidine 2x3x150 mg/kg s.c.	16/20	+ 3.2
3	MTX 2 x 200 mg/kg i.v. 3^h Interval	CF 2 x 10 mg/kg s.c. 18^h, 21^h	18^h 21^h	Cytidine 300 mg/kg s.c. Cytidine 3x100 mg/kg s.c.	7/15	+ 3.7
	Untreated Control					− 4.9

TABLE V. Overadditive Chemotherapeutic Action of Inosine-Bleomycin-Combinations on HRS-Reticulum-Cell-Sarcoma Implanted Intramuscularly in Swiss-Mice. (Onset of Treatment: 3 Days after Implantation. Average Weight of Tumor 1.1 Grams. Duration of Treatment: 4 Weeks).

Group Nr.	Scheme of Treatment (3 Doses or Combinations per Week in Time Intervals of 48 Hours)			Ratio of Cured Animals to Total Number	Average Weight of Tumor in Grams	Average of Body-Weight between Onset and End of Experiment (%)
	Antibiotic	Time Interval	Nucleoside			
1	Bleomycin 5 mg/kg i.v.	0	Inosine 3 x 200 mg/kg sc	9/15	0.7 ± 0.9	- 2.1
2	Bleomycin 5 mg/kg i.v.	6	Inosine 3 x 200 mg/kg sc	7/15	1.5 ± 1.7	- 2.2
3	Bleomycin 5 mg/kg i.v.	18	Inosine 3 x 200 mg/kg sc	6/15	1.6 ± 1.5	- 3.9
4			Inosine 3 x 200 mg/kg sc	0/15	6.4 ± 1.8	- 4.1
5	Bleomycin 5 mg/kg i.v.			0/15	5.1 ± 2.8	- 4.8
6	Untreated Control			0/15	8.9 ± 3.8	- 4.9

From reference 7 by permission of *Elsevier/North-Holland, Amsterdam.*

bleomycin were significantly diminished. Up to now, nearly 12 months after this combined course, no lung fibrosis in patients was detectable.

DISCUSSION

These various investigations (2-8) failed to allow us to predict which nucleosides were suitable to be combined with an antineoplastic agent. The results also did not allow us to predict the optimal interval or sequence of combination partners. In most of these combinations, the conversion of the nucleoside to the triphosphate derivative may be important. Not all combinations are effective. Some nucleosides, in combination with antineoplastic agents, diminished only the toxicity (methohexate-thymidine), without influencing the chemotherapeutic action, while others increased the toxicity and diminished or increased the chemotherapeutic action of combination. The most elegant hypothesis of a synchronizing effect does not allow us sufficient knowledge to support any firm conclusions (2-8). Perhaps an assumption that these combinations exert their influence on the repair mechanism might be a more tenable hypothesis.

REFERENCES

1. Brachetti, A., Leonhardt, A., Limburg, H., Osswald, H., and Schmähl, D., Klinische Erfahrungen mit einer Kombinationstherapie bei metastasierenden Genital- und Mammakarzinamen, *Munch. Med. Wsch. 116*, 2037-2042 (1974).
2. Osswald, H., Synergismus der chemotherapeutischen Wirkung einer Kombination von Thymidin und Endoxan (Cyclophosphamid) beim Ehrlich-Ascites-Tumor, *A. Krebsforsch. 74*, 374-382 (1970).
3. Osswald, H., Potenzierung der chemotherapeutischen Wirkung von Vinblastin durch Thymidin. *Drug Res. 22*, 1421 (1972).
4. Osswald, H., Überadditiver Synergismus einer Kombination von Cyclophosphamid mit Thymidin, Adenosin oder Uridin bei Transplantationstumoren, *Drug Res. 22*, 1184-1188 (1972).
5. Osswald, H., Überadditiver Synergismus der chemotherapeutischen Wirkung bei Kombination von Nukleosiden und antineoplastischen Chemotherapeutika, in *Aktuelle Probleme der Therapie maligner Tumores*, Thieme Verlag, Stuttgart, 1973, pp. 258-270.

6. Osswald, H. Potenzierung der chemotherapeutischen Wirkung
    von Adriamycin durch bestimmte Nukleoside, in *Ergebnisse
    der Adriamycin-Therapie,* Springer Verlag, Berlin-Heidel-
    berg-New York, 1975, pp. 24-31.
7. Osswald, H., Youssef, M, Potentiation of the chemotherapeutic
    action of bleomycin by combination with inosine on HRS-
    sarcoma, *Cancer Lett. 1,* 55-58 (1975).
8. Osswald, H., Leonhardt, A., Gründlagen und Erfuhrungen
    beider Kombinationstherapie von Tyrosta Zytostatika mit
    Nukleosiden, in Prophylaxe und Therapie von Behemdlüngs-
    folgen bei Karazinomen der Fran, Thieme Verlag, Stutt-
    gart, 1976, pp. 134-145.

COFORMYCIN AND DEOXYCOFORMYCIN:
TIGHT-BINDING INHIBITORS OF ADENOXINE DEAMINASE

R. P. AGARWAL, SUNGMAN CHA, G. W. CRABTREE,
R. E. PARKS, JR.

Section of Biochemical Pharmacology
Division of Biology and Medicine
Brown University
Providence, Rhode Island

ABSTRACT

Studies were conducted employing a series of adenosine deaminase
(ADA) inhibitors, varying from readily reversible to tight binding (K_i values 10^{-6} to $10^{-12}M$). Each class presents unique problems in kinetic analysis. The theoretical approaches described
here facilitate the determination of K_i values and other parameters, e.g., catalytic number, molar equivalency of the enzyme,
the velocity constants of the association and dissociation reactions between the enzyme and inhibitor, as well as the mechanism of inhibition, i.e., competitive, noncompetitive, and uncompetitive. The ADA inhibitors employed were coformycin (3-β-D-
ribofuranosyl-6,7,8-trihydroimidazo(3,4-d)(1,3)diazepin-8-(R)ol)
and deoxycoformycin (d-coformycin or (R)-3-(2-deoxy-β-D-erythro-
pento-furanosyl)-3,6,7,8-tetrahydroimidazo(4,5-d)(1,3)diazepin-
8-ol or Covidarabine[R] (tight-binding inhibitors); erythro-9-(2-
hydroxy-3-nonyl) adenine (EHNA), a semitight-binding inhibitor;
and 1,6-dihydro-6-hydroxymethyl purine ribonucleoside (DHMPR),
a readily reversible inhibitor. Application of the new theoretical approaches to the study of ADA enabled estimation of K_i values
of 2.5 x 10^{-12} M for deoxycoformycin, 1.2 x 10^{-10} to 1 x 10^{-11} M
for coformycin, 1.6 x 10^{-9} M for EHNA, and 1.3 x 10^{-6} M for
DHMPR. The inhibition of ADA by coformycin in intact human erythrocytes greatly enhanced the formation of analog nucleotides
of arabinosyl adenine, 2,6-diaminopurine ribonucleoside, 8-aza-
adenosine, N^6-hydroxyaminopurine ribonucleoside, and formycin A.
Some of the potential therapeutic uses of these inhibitors are
discussed.

I. INTRODUCTION

Recently, there has been increasing interest in analogs of adenosine as potential therapeutic agents, e.g., antineoplastic, antiviral, antiparasitic, and as blockers of blood platelet aggre-
gation (for recent reviews, see Refs. 1-4). The major biochemical mechanism that inactivates many of these analogs is deamination by the enzyme, adenosine deaminase (ADA, adenosine aminohydrolase, EC 3.5.4.4), which in the human is found in various
tissues, including erythrocytes. A typical reaction catalyzed·
by ADA is

$$\text{Adenosine} + H_2O \xrightarrow{\text{ADA}} \text{Inosine} + NH_3 \tag{1}$$

Thus an active antiviral compound, e.g., arabinosyl adenine (ara-
A), may be inactivated by erythrocytic ADA before reaching the
target site, i.e., a DNA virus. Therefore, it has become a
matter of some urgency to develop adenosine analogs that are resistant to this enzymic deamination, but which may still be converted to the nucleotide level by enzymes in the target cell.
Another approach to this problem has been the search for powerful
and specific inhibitors of ADA.

The present chapter deals with a series of investigations
carried out in this laboratory over the past few years: (a) the
development of new theoretical approaches and their application
to the study of tight-binding enzyme inhibitors, including the
inhibitors of ADA, coformycin, and deoxycoformycin (5-8); and
(b) the effect of these potent ADA inhibitors on incorporation
into intracellular nucleotide pools of several adenosine analogs.
The analogs that have been chosen for study are arabinosyl adenine, formycin A, 8-azaadenosine, N^6-hydroxyaminopurine ribonucleoside (HAPR), and 2,6-diaminopurine ribonucleoside (DAPR).
The chemical structures of several compounds discussed in this
manuscript are presented in Fig. 1.

II. THEORETICAL BASIS FOR THE STUDY OF TIGHT-BINDING ENZYME INHIBITORS

In kinetic studies of enzymes in the presence of potent inhibitors, e.g., adenosine deaminase inhibited by coformycin, the
classic experimental techniques based on steady-state kinetics
cannot be applied for the determination of the inhibition mechanisms or inhibition constants. The reasons are twofold. First,
under commonly employed experimental conditions, the transient
or non-steady-state phase of the interaction between the enzyme

SUBSTRATES

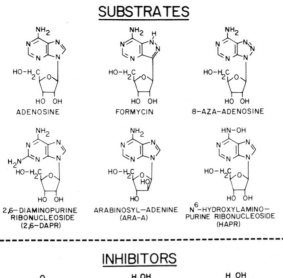

INHIBITORS

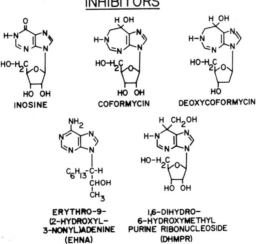

Figure 1. Structures of substrates and inhibitors of adenosine deaminase.

and inhibitor is so markedly prolonged that steady-state assumptions are not valid. Second, the optimal experimental inhibitor concentrations are so low that they are in the same order of magnitude as the molar concentration of the enzyme. Therefore, depletion of the free inhibitor by binding to the enzyme may not be neglected. In recent publications from this laboratory, the kinetic properties of enzymes in the presence of tight-binding inhibitors have been extensively examined (5-8).

A general noncompetitive inhibition mechanism is represented by

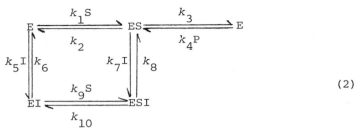

$$(2)$$

where E, I, P, and S are enzyme, inhibitor, product, and substrate, respectively. Kinetic constants are defined as $K_m = (k_2 + k_3)/k_1$; $K_m' = k_{10}/k_9$; $K_{is} = k_6/k_5$; $K_{ii} = k_8/k_7$; $V = k_3 E_t$. This generalized noncompetitive inhibition model is also applicable to other inhibition mechanisms: to competitive inhibition by setting $k_7 = 0$ $K_{ii} = \infty$, and $K_m' = \infty$; to uncompetitive inhibition by setting $k_6 = 0$, $K_{is} = \infty$, and $K_m' = 0$; and to classical noncompetitive inhibition by setting $K_{is} = K_{ii}$.

For enzymic reactions inhibited by a tight-binder, the following assumptions are reasonable:

(a) Steady-state conditions are reached instantaneously between ES, and EI and ESI.

(b) Prolonged non-steady-state conditions exist between E and EI, and between ES and ESI.

(c) The substrate concentration is much greater than the enzyme concentration so that depletion of free substrate by binding to the enzyme is negligible compared to the total substrate concentration, but depletion of free inhibitor by binding to the enzyme may or may not be negligible compared to the total inhibitor concentration.

(d) Experimental observations are made only while the effects of substrate depletion by conversion to product and the effects of the product (i.e., product inhibition and the reverse reaction are negligible.

(e) Unless stated otherwise, the enzymic reaction is started by the addition of enzyme to a reaction mixture containing the substrate and the inhibitor, i.e., no preincubation of the enzyme with the inhibitor.

A time-dependent rate equation has been derived (7) for enzymic reactions under the conditions assumed above:

$$v = \frac{v_s + (v_o(1 - \gamma) - v_s)e^{-\lambda t}}{1 - e^{-\lambda t}}$$

$$(3)$$

where v_s is the steady-state velocity; v_o the uninhibited velocity; λ the decay constant, a parameter that accounts for the
slow reaction of the inhibitor; and λ the depletion constant, a
number between 0 and 1, which accounts for the depletion of the
free inhibitor by binding to the enzyme.

If I_t (total inhibitor concentration) is much greater than
both E_t (total enzyme concentration) and K_i, both λ and v_s become
negligible, and Eq. (3) reduces to

$$\ln v = \ln v_o - \lambda t \tag{4}$$

which provides the basis for the determination of λ. A detailed
description of the development and use of these equations has
been recently reported (7).

The nature of the time-course of an enzymic reaction in the
presence of a tight-binding inhibitor is markedly affected by
the order of addition of reactants. If the enzyme is preincubated with the inhibitor and the reaction is started by addition
of the substrate, the velocity is initially slow and then gradually accelerates to a steady-state rate. If such preincubation
is carried out for a sufficiently long period of time, equilibrium is established between E, I, and EI. The EI complex thus
formed dissociates so slowly that the enzyme initially present
in the EI complex does not contribute significantly to the catalytic reaction. Therefore, the "initial velocity" observed
after the addition of the substrate represents the concentration
of the free enzyme (E) at the end of the preincubation. As time
goes on, however, the EI complex dissociates, although often
slowly, depending on the K_i value. The free enzyme liberated
from the EI complex has a better chance of reacting with a substrate, rather than an inhibitor molecule, because of the much
greater number of substrate molecules in the reaction mixture.
This process continues until new steady-state conditions are
reached among the enzymic species, i.e., E, ES, and EI. Therefore, the catalytic velocity gradually increases to a limiting
value, the steady-state velocity, as illustrated in Fig. 2.

Conversely, if the enzymic reaction is started by the
addition of the enzyme to a reaction mixture containing both
substrate and inhibitor, the initial velocity is similar to the
uninhibited rate because the association of the inhibitor with
the enzyme is slow due to the fact that there are fewer molecules
of the inhibitor than the substrate in the reaction mixture. As
increasing numbers of enzyme molecules are progressively removed
from the catalytic cycle, in the form of EI, the velocity gradually decreases until it reaches the steady-state level illustrated in Fig. 2.

This special behavior of enzymic reactions in the presence
of tight-binding inhibitors explains the earlier observations
of Easson and Stedman (9) and Werkheiser (10) that all types of

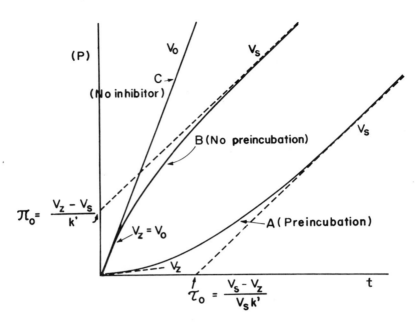

Figure 2. Schematic illustration of lag periods in the rate of formation of product in an enzymic reaction in the presence of a tight-binding inhibitor. Curve A, reaction curve for the case where the enzyme is preincubated with the inhibitor and the reaction is started with addition of the substrate; curve B, the reaction is started by adding the enzyme to the mixture containing both the substrate and inhibitor; and curve C, the control in the absence of the inhibitor. (From Cha (5))

"stoichiometric inhibitions are apparently non-competitive." The effect of preincubation with an inhibitor is as though the enzyme concentration were reduced by the amount of EI complex formed. Under these conditions the apparent K_m remains the same but the apparent V_{max} decreases in the presence of the in- hibitor, and the result resembles classical noncompetitive in- hibition. However, great caution must be employed in interpre- ting these data, as demonstrated earlier in studies of the inhi- bition of ADA by coformycin (6).

First of all, the apparent noncompetitive inhibition pattern, in which the straight lines of the Lineweaver-Burk plot intersect on the 1/S axis, would be observed only if the enzyme were pre- incubated with the inhibitor. Second, an apparent noncompetitive inhibition pattern of this type differs from that of classical noncompetitive inhibition. In classical noncompetitive inhibi- tion, the replots of both slope vs I and intercept vs I are

linear whereas in the case of inhibition by a tight binder the
replots are hyperbolas, regardless of the inhibition mechanism.
This subtle but important difference may serve as a diagnostic
test.   Thus the fact that the extrapolation of double-reciprocal
plots intersect at a common point on the $1/S$ axis may not be
regarded as indicative of noncompetitive inhibition.

A. STEADY-STATE RATE EQUATION

If t is set at infinity, Eq. (3) reduces to $v = v_s$ and the
resulting expression of the steady-state velocity becomes equivalent to the equation derived by Henderson (11):

$$\frac{I_t}{1 - (v_i/v_o)} = E_t + \frac{S + K_m}{(K_m/K_{is}) + (S/K_{ii})} \frac{V_o}{V_i} \tag{5}$$

B. SIGNIFICANCE OF I_{50}

If the enzymic reaction is allowed to proceed in the presence
of both substrate and inhibitor, and the I_{50} values (total inhibitor concentration at which the enzymic reaction is 50% of the
uninhibited reaction) are determined, the expressions of I_{50} derived from Eq. (5) are different depending on the inhibition
mechanism (5).   However, experiments based on these different expressions of I_{50} do not provide accurate results if the non-
steady-state phase of the enzyme-inhibitor interaction is too
long; i.e., too much substrate may be converted to product before steady-state conditions are reached.   Thus, it is preferable
to preincubate the enzyme with the inhibitor for a sufficiently
long time, with subsequent measurement of the initial velocity
after addition of the substrate.   In such cases, the various expressions of I_{50}, based on steady-state rate equations (Eq. (5))
no longer apply.   Rather, the following equation may be employed
for both competitive and noncompetitive inhibitions:

$$I_{50} = (E_t/2) + K_i \tag{6}$$

From this equation, the K_i value and the molar equivalency of
the enzyme concentration can be determined (5).

C. ACKERMANN-POTTER PLOT

The plot of velocity vs enzyme concentration has been commonly used to demonstrate "stoichiometric" inhibition (12).   In this

procedure, varying amounts of the enzyme and inhibitor are pre-incubated before measurement of the initial velocities. The plot of v vs E_t is a hyperbola having an asymptote (5) expressed by

$$v = \frac{k_3 S}{K_m + S} E_t - \frac{k_3 I_t S}{K_m + S} \qquad (7)$$

As illustrated in Fig. 3, the molar equivalency of the enzyme concentration may be estimated from the intercepts of the asymptotes on the E_t axis, and the catalytic number (k_3) may be calculated from the slopes or from the v-axis intercept of the asymptotes and the known values of K_m, S, and/or I_t. This graphical analysis does not permit direct evaluation of K_i. However, both methods of analysis, i.e., I_{50} and Ackermann-Potter plot, can be carried out on data from the same set of experiments.

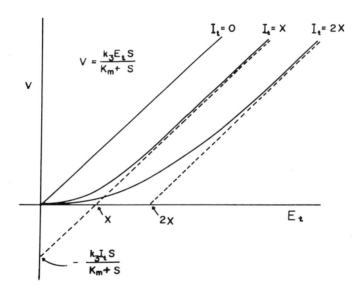

Figure 3. Schematic illustration of Ackermann-Potter Plot according to Eq. (7). (From Cha (5).)

D. DETERMINATION OF INHIBITION MECHANISM

As demonstrated by the inhibition of human erythrocytic ADA by coformycin (6) and as explained above, the familiar methods of analysis of steady-state kinetic data can readily lead to erroneous conclusions about the inhibition mechanism and K_i values. However, the inhibition mechanism, i.e., competitive or noncompetitive, can be elucidated by an entirely different approach. In this new method, the manner in which the substrate interferes with the binding of the inhibitor to the enzyme is determined (7). The procedure involves the determination of the pseudo-first-order rate constants (λ) according to Eq. (4) at various concentrations of substrate and inhibitor. Under the specific conditions where Eq. (4) is applicable, the value of λ may be expressed as

$$\lambda = \alpha + \beta I_t \tag{8}$$

where

$$\alpha = \frac{k_6 + (k_8 S/K'_m)}{1 + (S/K'_m)} \tag{9}$$

$$\beta = \frac{k_5 + (k_7 S/K_m)}{1 + (S/K_m)} \tag{10}$$

Let $\alpha_0 = k_6$, $\beta_0 = k_5$, i.e., the values of α and β measured in the absence of the substrate; $\Delta\alpha = \alpha_0 - \alpha$; $\Delta\beta = \beta_0 - \beta$; $\Delta\alpha_{max} = k_6 - k_8$; and $\Delta\beta_{max} = k_5 - k_7$. Then Eqs. (9) and (10) can be transformed to

$$\Delta\alpha = \frac{\Delta\alpha_{max} S}{K'_m + S} \tag{11}$$

$$\Delta\beta = \frac{\Delta\alpha_{max} S}{K_m + S} \tag{12}$$

From the double reciprocal plots of these equations, k_5, k_6, k_7, k_8, K_m, and K'_m, as well as the inhibition mechanism, may be determined as demonstrated in a recent study of calf intestinal ADA inhibited by coformycin (7).

E. CALCULATION OF INHIBITION CONSTANTS (K_i) BY DIRECT DETERMIN-
ATION OF INDIVIDUAL VELOCITY CONSTANTS OF THE ASSOCIATION
AND DISSOCIATION OF THE EI COMPLEX

For the simple case of uncomplicated competitive inhibition
by a tight-binding inhibitor, the inhibition constant K_i, according to Eq. (2), may be represented as

$$K_i = k_6/k_5 \tag{13}$$

Under favorable experimental circumstances (e.g., stable enzymes and inhibitors, effective methods for separating the inhibitor from the enzyme (both E and EI forms), rapid and sensitive enzymic assays) it may be possible to measure directly the
individual velocity constants for the association and dissociation of the EI complex. With a typical tight-binding inhibitor
k_6, the first-order dissociation rate constant, is very small,
with a $t_{1/2}$ (= ln $2/k_6$) in the order of many minutes to hours.
For this reason, when the free enzyme reacts with the inhibitor
it may become "trapped" in the form of the EI complex, permitting
direct assay of the remaining unreacted free enzyme unaffected
by continuing release of the free enzyme (E) from the EI complex.
Thus, if one can cleanly separate the EI complex from unreacted
inhibitor it may be possible to measure the rate of liberation
of free enzyme, i.e., the first-order dissociation constant k_6.
Similarly, one can measure the rate of interaction between the
free enzyme and inhibitor if one uses a sufficient excess of the
inhibitor (molar concentration) over that of the enzyme. The
rate of formation of the EI complex may be sufficiently slow to
permit direct measurement of the rate of inactivation of the
free enzyme and therefore to determine the second-order association velocity constant k_5 in terms of pseudo-first-order velocity
constants k_5I. Through determination of several pseudo-first-
order velocity constants at different concentrations of the inhibitor, one may readily calculate k_5. For further considerations of the theoretical and practical aspects of these methods
see Ref.5-8 and the examples described below (Fig. 7 and 8).
It is apparent that K_i, as described in Eq. (13), is the
dissociation constant (K_d) of the EI complex. Therefore, under
favorable experimental conditions this value may be determined
by alternative methods, e.g., equilibrium dialysis, with estimation of K_d from the slope of a Scatchard plot. However, such
a procedure usually requires the availability of a radioactively
labeled inhibitor of sufficiently high specific activity to make
such experiments practical. A recent example of the application
of one of these methods is the binding of tetrahydrouridine to
cytidine deaminse (Stoller et al., personal communication).
When suitably labeled coformycin and deoxycoformycin become

available it will be important to determine the K_d values (and therefore the validity of the K_i values estimated above) by one or more of these alternative procudures.

F. *PRELIMINARY ESTIMATION OF TIGHTNESS OF BINDING OF INHIBITOR TO ENZYME*

It should be apparent from the above discussion that kinetic studies of tight-binding inhibitors require theoretical bases and experimental approaches different from those of rapidly reversible inhibitors. Therefore, in the initial study of an enzyme inhibitor it is important to apply a simple "diagnostic test" to determine the class of inhibitor: readily reversible, semitight-binding or tight-binding. Such a procedure is as follows: determine the time course of the enzymic reaction with and without preincubation in the presence of inhibitor. If the results are identical (Fig.4A), the inhibitor is rapidly reversible and classical steady-state methods of analysis are applicable. If the results are different and show a pattern similar to that illustrated in Figs 2 and 4B, the inhibitor is classified as a semitight-binder. Both the steady-state kinetics and the present methods for tight-binders (as described above, e.g., analyses of I_{50} and λ) may be employed. However, in order to apply steady-state methods to semi-tight-binding inhibitors, the steady-state regions, i.e., linear portions, of the time-course curves must be used instead of the true initial velocities. It is also advantageous to preincubate the enzyme with the inhibitor, because less substrate is converted to the product prior to the attainment of steady-state conditions. For the I_{50} analyses the enzyme must be preincubated for a sufficiently long period of time, and the initial velocities and Eq. (6) must be used. Finally, if the time course of the reaction after preincubation shows marked inhibition with an extremely slow or negligible increase in the velocity (whereas in the absence of preincubation the initial enzymic velocity would be close to the uninhibited velocity) and a slowly progressive inhibition becoming apparent only after many minutes of reaction, the inhibitor is a tight-binder (Fig.4C). Of course, in order to observe this slowly progressing inhibition it is necessary to employ very low, but effective concentrations of the inhibitor since the slow onset of inhibition is largely due to the infrequent collisions between the enzyme and inhibitor in the reaction mixture. Methods such as I_{50}, the Ackermann-Potter plot, and direct determination of velocity constants should be employed in the study of such tight-binding inhibitors.

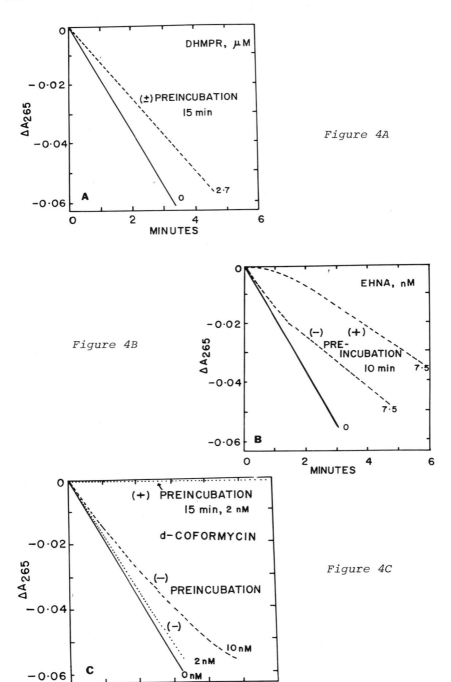

Figure 4A

Figure 4B

Figure 4C

III. APPLICATION OF TIGHT-BINDING INHIBITOR THEORY TO THE EVALUATION OF ADENOSINE DEAMINASE INHIBITORS

A. CLASSIFICATION OF ADA INHIBITORS ACCORDING TO TIGHTNESS OF BINDING

As shown in Figure 4A, when the inhibitor DHMPR (see Fig. 1) was examined with and without preincubation for 15 min with human erythrocytic ADA, identical reaction velocities were observed over a 5-min incubation period. This indicated that DHMPR is a readily reversible inhibitor with rapid establishment of equilibrium between E, ES, and EI, i.e., within a few seconds. Therefore, it was possible to employ the initial velocities to determine both the inhibition mechanism and K_i value by use of Lineweaver-Burk and Dixon analyses. In this case the double-reciprocal plot yielded a classical pattern of competitive inhibition and the replot gave a K_i value of 1.3×10^{-6} M (8).

In contrast to the results with DHMPR, the semitight-binding inhibitor EHNA (see Fig. 1) yielded the time-course patterns shown in Fig. 4B. In this case, the sequence of addition of components to the reaction mixture played a crucial role in the results obtained. When the reaction was initiated by addition of the enzyme (without preincubation with the inhibitor), little inhibition was seen during the first minute. Thereafter, progressive inhibition became apparent, with the steady-state rate occurring after two or three minutes. On the other hand, if EHNA was preincubated for 10 min with the enzyme and the reaction initiated by the addition of adenosine, profound inhibition was observed initially with gradual release of the inhibition becoming apparent after about one minute. Again, steady-state

Figure 4. Spectrophotometric tracings of the reaction of human erythrocytic adenosine deaminase in the presence of DHMPR (frame A), EHNA (frame B), and deoxycoformycin (frame C). Reaction mixtures in a total volume of 1.0 ml contained: phosphate buffer, 50 mM, pH 7.5; adenosine, 0.1 mM; human erythrocytic ADA, 0.0033 units and inhibitors (concentration as indicated). Where inhibitors were used two methods were employed: preincubation of enzyme and inhibitor (+), here the enzyme and inhibitor were preincubated for the time indicated in 0.99 ml phosphate buffer at room temperature and the reaction was started by addition of substrate, adenosine (10 μl); without preincubation (-), here the reaction was started by addition of ADA (10 μl) to a mixture of adenosine and inhibitor in phosphate buffer. The reactions were followed at room temperature by measuring decrease in absorbance at 265 nm in a Beckman spectrophotometer equipped with a Gilford recorder (From Agarwal et al. (8).)

velocities were attained after two to three minutes, at which
time the velocity paralleled that seen without preincubation.
Therefore, it is obvious that one may *not* employ initial velocity measurements for the determination of inhibition constants
in such cases. Furthermore, if one is to employ the steady-
state velocity it should be appreciated that excessive consump-
tion of substrate prior to the attainment of steady-state conditions can lead to erroneous results. A method of minimizing
this complication is to preincubate the inhibitor and enzyme
and initiate the reaction by addition of the substrate. In this
case, less substrate depletion occurs than if the reaction is
initiated by the addition of enzyme (see Fig. 4B). When steady-
state velocities were employed for the determination of the K_i
value and the reaction mechanism of EHNA for human erythrocytic
ADA, the Lineweaver-Burk plot yielded a classical pattern of
competitive inhibition and a K_i value of 1.6 x 10^{-9} M (8). When
a similar analysis was performed on the inhibition of calf duo-
denal ADA by EHNA, time-course patterns essentially identical to
those of Fig. 4B were found. By use of the I_{50} method, a K_i
value of 6.5x$10^{-9}M$ was obtained (8).

As seen in Fig. 4C, markedly different results were obtained
with the tight-binding inhibitor, deoxycoformycin. When human
erythrocytic ADA was preincubated for 15 min with 2 x $10^{-9}M$ deoxycoformycin and the enzymic reaction was initiated by the
addition of adenosine, almost complete inhibition was observed
during the 6-min observation period. On the other hand, when
the reaction was initiated by addition of the enzyme to a reaction mixture containing 2 x $10^{-9}M$ deoxycoformycin and adenosine,
negligible inhibition was observed. At a higher concentration
of deoxycoformycin (10 x $10^{-9}M$) (Fig. 4C), little inhibition was
initially seen. However, inhibition became clearly apparent
after two to four minutes. Obviously, under such conditions
neither initial nor steady-state velocities may be employed for
the determination (by conventional methods) of either the K_i
value of the inhibition mechanism. In fact, attempts to apply su
data to determine either the K_i or the inhibition mechanism have
in the past lead to much controversy and grossly erroneous inter-
pretations (e.g., see Ref. 6). However, other methods such as
the I_{50} procedure, the Ackermann-Potter analysis, and the determination of the association and dissociation velocity constants
may be successfully employed in the study of a tight-binding inhibitor such as deoxycoformycin.

B. **APPLICATION OF ACKERMANN–POTTER ANALYSIS TO COFORMYCIN AND ADENOSINE DEAMINASE**

As noted above (see Ref. 5,6 and 8), the application of the Ackermann-Potter analysis to the study of a tight-binding enzyme inhibitor may yield important information on the properties of the enzyme, e.g., the molar equivalency of the enzyme and the catalytic number k_3, but does not permit direct calculation of the K_i value. Figure 5 presents an Ackermann-Potter analysis of human erythrocytic ADA preincubated for 50 min with coformycin. As seen in Fig. 5A, extrapolation of the asymptotes yielded a family of parallel lines, which permitted replotting of the E_t and v intercepts against the coformycin concentration (see Fig. 5B and 5C). Both replots were linear. From the slope of the line in Fig. 5B, it was estimated that one unit per liter of enzyme corresponds to about $1.0 \times 10^{-10}M$. From the plot of Fig. 5C, the catalytic number k_3 was estimated to be 0.995×10^4 min^{-1}. When a similar Ackermann-Potter analysis was performed with deoxycoformycin and erythrocytic ADA, very similar graphs were obtained (8). The molar equivalency of one unit of enzyme per liter was estimated to be $1.3 \times 10^{-10}M$ and the catalytic number k_3 was estimated at 0.8×10^4 min^{-1} (8). Thus, by the use of two different inhibitors, reasonably similar values were obtained for these parameters, the molar equivalency of the enzyme, and the catalytic number.

C. **APPLICATION OF THE I_{50} ANALYSIS TO COFORMYCIN AND DEOXYCO-FORMYCIN**

As discussed above and in earlier publications (5,6,8), the same data employed in the Ackermann-Potter analysis may be used by the I_{50} method to estimate the K_i value and the molar equivalency of the enzyme. Figure 6 illustrates the application of this procedure to the study of the interaction between coformycin and human erythrocytic ADA. Figure 6A illustrates the method employed to determine the I_{50} values. When these values were replotted against the enzyme concentration (as in Fig. 6B) according to Eq. (5) the intercept on the I_{50} axis equaled the K_i value. In this case, when a 50-min. preincubation was performed, the K_i was $1.2 \times 10^{-10}M$. However, as shown in Fig. 6B, a similar I_{50} determination performed after 10 hr of preincubation of coformycin and ADA yielded a straight line that intercepted the I_{50} axis very close to the origin. It was not possible to estimate precisely the K_i value from this line. However, it is apparent that the inhibition of ADA by coformycin markedly increases with prolonged incubation, perhaps as a result of pro-

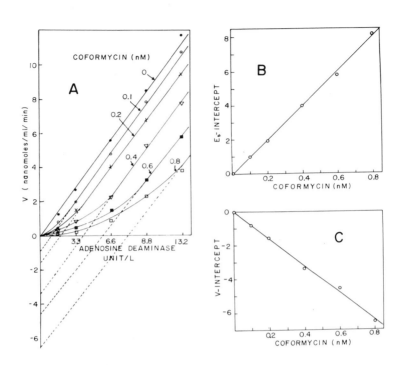

*Figure 5. Ackermann-Potter plot of adenosine deaminase pre-
incubated with coformycin. Various amounts of adenosine deaminase were incubated in a total volume of 0.99 ml containing 50 mM
of potassium phosphate buffer, pH 7.5, and varying concentrations
of coformycin. After incubation for 50 min at room temperature,
the enzymic reaction was started by the addition of 10 μl of 10
mM adenosine (final concentration 0.1 mM). The reaction was followed by measuring the decrease in absorbancy at 265 nm at room
temperature. Frame A: plot of the enzymic velocities (nmoles/
ml/min) vs the enzymic concentration (units/liter). Frame B:
Plot of E_t intercept (from frame A) vs coformycin concentration.
From this plot the molar equivalency of the enzyme was calculated
to be 1 unit/liter, corresponding to about 1.0×10^{-10} M. Frame
C: plot of v intercept (from frame A) vs coformycin concentration.
From the slope of the line the k_3 value was calculated to be
0.995×10^4 min^{-1}. (From Cha et al. (6).)*

gressive conformational changes (see also Ref. 6). When an
attempt was made to apply the I_{50} method to study the inhibition
of human erythrocytic ADA by deoxycoformycin, the results obtained
resembled those seen with coformycin after 10 hr of incubation,
i.e., the line intersected the I_{50} axis near the origin. Although
it was appreciated that this parameter would be subject to significant error, a K_i value of $1.5 \times 10^{-11}M$ was estimated. This

illustrates certain limitations of the I_{50} procedure. If the molar equivalency of the enzyme employed in the experiment is significantly less (tenfold or smaller) than the K_i, the plot of I_{50} vs. the enzyme concentration will yield an apparently horizontal line and the K_i value will be very close to the I_{50} values determined at each enzyme concentration. On the other hand, if one studies a very potent enzyme inhibitor, it may be necessary to employ enzyme concentrations that exceed the K_i value by tenfold or more. In such cases, e.g., as seen with deoxycoformycin and with coformycin after 10 hr of preincubation, the plot of I_{50} vs enzyme concentration yields a line that intersects the I_{50} axis near the origin and is of minimal value for the calculation of either the molar equivalency of the enzyme or the K_i value. To apply the I_{50} method effectively in such instances it would be necessary to devise an enzymic assay that permitted accurate determination of individual I_{50} values with the use of enzyme molar equivalencies in the same order of magnitude as the K_i value. It should also be appreciated that one may effectively apply the I_{50} method to the study of semitight-binding inhibitors, i.e., K_i values in the range 10^{-7} to 10^{-9} M,

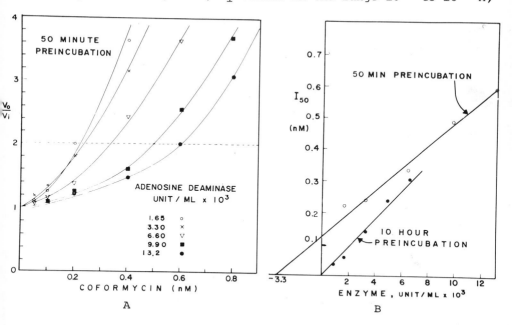

Figure 6. Determination of I_{50}. Frame A: a plot of v_o/v_i vs coformycin concentration. The same data as in Fig. 5 were used. Frame B: I_{50} values were estimated from frame A and plotted against the enzyme concentration (o). The I_{50} values from a similar experiment in which the enzyme was preincubated for 10 hr are also presented (●). (From Cha et al. (6).)

if one decreases the enzymic reaction velocity by modifying assay
conditions or through the use of a substrate with a low V_{max}
value. The low V_{max} for the substrate would permit the use of a
high concentration of enzyme, i.e., in the same order of magnitude as K_i. However, in this case both association and dissociation of inhibitor would be quite rapid. Therefore, the expression of I_{50} as presented in Eq. (6) would not apply. Instead,
one should employ different expressions of I_{50} values, depending
on the inhibition mechanism, which are based on the following
steady-state kinetics (5). Of course, there is usually little
merit in employing such maneuvers for the determination of K_i
values with semitight-binding inhibitors, where simpler and less
wasteful methods are available. However, this approach might
prove of value for the estimation of molar equivalencies of enzymes when studying partially purified enzyme preparations.
Furthermore, if the Ackermann-Potter analysis is applied to data
obtained in such a manner, i.e., with a semitight-binding inhibitor, it may be possible to estimate the catalytic number k_3.

$$I_{50} = \tfrac{1}{2}E_t + \frac{S + K_m}{K_m/K_{is} + S/K_{ii}} \qquad \text{(noncompetitive)} \qquad (15)$$

$$I_{50} = (\tfrac{1}{2}E_t + K_i) + (K_i S/K_m) \qquad \text{(competitive} \qquad (16)$$

$$I_{50} = (\tfrac{1}{2}E_t + K_i) + (K_i K_m/S) \qquad \text{(uncompetitive)} \qquad (17)$$

D. *DETERMINATION OF VELOCITY CONSTANTS OF THE ASSOCIATION AND
    DISSOCIATION REACTIONS BETWEEN ADENOSINE DEAMINASE AND
    COFORMYCIN OR DEOXYCOFORMYCIN*

As discussed above and previously (5-8), under favorable
experimental conditions it may be possible to measure directly
the individual rate constants for the association and dissociation of the EI complex. Such conditions can be fulfilled with
adenosine deaminase and the coformycin-type inhibitors. For
example, both the human erythrocytic and calf intestinal enzymes
are easily obtained in sufficient quantity and at a state of
purity satisfactory for kinetic analysis (13, 14). The enzymes
are relatively stable and may be maintained for prolonged periods at room temperature even when stirred in the presence of
charcoal. The enzymic reaction catalyzed is a simple displacement of an amino group from adenosine by a hydroxyl group and
only one substrate is involved (in addition to H_2O). Furthermore
the enzyme is readily and sensitively assayed by a variety of
methods. A number of substrates are also available that vary
markedly in both K_m and V_{max} values. Thus, the interaction be-

tween ADA and its inhibitors offers an ideal model system for the elucidation of inhibition mechanisms.

As shown in Fig. 7, when various concentrations of deoxycoformycin were incubated with human erythrocytic adenosine deaminase, the semilogarithmic plot of remaining enzymic activity against time yielded a family of straight lines from which the pseudo-first-order velocity constants, i.e., $k_5(I_t)$, for the association reaction could be calculated. From these values the second-order rate constant k_5 of the association reaction was readily obtained by mathematical or graphical methods (8). The value of this association rate constant k_5 was estimated at

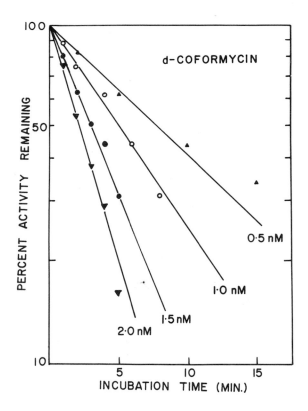

Figure 7. Rate of association of human erythrocytic adenosine deaminase and deoxycoformycin. ADA (0.0033 units) was incubated with various concentrations of deoxycoformycin in 0.99-ml phosphate buffer. After incubation at room temperature for various time intervals, the reactions were started by addition of 10 µl of 10 mM adenosine (final concentration 0.1 mM). (From Agarwal et al. (8))

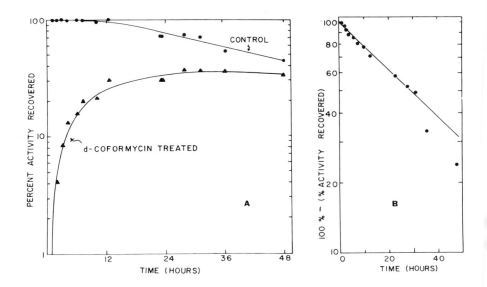

Figure 8.   Dissociation of adenosine deaminase - deoxycofor-mycin complex.   In a reaction mixture of 0.5 ml, human erythro-cytic ADA (0.13 units), deoxycoformycin (4 x 10^{-11} moles), and phosphate buffer (25 μmoles, pH 7.5) were incubated at room temp-erature for 1 hr.   Deoxycoformycin was replaced by an equal vol-ume of water in control experiments.   After incubation, 0.4 ml of hemoglobin-coated charcoal in 50 mM phosphate buffer, pH 7.5, was added.   After stirring occasionally for 5 min the charcoal was removed by centrifugation.   Immediately, the supernatant fluids were diluted fifty-fold with 50 mM phosphate buffer, pH 7.5, containing charcoal (0.5 ml), Na_2EDTA (0.5 mM) and bovine plasma albumin (0.03%, wt./vol.).   The mixtures were stirred at room temperature using a magnetic stirrer. Aliquots (1.2 ml) of these suspensions were withdrawn at different time intervals (time immediately after dilution was recorded as zero time), cen-trifuged for 5 min, and 1.0 ml of the supernatants were assayed at room temperature for enzymic activity in the presence of 0.1 mM adenosine.   Frame A: recovery of activity from enzyme-deoxy-coformycin (EI) complex with time.   The percent of activity re-covered was calculated on the basis of the activity of control at zero time.   Frame B: semilog plot of 100% minus the percent of activity recovered (percent of enzyme remaining as EI complex) vs time.   The percent of activity recovered was calculated on the basis of the activity of control sample at each particular time taken as 100%.   From this plot the $t_{\frac{1}{2}}$ value was estimated at about 29 hr.   (From Agarwal et al. (8).)

$2.6 \times 10^6 \ M^{-1} \ \text{sec}^{-1}$ (8) which is similar to the value $(2.1 \times 10^6 \ M^{-1} \ \text{sec}^{-1})$ obtained with coformycin (6). The similarity in these values is not unexpected since both experiments employed the same enzyme and two inhibitors that resemble each other closely in structure and in physical properties. Also, it may be recalled that the second-order rate constant k_5 is markedly affected by the collision frequencies and rates of diffusion of two interacting molecules, factors that should be similar with both inhibitors. When similar studies were performed with calf intestinal ADA and coformycin, the second-order rate constant was $1.01 \times 10^6 \ M^{-1} \ \text{sec}^{-1}$ (7). This smaller value most likely reflects subtle differences between the two enzymes studied. When similar measurements were made with EHNA and calf intestinal ADA, a value of $0.7 \times 10^6 \ M^{-1} \ \text{sec}^{-1}$ (8) was found, which is close to the value obtained for coformycin (7).

The excellent stability of erythrocytic ADA has made it possible to measure directly the first-order rate constants k_6 (see Eqs. (2) and (13)) for the dissociation of EI complexes with either coformycin or deoxycoformycin. Figure 8 illustrates the results of such an experiment performed with deoxycoformycin (8). Erythrocytic ADA was incubated at room temperature for one hour with an excess of deoxycoformycin. At the end of this period, no free enzyme could be detected. The unreacted deoxycoformycin was removed from the reaction mixture by treatment with hemoglobin-coated charcoal (6,15). The solution containing the EI complex was then diluted fifty-fold and stirred at room temperature in the presence of charcoal. Periodically, aliquots were removed and assayed for free enzyme. As shown in Fig. 8 progressive release of free enzyme was observed, which yielded a linear monophasic graph when the logarithm of the remaining EI complex was plotted against time. From this latter graph, the $t_{\frac{1}{2}}$ was estimated as about 29 hr. From the slope of this line, as well as the relationship

$$k_6 = 0.693/t_{\frac{1}{2}} \tag{14}$$

(with t given in seconds) the value of k_6 was estimated to be $6.6 \times 10^{-6} \ \text{sec}^{-1}$.

When a similar experiment was performed with coformycin and erythrocytic ADA, a biphasic semiogarithmic plot was obtained, which yielded two $t_{\frac{1}{2}}$ values, 75 min and 7.9 hr, and velocity constants of 1.54×10^{-4} and $2.44 \times 10^{-5} \ \text{sec}^{-1}$, respectively (6). The value of k_6 for the dissociation of the coformycin-calf intestinal ADA complex, estimated by an entirely different method, was $2.2 \times 10^{-4} \ \text{sec}^{-1}$ (7).

From these values of k_5 and k_6, the K_i values for human erythrocytic ADA were calculated according to Eq. (13). The K_i of deoxycoformycin was $2.5 \times 10^{-12} \ M$ and the values for coformycin were $1.7 \times 10^{-10} \ M$, calculated for the 75-min $t_{\frac{1}{2}}$ component, and $1.1 \times 10^{-11} \ M$ for the 7.9-hr $t_{\frac{1}{2}}$ component of the dissociation curve (6).

     It is of considerable interest that although the k_5 values
of coformycin and deoxycoformycin were similar, the values of k_6
differed by fourfold or greater; thus, the greater potency (at
least fourfold) of deoxycoformycin as an inhibitor of erythrocytic
ADA results primarily from its smaller k_6 value and the slower
rate of dissociation of the EI complex. These observations are
consistent with earlier findings (13) that the K_m or K_i values
of deoxynucleoside substrates or inhibitors are two- to fourfold
lower than the values of the corresponding ribonucleoside derivatives, as shown in Table I. This observation indicates that

TABLE I   Kinetic Constants of Inhibitors and Substrates of
          Adenosine Deaminases

Compound	K_m $(10^{-6}\,M)$	K_i (M)	k_5^a $(10^6\,M^{-1}\,sec^{-1})$	k_6^a $(10^{-6}\,sec^{-1})$	Ref.
Human Erythrocytic Adenosine Deaminase					
Adenosine	25	--	--	--	(13)
Deoxyadenosine	7	--	--	--	(13)
Inosine	--	60×10^{-6}	--	--	(13)
Deoxyinosine	--	19×10^{-6}	--	--	(13)
DHMPR	--	1.3×10^{-6}	--	--	(8)
EHNA	--	1.6×10^{-9}	--	--	(8)
Coformycin	--	10×10^{-12}	2.1	24	(6)
Deoxycoformycin	--	2.5×10^{-12}	2.6	6.6	(8)
Calf Intestinal Adenosine Deaminase					
Adenosine	36	--	--	--	(7)
Coformycin	--	220×10^{-12}	1.0	220	(7)
EHNA	--	6.5×10^{-9}	70.0	4500	(8)

a In references 6 and 8 k_5 and k_6 are called k_1 and k_2,
respectively.

the 2'-position on the ribose moiety contributes a factor of
about fourfold to the binding of this class of compounds to the
active site of the enzyme.

E. DETERMINATION OF INHIBITION MECHANISM FROM PSEUDO-FIRST-ORDER RATE CONSTANTS

The mechanism of inhibition of ADA from calf intestinal mucosa by coformycin was determined by a new principle (7). If the inhibition is competitive, only the free enzyme (E) of the uninhibited enzyme species (E and ES) is available for the binding of the inhibitor in the presence of the substrate. Therefore, the effect of the presence of the substrate on the rate of formation of the EI complex is as though the concentration of the enzyme was reduced by a factor of $1/(1 + S/K_m)$, which is equivalent to the ratio $(E)/((E) + (ES))$, and the apparent second-order rate constant became $k_5/(1 + S/K_m)$ instead of k_5. Thus, in the presence of saturating concentrations of the substrate, the binding of the inhibitor could be completely blocked and the apparent second-order rate constant would become zero. In contrast, in noncompetitive inhibition, even if the system is saturated with the substrate, the binding of the inhibitor to the ES complex would proceed with a rate constant k_7. In the presence of nonsaturating concentrations of the substrate the rates of association of the inhibitor with E and ES would be $k_5/(1 + S/K_m)$ and $k_7(S/K_m)/(1 + S/K_m)$, respectively, and the sum of these rates would be given by β, as in Eq. (10). Note that the factor $(S/K_m)/(1 + (S/K_m))$ is equivalent to $(ES)/(E) + (ES))$ when the steady-state conditions are reached between E and ES. Thus the apparent rate constant will change from k_5 to k_7 as the concentration of S changes from 0 to ∞. The value of β at any finite concentration of the substrate is determined by Eq. (10).

Figure 9 demonstrates that the plot of $1/\beta$ vs S is a straight line. Therefore, k_7 must be zero and coformycin must compete with adenosine (for experimental details see Ref. 7).

IV. EFFECT OF ADENOSINE DEAMINASE (ADA) DEFICIENCY OR INACTIVATION ON THE INCORPORATION OF ADENOSINE ANALOGS INTO INTRACELLULAR NUCLEOTIDE POOLS

It has recently become possible to examine directly the role of ADA in the regulation and incorporation of adenosine and its analogs into intracellular nucleotide pools. This results from two important new developments: first, the recent identification of potent inhibitors of ADA, such as EHNA, coformycin, and deoxycoformycin (see Fig. 1), which can function intracellularly as well as with purified enzymes; second, a newly identified disease syndrome first described in 1972 (16), referred to as severe combined immunodeficiency disease (SCID),-adenosine deaminase deficient (17). Children born with this genetic disorder lack both T and B lymphocytic functions, often have bony abnormalities, and usually succumb to infection early in life. Fortunately,

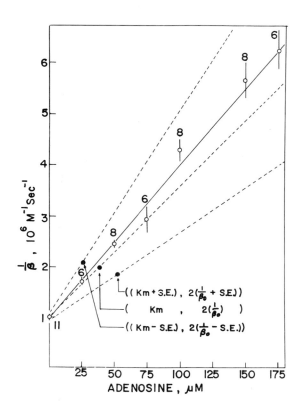

Figure 9. Plot of the reciprocal of the apparent second-order rate constant vs adenosine concentration according to Eq. (10). The circles and the perpendicular bars indicate the reciprocals of apparent second-order rate constants and their standard errors of estimation. The number on each datum point indicates the number of replicates. The middle dashed line indicates the theoretical expectation on the basis of competitive inhibition (i.e., $k_7 = 0$), and the values of k_5 and K_m being 1.009×10^6 M^{-1} sec^{-1} and 38.8×10^{-6} M, respectively. The upper and lower dashed lines represent the cases where the true values of $1/k_5$ and K_m are $(1/k_5) \pm S.E.(1/k_5)$ and $K_m \pm S.E.(K_m)$, respectively. The coordinates of one point on each of these three theoretical lines are given in the figure. (From Cha (7).)

several children have been successfully maintained in aseptic environments, and at least two patients have developed immunocompetence and have survived for several years following transplantation with sibling bone marrow (18). Another child has responded to frequent transfusions with normal irradiated ADA-containing erythrocytes (19). Through the generous collaboration of clinical colleagues, it has been possible for us to examine the ADA-deficient erythrocytes of several of these surviving children (20).

It was learned, using erythrocytes from SCID-ADA-deficient patients, that after exposure to adenosine there occurred a rapid rise (of 130% to 150%) in the intraerythrocytic ATP levels (20). In normal subjects, with erythrocytes that contain ADA (normally in the range of 0.2 to 0.4 E.U./ml of cells), most of the added adenosine is deaminated and is eliminated from the cell or enters the nucleotide pools as inosinate, rather than as adenine nucleotides (20). Figure 10 shows the result of incubating normal erythrocytes and SCID-ADA-deficient erythrocytes with the adenosine analog, formycin A (see Fig. 1). This compound was selected for study because it is deaminated about eight times more rapidly than adenosine by ADA (13) and enters the nucleotide pools of normal erythrocytes only in small quantity (21). However, as seen in Fig. 10, after four hours of incubation with formycin A, large new peaks of formycin triphosphate, diphosphate, and monophosphate were readily identified by high-pressure liquid chromatography. Almost identical nucleotide profiles have been obtained with normal erythrocytes incubated with formycin in the presence of the tight-binding ADA inhibitor, coformycin (unpublished results).

On the basis of these observations the hypothesis was offered that a primary function of the enzyme ADA is the protection of cells against rapid surges in the concentrations of ATP following exposure to relatively large amounts of adenosine, as might occur in the presence of tissue damage or after a purine-rich meal. It is well known that many metabolic reactions such as glycolysis and PRPP synthetase are subject to allosteric inhibition by high concentrations of ATP (see discussion, Ref. 20). Of course, there are many other possible explanations of the cytotoxicity of adenosine to be explored in the future.

Figure 11 illustrates the effect of the tight-binding inhibitor, coformycin, on intraerythrocytic ADA. This experiment employed an assay devised in this laboratory that makes possible the measurement of ammonia liberation from adenosine or adenosine analogs in intact cells (20, 22). Coformycin, at a concentration of 2 µg/ml of reaction mixture, abolished the liberation of ammonia from adenosine. Thus, this concentration of coformycin was employed in continuing studies of the effects of ADA inhibition on the incorporation of a variety of adenosine analogs into the nucleotide pools of human erythrocytes. The latter studies employed the technique of high-pressure liquid chromatography (HPLC) with monitoring of the effluent at two wavelengths, i.e., 254 nm, which measures the natural nucleotides, and a second wavelength of choice, usually the λ_{max} of the analog nucleoside under

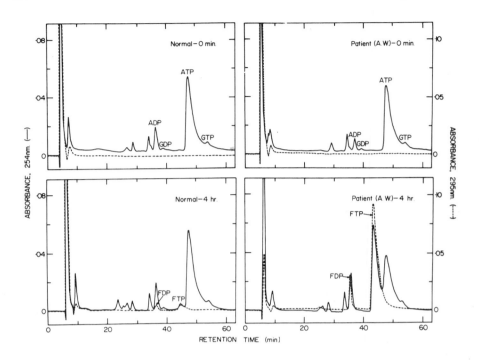

Figure 10. *High-pressure liquid chromatographic profiles of erythrocytes from normal donors and from patients with SCID-ADA deficiency incubated with formycin A. Erythrocytic suspensions (15-20% in medium consisting of potassium phosphate buffer, 50 mM, pH 7.5; MgSO$_4$, 2 mM; NaCl, 75 mM; glucose, 10 mM; penicillin, 10 units/ml, and streptomycin, 10 μg/ml) were incubated at 30°C in a shaking water bath with 1.0 mM formycin A. Aliquots (200 μl) were removed at various time intervals, after addition of formycin, and added to 100 μl of cold 12% perchloric acid. After thorough mixing, the samples were centrifuged and the supernatant solutions were neutralized with KOH. After removal of KClO$_4$ by centrifugation, 20 μl of the neutralized extracts were chromatographed on a Varian LCS-1000 liquid chromatograph equipped with a Reeve-Angel AS Pellionex-SAX column (3 m x 1 mm). A linear elution gradient was used, which consisted of 0.002 M potassium phosphate, pH 4.5 as the low-concentrate buffer and 0.5 M potassium phosphate, pH 4.5 in 1.0M KCl as the high-concentrate buffer. The starting volume was 40 ml and the flow rates were 14 and 7 ml/hr for the column and gradient, respectively. Column effluents were monitored at both 254 and 295 nm (λ_{max} for formycin A). "FDP" and "FTP" refer to the di- and triphosphate nucleotides of formycin A. (From Agarwal et al. (20).)*

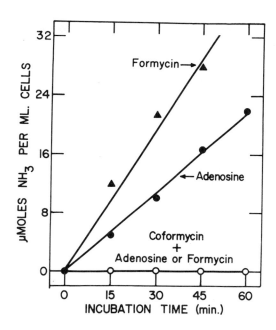

Figure 11. Inhibition of adenosine deaminase by coformycin in intact normal human erythrocytes. Suspensions (2%) of erythrocytes were incubated for 30 min at 30°C in a medium described in Fig. 10 containing coformycin, 2µg/ml (where indicated). Formycin A or adenosine was added at the concentration of 1 mM (where indicated), and samples were withdrawn for measurement of ammonia at various time intervals after incubation at 30°C in a shaking water bath.

study. The following presents several experiments designed to demonstrate the role of inhibition of ADA on the incorporation of several representative adenosine analogs into the nucleotide pools. The compounds chosen to illustrate these effects are the promising antiviral agent, arabinosyl adenine (ara-A), 2,6-diaminopurine ribonucleoside (DAPR), N^6-hydroxyaminopurine ribonucleoside (HAPR) and 8-azaadenosine, the structures of which are presented in Fig. 1.

As shown in Fig. 12, when normal erythrocytes were incubated with arabinosyl adenine, little analog nucleotide was detectable over a four-hour period. However, in the presence of coformycin, the incorporation of arabinosyl adenine into nucleotides occurred progressively and after four hours, the peak of araATP approximated the size of the natural ATP peak. These observations are consistent with recent findings, which demonstrated a striking

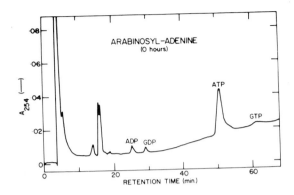

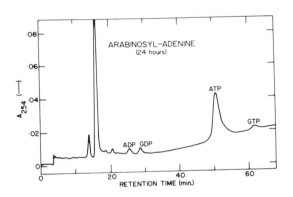

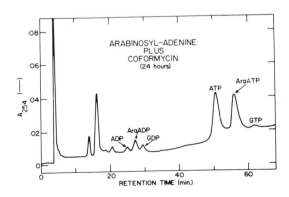

Figure 12.

potentiation in the antitumor action of arabinocyl adenine by coadministration of deoxycoformycin (23-26).

Figure 13 illustrates the effect of coformycin treatment on the incorporation of 2,6-diaminopurine ribonucleoside into the nucleotide pools of human erythrocytes. As described earlier (22), when normal erythrocytes were incubated with DAPR, new peaks appeared in the nucleotide profiles with retention times identical to those of the guanine nucleotides. A small peak also appeared just before the ATP peak. This indicated that DAPR was predominantly deaminated in the 6 position, forming guanosine, which is consistent with observations made with isolated ADA (13). As shown elsewhere, incubation of human erythrocytes with guanosine leads to the rapid formation of large amounts of guanine nucleotides (22). In contrast, when DAPR and coformycin were incubated with erythrocytes there was no evidence of the formation of guanine nucleotides. Rather, large new peaks appeared with absorbancy maxima at 295 nm, which represent the polyphosphate nucleotides of DAPR. With the column employed in this HPLC, the DAPR nucleotide peaks coincided with the adenine nucleotide peaks but were readily distinguishable by their absorption at 295 nm (see Fig. 13).

Analogs of the 8-azapurine class have been extensively studied since their introduction in 1945 by Roblin and associates (27). Most experiments have involved the carcinostatic agent, 8-azaguanine (28) whereas 8-azaadenine and its derivatives have received less attention, primarily because they displayed inferior therapeutic indices in chemotherapeutic trials. In studies with partially purified human erythrocytic ADA it was learned that 8-azaadenosine is deaminated about three times more rapidly than adenosine (13), which probably accounts for its failure to form analog adenine nucleotides or to enter the nucleic acids of cells and display significant chemotherapeutic activity. There-

Figure 12. Effect of coformycin on the incorporation of arabinosyl adenine (Ara-A) into the nucleotide pools of normal human erythrocytes. Experimental details were similar as described in Fig. 10 except that the cells were preincubated with coformycin (2 µg/ml) for 15 min at 37°C, and arabinosyl adenine (1.0 mM) was used instead of formycin. In addition, the column used for HPLC was a Reeve-Angel Partisil-10 SAX (25 cm x 4.6 mm) column. Low-concentrate eluant was 0.002 M KH_2PO_4 (pH 4.5) and the high-concentrate eluant was 0.5 M KH_2PO_4 (pH 4.5). Starting volume was 25 ml and flow rates for both column and gradient were 40 ml/hr. Eluants were monitored at 254 nm.

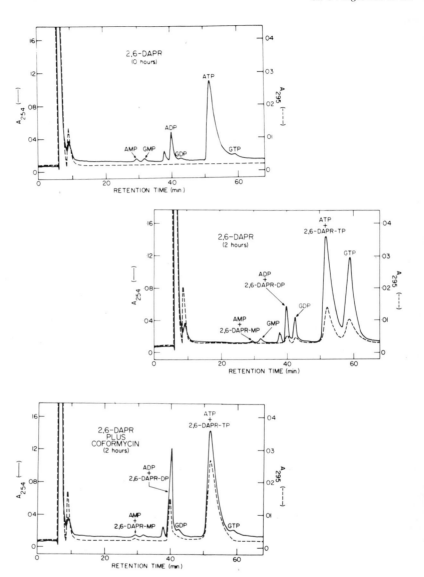

Figure 13. Effect of coformycin on the incorporation of
*2,6 diaminopurine ribonucleoside (DAPR) into the nucleotide pools
of normal human erythrocytes. Experimental conditions were similar as described in Figs. 10 and 12 except that DAPR (1.0
mM) was used here, and the effluents were monitored at 254 and
295 nm. HPLC columns and eluting conditions were as in Fig. 10.
2,6-DAPR-MP, 2,6-DAPR-DP, and 2,6-DAPR-TP are the mono- di-, and
triphosphate nucleotides of DAPR, respectively.*

fore, experiments were performed in which 8-azaadenosine was incubated with human erythrocytes in the presence and absence of
coformycin (1 µg/ml), the results of which are shown in Fig. 14.
When normal erythrocytes were incubated for 4 hr with 8-azaade-
nosine (minus coformycin), small new peaks were found following
the ATP and ADP peaks that absorbed more strongly at 271 nm
(the λ_{max} of 8-azaadenine nucleotide) than at 254 nm (Fig. 14
middle frame). On the other hand, when similar incubations were
performed in the presence of coformycin, the nucleotide profile
of Fig. 14 (bottom frame) was obtained. Here, large peaks with
high absorbance at 271 nm emerged after the peaks of ATP and ADP
and a large 271-nm absorbing peak appeared in the monophosphate
nucleotide region of the profile. Of interest is the observation
that in the zero time sample the initial peak, which emerged with
the void volume (Fig. 13 top frame), had markedly decreased in both
experiments at 4 hr. This indicates that in the presence of coformycin there is almost total conversion of the added 8-azaaden-
osine into intracellular 8-azaadenine nucleotides. Consistent
with earlier observations with human erythrocytes (22), the concentrations of the normal adenine nucleotides ADP and ATP were
unaffected, despite the synthesis of very large amounts of
analog nucleotides. Since most purine analogs display cytotoxic
activity only after conversion to the nucleotide level, the present observations suggest the possibility that adenosine analogs
such as 8-azaadenosine, that are rapidly deaminated (usually to
an inactive form) by ADA, may become cytotoxic if employed in the
presence of a potent ADA inhibitor. Therefore, it is highly advisable that the chemotherapeutic activity of 8-azaadenosine be
reevaluated in biological systems in which the adenosine deaminase activity is inhibited, e.g., by an inhibitor such as EHNA or
deoxycoformycin.

For a number of years the adenine analog N^6-hydroxyamino-
purine and its ribonucleoside derivatives have received major
attention as potential chemotherapeutic agents (see Ref. 29).
Interestingly, N^6-hydroxyamino-9-β-D-ribofuranosylpurine (HAPR)
displayed activity against a number of murine tumors, including
sublines resistant to some of the classic agents such as 6-mer-
captopurine, methotrexate, vincristine, and arabinosyl cytosine
(30). Therefore, HAPR was considered to have sufficient promise
to warrant preliminary pharmacologic studies and clinical trials
in patients with far advanced neoplastic diseases (29). Unfortunately, it was learned that HAPR reacts with ADA resulting in
the formation of inosine and hydroxylamine. Hydroxylamine is a
potent methemoglobin former in human erythrocytes and, during
these preliminary clinical trials, acute hemolysis was detected at
HAPR doses far below those expected to display therapeutic activity. Furthermore, treatment of patients with the nucleoside
transport inhibitor, dipyridamole, did not prevent the occurence
of hemolytic episodes after treatment with HAPR (29). Therefore,

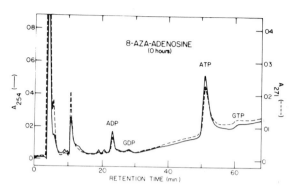

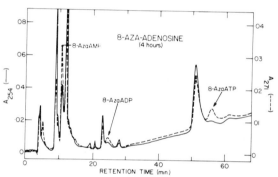

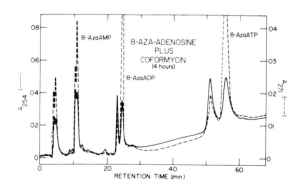

Figure 14.

further clinical studies with this compound were abandoned. However, it is possible that this group of potentially useful chemotherapeutic agents might be salvaged if one could prevent the
intraerythrocytic liberation of hydroxylamine through inactivation of the adenosine deaminase. Therefore, in collaboration
with a Brown University medical student, Mr. Jonathan Fleisch-
mann, and Dr. Leroy Townsend and his colleagues from the University of Utah, we have examined HAPR in human erythrocytes in the
presence and absence of deoxycoformycin (4 µg/ml). In addition
to studies of incorporation of HAPR into the nucleotide pools,
the methemoglobin levels of the erythrocytes were examined. In
the presence of 1 mM HAPR, the methemoglobin level rose from control values of about 5% to about 20% after 30 min of incubation.
Some decline in the methemoglobin levels occured after 1 hr and
methemoglobin levels were maintained at approximately 15% through
the first 8 hr of the experiment. When the erythrocytes were
preincubated for 15 min with deoxycoformycin (4 µg/ml) prior to
the addition of HAPR, the methemoglobin levels throughout the 8
hr of the experiment did not vary significantly from those of
the untreated controls. However, after a prolonged delay (about
24 hr), significant methemoglobin formation was produced even in
the presence of deoxycoformycin. Figure 15 presents nucleotide
profiles of extracts of human erythrocytes incubated for 4 hr
with HAPR in the presence and absence of coformycin. In this
experiment the column eluents were measured at both 254 and 265
nm (the λ_{max} of HAPR nucleotides). Interestingly, in erythrocytes
incubated with HAPR alone some increases in nucleotide peaks were
observed in positions normally occupied by GDP and GTP. These
peaks had greater absorbancy at 265 than at 254 nm and probably
represent the formation of the di- and tri-phosphate nucleotides
of HAPR. Furthermore, a sizable new peak with greater absorbancy
at 254 than at 265 nm was observed in the monophosphate region
(retention time about 12 min). This probably represents the formation of IMP since similar IMP peaks are usually observed when
normal erythrocytes are incubated with adenosine (21). When the
HAPR was incubated in the presence of coformycin, strikingly
different nucleotide profiles were observed. Substantially larger
peaks emerged with retention times of about 30 and 68 min, which
are consistent with the formation of relatively large amounts of
the di- and triphosphate nucleotides of HAPR. Furthermore, the
large 254-nm absorbing peak in the mononucleotide region (presumably IMP) did not appear in the presence of coformycin. Rather,
a peak was found in this region with greater absorbancy at 265
than at 254 nm, which would be consistent with the formation of

*Figure 14. Effect of coformycin on the incorporation of
8-azaadenosine into the nucleotide pools of normal human erythrocytes. Experimental conditions were similar as described in
Figs. 10 and 12, except that 8-azaadenosine (1.0 mM) was used
here and the eluants were monitored at 254 and 271 nm. HPLC
columns and eluting conditions were the same as in Fig. 12.*

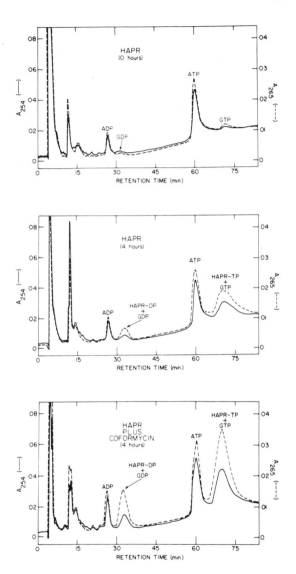

Figure 15. Effect of coformycin on incorporation of N[6]-hydroxyaminopurine ribonucleoside (HAPR) into the nucleotide pools of normal human erythrocytes. Experimental conditions were similar as described in Figs. 10 and 12, except that HAPR (1.0 mM) was used here and the eluants were monitored at 254 and 265 nm. HPLC columns and eluting conditions were the same as in Fig. 12.

the 5'-monophosphate nucleotide of HAPR. It will be of interest
to learn whether the severe hemolytic toxicity of HAPR found
in initial clinical trials can be prevented through the stategy
of inactivating the erythrocytic adenosine deaminase by a drug
such as deoxycoformycin.

V. COMMENTS

This chapter describes studies performed in our laboratory
over the past few years on the development and application of
new theoretical approaches to the study of tight-binding enzyme
inhibitors. These investigations have taken advantage of the
recent availability of a series of inhibitors of the enzyme
adenosine deaminase that vary in inhibitory potency, with K_i
values ranging from about $10^{-6}M$ to $10^{-12}M$. We are confident that
many of the concepts that have been developed during these studies
will prove of value in future investigation of tight-binding inhibitors of enzymes or other macromolecules. Hopefully, some of
these observations will result in useful therapeutic applications.
Perhaps the most important theoretical generalization advanced by these studies is the concept that tight-binding inhibitors
may interact with a macromolecule such as an enzyme very much
more slowly than the usual interaction between the macromolecule
and its natural ligand. This concept indicates that if one pro-
poses to apply kinetic analyses to the study of a tight-binding
enzyme inhibitor, it is essential to recognize that a combination
of steady-state kinetics (in the interaction between the enzyme
and the substrate) and non-steady-state kinetics (in the interaction between the enzyme and the inhibitor) may exist. This
work and related publications offer a number of new, practical
approaches to the investigation of such situations (5-8).
Over the past few years it has been our good fortune to gain
access to a series of substrates and inhibitors of adenosine
deaminase that have yielded many insights into the structure-
activity relationships of the enzymic reaction. First of all,
as demonstrated in Table I, it is apparent that the 2' position
of the sugar moiety contributes to the binding by a factor of
about fourfold, e.g., the K_m value of deoxyadenosine is about
fourfold lower than that of adenosine and the K_i values of coformycin and deoxycoformycin differ by a factor of about fourfold.
As demonstrated through our studies of the individual rate constants, it is also apparent that the dissociation of the EI complex (of the ADA reaction) is the factor responsible for this
difference in binding, rather than the rate of association between
the enzyme and inhibitor or substrate. Another important point
is that it is possible to retain affinity by replacing the ribose

moiety entirely by a large aliphatic alcohol, such as the ery-
thrononyl moiety of EHNA (31). This observation strongly suggests that there is a hydrophobic group on the ADA protein at
the substrate binding site.

Among the most interesting and unique enzyme inhibitors
yet discovered are coformycin and deoxycoformycin. As we have
discussed earlier (6), a strong case may be made that these compounds represent "transition state" analogs of the type originally predicted by Pauling about 30 years ago (32). These inhibitors may be viewed as structural analogs of the physiological
products of the ADA reaction, inosine and deoxyinosine. However,
these natural products have K_i values in the range of 10^{-4} to
10^{-5} M (Table I) whereas the K_i values of coformycin and deoxycoformycin are of the order of 10^{-10} to 10^{-12} M, i.e., more than
one million times more potent. As seen in Fig. 1, coformycin
differs from inosine by the interposition of a methylene group
between N-1 and C-6 of the purine ring. Thus, coformycin has a
seven-membered diazepin ring in the place of the six-membered
pyrimidine ring of inosine. With coformycin, therefore, the
usual keto-enol tautomerism does not occur, and the hydroxyl
group is in fact a secondary alcohol. Furthermore, the planarity
of the ring is altered and the coformycin structure is puckered
with a disruption of the aromaticity found in the purine ring.
These presumably minor structural changes, however, result in a
remarkable increase in the binding affinity of the coformycin-type
molecule to the active site of adenosine deaminase. As commented
upon elsewhere (see Ref. 8), an exciting challenge to the medicinal chemist is that of adding a structure such as the erythro-
nonyl group of EHNA to the ring-structure of coformycin (see
Fig. 1), thus combining the features of a transition-state analog
with the properties of enhanced hydrophobic bonding. It is in-
triguing to speculate that such a compound might have a K_i value
in the range of 10^{-15} to 10^{-16} M, i.e., in the order of magnitude
of a covalent bond.

The availability of new ADA inhibitors of widely different
potencies offers promising therapeutic opportunities. A number
of adenosine analogs, which have shown promise as chemotherapeutic agents, e.g., formycin A, arabinosyl adenine, and HAPR, are
excellent substrates for ADA from erythrocytes and other tissues.
The ADA reaction usually renders these analogs biologically inactive. The use of ADA inhibitors might be expected to produce
drastic modifications in the activity, and perhaps the toxicity,
of various adenosine analogs currently under study. For example,
one possibility might be to introduce a tight-binding inhibitor
such as deoxycoformycin into the erythrocytes by means of a dialysis device, e.g., an "artificial kidney" (33). Thus it may be
possible to inactivate specifically the erythrocytic ADA without
affecting the enzyme in other vital organs such as the liver and
gastrointestinal tract. However, it will be important to examine

the rates of inactivation and reactivation of ADA in intact erythrocytes and to compare these rates with those seen with the
isolated enzyme. An important conclusion that can be drawn from
our kinetic studies of ADA inhibitors is that the tight-binding
inhibitor must be administered before (one to two hours in the
case of coformycin) the main drug whose activity one wishes to
potentiate by inhibiting ADA.

Another area of current interest is the possible use of potent ADA inhibitors as immunosuppressive agents (13,20,22,34).
Certain recent studies indicate that coformycin can block lym-
phatic stem-cell maturation (34) and that EHNA can potentiate
the ability of adenosine to inhibit lymphocytic blastogenesis
(35) and lymphocyte-mediated cytolysis (36). In addition, the
finding that human and murine leukemic cells have much greater
ADA activity than their normal cellular counterparts, e.g., normal peripheral lymphocytes (37), suggests a possible direct role
for tight-binding ADA inhibitors in chemotherapy.

There are likely to be instances in which ADA blockade of
long duration, as seen with coformycin or deoxycoformycin, might
be undesirable, so that a weaker inhibitor with a shorter duration
of action such as EHNA might be preferred. Obviously, the subject of ADA inhibitors is one of great current interest, both for
the provision of potent new tools for biochemical studies and for
the development of possible therapeutic agents.

ACKNOWLEDGMENTS

The unpublished experimental work from this laboratory presented here and the preparation of this chapter have been supported by USPHS Grants CA 07340, CA 13943, and CA 12531. For gifts
of compounds described in this article, the authors wish to thank
the following persons: Dr. H. Umezawa of the Institute of Microbial Chemistry, Tokyo, for coformycin and formycin A; Dr. H. W.
Dion of the Parke, Davis and Company, Detroit, Michigan, for de-
oxycoformycin; Dr. R. V. Wolfenden of the University of North
Carolina, Chapel Hill, for DHMPR; Dr. G. B. Elion of the Wellcome
Research Laboratories, Research Triangle Park, North Carolina for
EHNA; Dr. L. B. Townsend of the University of Utah, Salt Lake
City, for HAPR; and Dr. Harry B. Wood Jr. of the Division of Drug
Research and Development, National Cancer Institute, Bethesda,
Maryland, for arabinosyl adenine and 8-azaadenosine.

REFERENCES

1. Bloch, A. (ed.) Chemistry, biology, and clinical uses of
    nucleoside analogs, *Ann. N Y Acad. Sci. 255* (1975).
2. Antineoplastic and Immunosuppressive Agents II in *Handbook
    of Experimental Pharmacology,* Vol. XXXVIII/2 (A.C. Sartor-
    elli and D. G. Johns, ed.), Springer-Verlag, New York,
    1975.
3. Crabtree, G. W., and Senft, A. W., *Biochem. Pharmacol. 23,*
    649 (1974).
4. Agarwal, K. C., and Parks, R. E., Jr., *Biochem. Pharmacol.
    24,* 2239 (1975).
5. Cha, S., *Biochem. Pharmacol. 24,* 2177 (1975).
6. Cha, S., Agarwal, R. P., and Parks, R. E., Jr., *Biochem.
    Pharmacol. 24,* 2187 (1975).
7. Cha, S., *Biochem. Pharmacol. 25,* 2695 (1976).
8. Agarwal, R. P., Spector, T., and Parks, R. E., Jr., *Biochem.
    Pharmacol. 26,* 359 (1977).
9. Easson, L. H., and Stedman, E., *Proc. R. Soc. London Ser. B.
    112,* 142 (1936).
10. Werkheiser, W. C., *J. Biol. Chem. 236,* 888 (1961).
11. Henderson, P. J. F., *Biochem. J. 127,* 321 (1972).
12. Ackermann, W. W., and Potter, V. R., *Proc. Soc. Exp. Biol.
    Med. 72,* 1 (1949).
13. Agarwal, R. P., Sagar, S. M., and Parks, R. E., Jr., *Biochem.
    Pharmacol. 24,* 693 (1975).
14. Brady, T. G., and O'Connell, W., *Biochim. Biophys. Acta 62,*
    216 (1962).
15. Waxman, S., Schreiber, C., and Herbert, B., *Blood 38,* 219
    (1971).
16. Giblett, E. R., Anderson, J. E., Cohen, F., Pollara, B.,
    and Meuwissen, H. J., *Lancet 2,* 1067 (1972).
17. Meuwissen, H. J., Pollara, B., and Pickering, R. J., *J. Ped-
    iatr. 86,* 169 (1975).
18. Parkman, R., Gelfand, E. W., Rosen, F. S., Sanderson, A.,
    and Hirshhorn, R., *N. Engl. J. Med. 292,* 714 (1975).
19. Polmar, S. H., Stern, R. C., Schwartz, A. L., Wetzler, E.
    M., Chase, P. A., and Hirschhorn, R. *N. Engl. J. Med.
    295,* 1337 (1976).
20. Agarwal, R. P., Crabtree, G. W., Parks, R. E., Jr., Nelson,
    J. A., Keightley, R., Parkman, R., Rosen, F. S., Stern,
    R. C., and Polmar, S. H., *J. Clin. Invest. 57,* 1057 (1976)
21. Parks, R. E., Jr., and Brown, P. R., *Biochemistry 12,* 3294
    (1973).
22. Parks, R. E., Jr., Crabtree, G. W., Kong, C. M., Agarwal, R.
    P., Agarwal, K. C., and Scholar, E. M., *Ann. NY Acad. Sci.
    255,* 412 (1975).

23. Schabel, F. M., Jr., Trader, M. W., and Laster, W. R., *Proc. Am. Assoc. Cancer Res. 17*, 46 (1976).
24. Brockman, R. W., Shaddix, S. C., Rose, L. M., and Carpenter, *J. Proc. Am. Assoc. Cancer Res. 17, 52* (1976).
25. Lepage, G. A., Worth, L. S., and Kimball, A. P., *Cancer Res. 36*, 1481 (1976).
26. Cass, C. E., and Au-Yeung, T. H. *Cancer es. 36*, 1486 (1976).
27. Roblin, R. O., J., Lampen, J. O., English, J. P., Cole, Q. P., and Vaughan, J. R., *J. Am. Chem. Soc. 67*, 290 (1945).
28. Parks, R. E., Jr., and Agarwal, K. C., in *Handbook of Experimental Pharmacology, Vol. XXXVIII/2*, (A. C. Sartorelli and D. G. Johns, eds.) Springer-Verlag, New York, 1975, P. 458.
29. Dollinger, M. R. and Krakoff, I. H., *Clin. Pharmacol. Ther. 17*, 57 (1975).
30. Burchenal, J. H., Dollinger, M., Butterbaugh, J., Stoll, D., and Giner-Sorolla, A., *Biochem. Pharmacol. 16*, 423 (1967).
31. Schaeffer, H. J., and Schwender, C. F., *J. Med. Chem. 17*, 6 (1974).
32. Pauling, L., *Am. Sci. 36*, 58 (1948).
33. Richardson, P. D., Galleti, P. M., and Born, G. V. R., *Proc. Am. Soc. Artificial Organs 5*, 67 (1976).
34. Ballet, J. J., Insel, R., Merler, E., and Rosen, F. S., *J. Exp. Med. 143*, 1271 (1976).
35. Carson, D. A. and Seegmiller, J. E., *J. Clin. Invest. 57*, 274 (1976).
36. Wolberg. R., Zimmerman, T. P., Hiemstra, K., Winston, M., and Chu, L. C., *Science 187*, 957 (1975).
37. Smyth, J. F., *Proc. Am. Assoc. Cancer Res. 17*, 59 (1976).

SYNTHESIS AND CHEMISTRY OF A NEW ANTIVIRAL-
ANTIBIOTIC NUCLEOSIDE IN THE ANHYDRO FORM:
APPLICATION OF A NOVEL 1,3-ELECTROPHILIC
(C-N-C) ANNULATION REAGENT

WENDELL WIERENGA AND JOHN A. WOLTERSOM

ABSTRACT

A novel 5-aza-nucleoside has recently been isolated from
Streptomyces platensis var. *clarensis* and assigned the unusual
structure of dihydro-5-azathymidine. Because of its interesting antibiotic-antiviral properties and relatively low fermentation yields, we have explored several chemical synthetic
routes for the preparation of this new nucleoside ring system
and related analogs. Utilization of a new annulation reagent
in several chemical approaches, the successful application of
lanthanide-induced shift studies and ^{13}C-NMR for structure determinations, as well as relevant pharmacology, will be discussed.

An unusual nucleoside was recently detected by its antibiotic
activity and subsequently isolated from *streptomyces platensis*
(var. *clarensis)*. This nucleoside was shown (1) to have the
structure (1). This deoxyribonucleoside, dihydro-5-azathymidine,
also exhibited (2) antiviral activity against Herpes simplex
type 1 (HSV-1), somewhat less activity against type 2, and
slight activity against vaccinia and pseudorabies. It was
found to be inactive against RNA viruses, nonvirucidal, and
topically inactive against HSV-2, however, antiviral activity
was demonstrated in mice against HSV-1 even when therapy was
initiated three days after infection. Even though dihydro-5-
azathymidine was less cytotoxic than most other antiviral nucleosides and no toxicity was noted in rats or mice, aplastic

anemia was seen in dogs on a chronic toxicology study and also
marrow suppression in cats and rabbits. Its mode of action
appears to occur by an inhibition of thymidine kinase since,
among other reasons, thymidine competitively inhibits antiviral
and antibacterial activity.

A chemical synthesis of this novel nucleoside and congeners
was desirable due to the aformentioned interesting biological
properties and low formation yields. Of several possible approaches, one might envision either a classical condensation route
(1,2) or, alternatively, a route employing the 2,2'-anhydro isomer
(3) as a key target for potential conversion to the deoxy nucleoside (4) and other analogs.

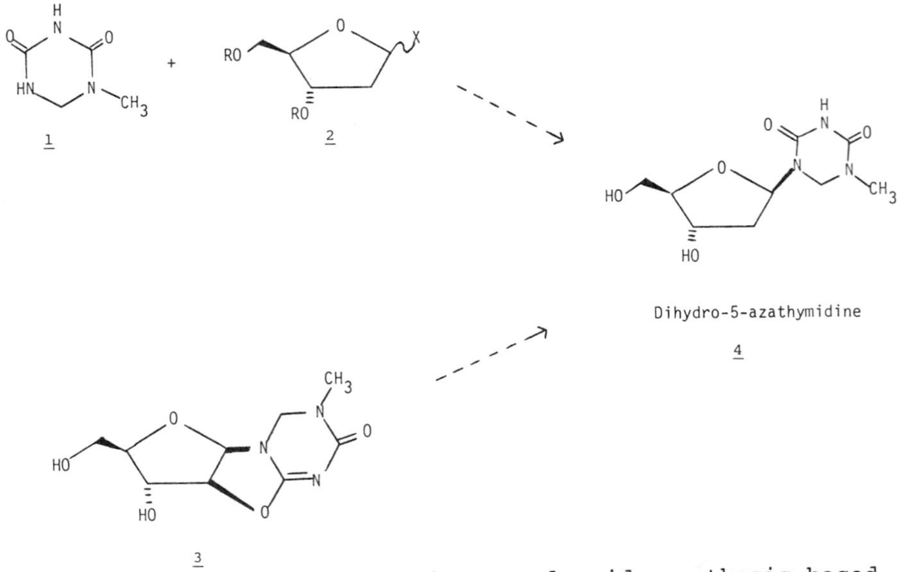

Dihydro-5-azathymidine

4

The utility of an approach to nucleoside synthesis based
on a 2,2'-anhydro synthon derived from the "Salk intermediate,"
the aminooxazoline of arabinose (5), has been well documented
and is illustrated (3) by the formation of 6 and 7. The "Salk
intermediate" not only is readily available, but also incorporate
3 elements common to the pyrimidine ring system bases, with the
correct sugar-ring stereochemistry.

The utility of this intermediate (Ref. 4,5) could be substantially enhanced if suitable, selective protection were available for the 3',5'-diol grouping. As shown below, we found that
the t-butyldimethylsilyl group(5) fit our requirements of selectivity, "base" stability (could not employ acid-deprotection
either), and enhanced lipophilicity for chemical manipulability
and analysis(9 and 10).

Since the desired triazine ring system is unsymmetrical
with regards to substitution, one could envision several routes
depending on the selectivity, if any, inherent in the amidine

grouping. For example, the reaction of 1̲0̲ with an appropriate isocyanate-type electrophile could lead, in a two-step process, to the desired 2,2'-anhydro nucleoside (1̲2̲).

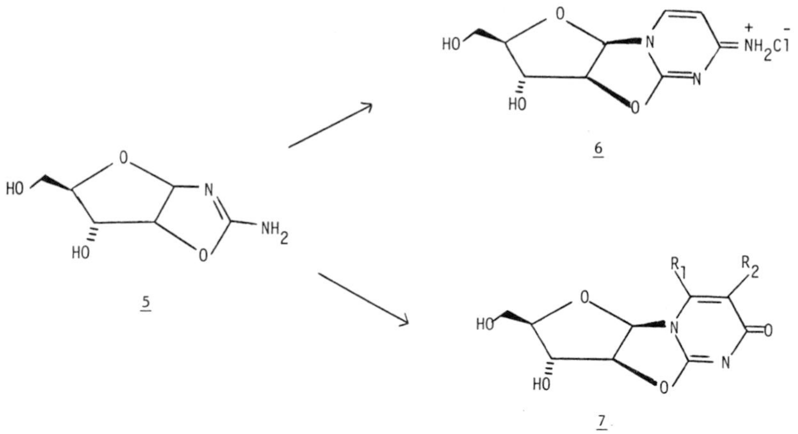

6̲

5̲

7̲

for example, A. Holy, Coll. Czech. Chem. Commun., 3̲9̲
3177 (1974) and ref. cited therein.

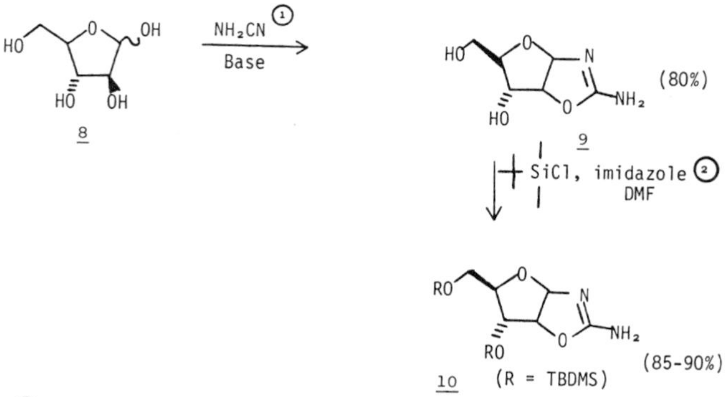

8̲

9̲ (80%)

1̲0̲ (R = TBDMS) (85–90%)

① (R.A. Sanchez and L.E. Orgel, J. Mol. Biol., 4̲7̲, 531, 1970)

② see K.K. Ogilvie and D.J. Iwacha, Tet. Letters, 317 (1973) for application to nucleosides.

We employed methyl isothiocyanate as a means of obtaining the desired dihydrotriazine ring system. However, acylation was nonselective and both isomers (1̲3̲ and 1̲4̲) were isolated in high yields as a 1:1 mixture. A clean separation was effected by HPLC (high-pressure liquid chromatography). Even more problematic was the fact that only the "wrong isomer" (1̲3̲, conjugated thione) was readily desulfurized. The use of more forcing desulfurization methods with Raney nickel or other reagents (e.g., phosphites) still proved unsuccessful in producing the desired 2,2'-anhydro isomer (1̲6̲).

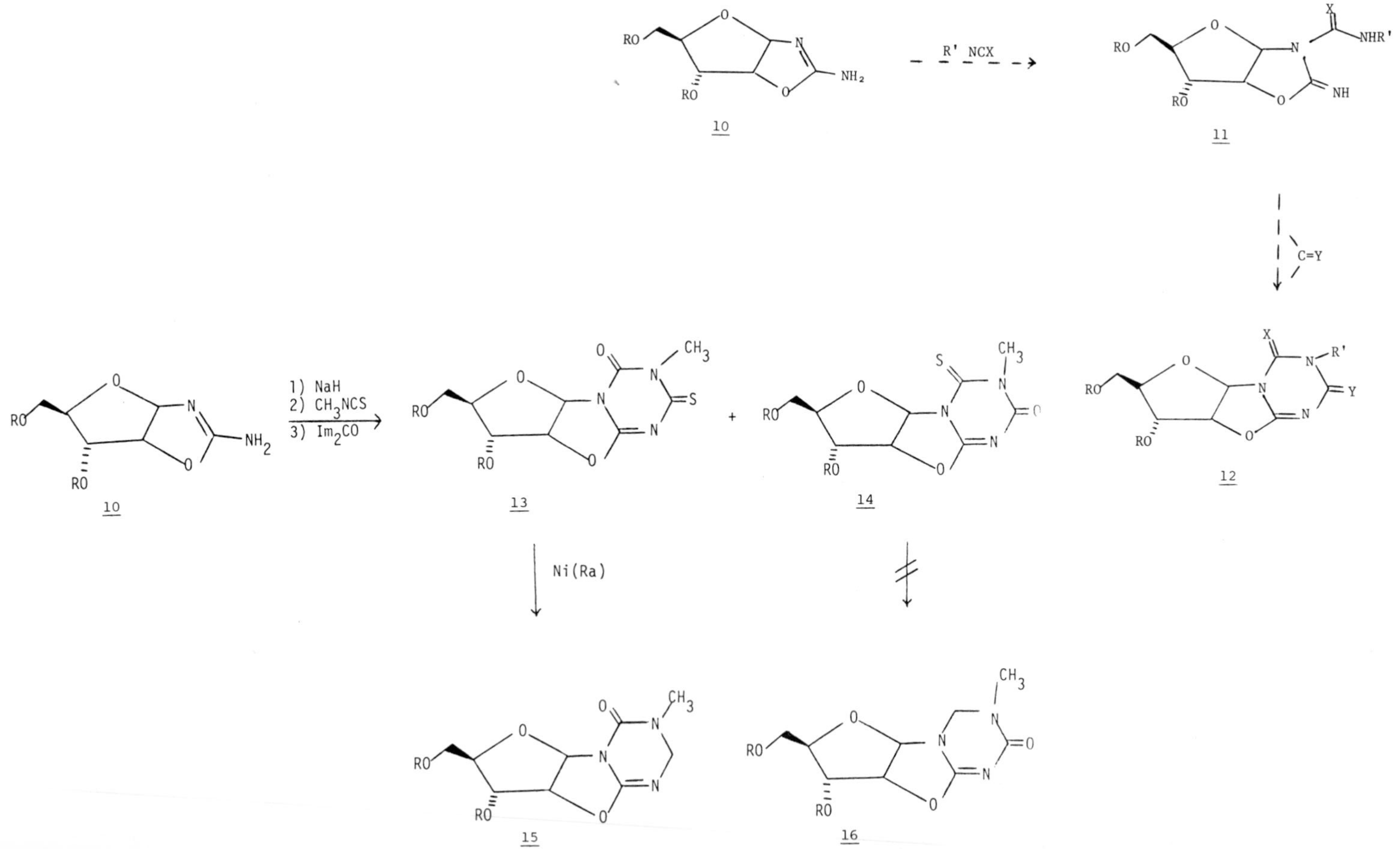

R' NCX
C=Y
1) NaH
2) CH3NCS
3) Im2CO
Ni(Ra)
10
11
12
13
14
15
16

As a means of circumventing these problems, we devised an annulation reagent that exhibited the selective alkylation required by the unsymmetrical amidine grouping of the aminooxazoline, had the desired oxidation state, and was readily available.

We initially attempted to generate an N-carbethozy-N-methyl imminium species via hydride abstraction on N,N-dimethyl carbamate with trityl fluoroborate. Even though a high yield of triphenyl methane was realized, the presumed iminium species proved too reactive to utilize directly. We then turned to a less reactive system involving an X-methylene carbamate, where X was a good leaving group. The bromomethylene derivative(17) was prepared as depicted and indeed afforded only ring-nitrogen alkylation of the sodium salt of the aminooxazoline.

Although the annulation reagent is a potential 1,3-electrophile, the carbamoyl carbonyl proved resistant to intramolecular nucleophilic attack to effect ring closure to the desired 2,2'-anhydro triazine nucleoside. As shown, several different leaving groups and numerous conditions including acid, base, neutral, and thermal were found to be ineffectual.

An alternate approach utilizing the capabilities of a thiocarbamate system to be modified by mild oxidation after the initial alkylation of the aminooxazoline substrate proved successful. The bromomethylene of the thiocarbamate was prepared analogously, alkylated by the aminooxazoline to afford 18 and then

cyclized with two equivalents of m-chloro-perbenzoic acid at
0 C in methylene chloride. Attempts to effect this cyclization
with base or metal catalysis (Hg^{2+}, Cu^{2+}) were not successful.
The remarkable ease of this transformation under oxidative conditions (presumably proceeding through the α-keto sulfone(6))
is indeed surprising in light of the previously mentioned attempts.

$R' = CH_3CH_2$, CH_3, $CH_2\emptyset$ ($\emptyset$, $p-NO_2\emptyset$)

Depending on the mode of preparation of tetra-n-butyl ammonium fluoride, we could obtain either the arabinoside deriv-
ative(19) or a mixture of 19 and 3.
      The 2,2'-anhydro-dihydro-5-azathymidine(3) exhibited the
usual chemical properties of facile base hydrolysis to afford
the arabinoside derivative(19). However, due to the very acid
labile nature of the C-6 masked formaldehyde carbon of this 5,
6-dihydro pyrimidine, we have been unable to directly convert
3 to either the riboside or 2'-deoxyriboside derivatives.

Since we had both of the 2,2'-anhydro isomers in hand, a comparison of the *nmr*, ir, and uv data strongly supported the assigned structures. However, to obtain more confirmatory data we turned to lanthanide shift studies. Shift studies have seen very limited application in nucleoside structure analysis due to limited solubility in noncomplexing solvents, multiple sites for metal interaction, and glycosidic rotation. However, we anticipated an application in this case because of the rigid structure, solubility in nonpolar solvents, and masking of *OH* complexation, both due to the t-butyldimethylsilyl groups, and finally based on the empirical relationship of amine > alcohol > ketone, am-

Dihydro-5-azathymidine

ide > ester > nitrile > ether complexation equilibria (steric
factors being equal).   In fact, as seen in Figs. 2-5, the experimental data fit well with the assigned structures and with the
calculated values arising from an LISC program generator (based
on coordinates derived from x-ray data of dihydro-5-azathymidine
(unpublished)).

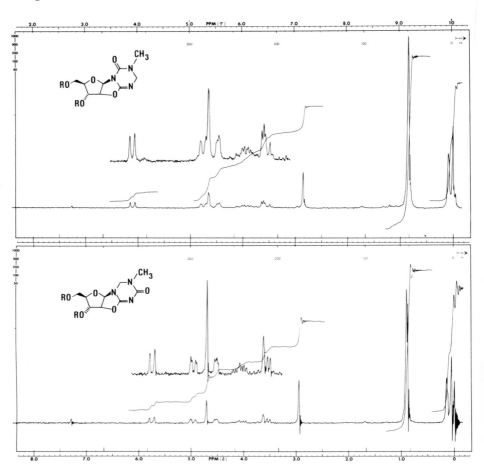

Figure 1.

As indicated in Fig. 6, the application of [13]C *nmr* proved
very useful in predicting and confirming structures of some
other symmetrical triazines synthesized.   The predicted trend
of downfield shift correlating with conjugation of the C=X
grouping holds in every case seen here.

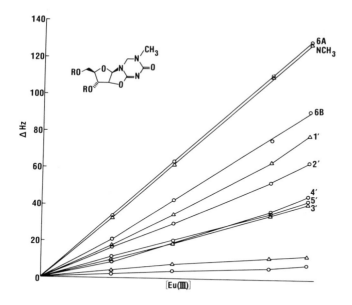

Figure 2.

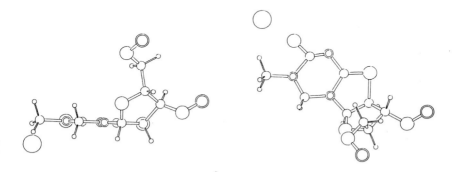

R- factor = 0.079

Figure 3.

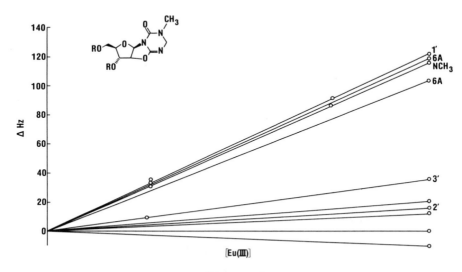

Figure 4.

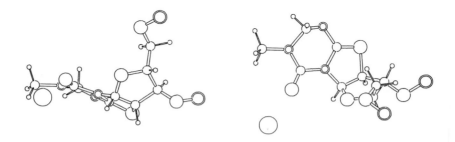

R- factor = 0.072

Figure 5.

C-2	152.8	157.2	155.5	153.0	155.9
C-4	64.8	162.8	161.2	159.3	187.9
C-6	149.8	59.4	147.4	174.1	145.8

Finally, we were able to prepare the alternate N-2,2'-anhydro isomer(15) by independent chemistry involving the diphenyl carbonate-mediated (7) dehydration of the corresponding N_3-riboside. It was shown to be identical by tlc, glpc, mass spectral, and *nmr* comparison to 15 prepared by the previous procedure.

ACKNOWLEDGMENTS

The authors thank Mr. S. A. Mizsak for performing some of the *nmr* shift study and processing the data, Drs. G. Slomp and B. V. Cheney for the use of their LISC program, Mr. H. I. Skulnick for a sample of the N_3-riboside of the title compound, and Dr. J. C. Babcock for helpful discussions.

REFERENCES

1. C. DeBoer and B. Bannister, US Patent Nos. 3,907,643 and 3,907,779, 1975.
2. H. E. Renis, B. A. Court, and E. E. Edison, Abstracts of the ASM Meeting, Atlantic City, N.J., 1976.

3. A. Holy, *Coll. Czech. Chem. Commun. 39,* 3177 (1974); and references cited therein.
4. R. A. Sanchez and L. E. Orgel, *J. Mol. Biol. 47,* 531 (1970).
5. K. K. Ogilvie and D. J. Iwacha, *Tetrahedron Lett.* 317 (1973).
6. T. Kumamoto and T. Mukaiyama, *Bull. Chem. Soc. Japan 41,* 2111 (1968); D. H. R. Barton, D. D. Manly, and D. A. Widdowson, *J. Chem. Soc., Perkin Trans. 1* 1968 (1975).
7. A. Hampton and A. W. Nichol, *Biochemistry 5,* 2076 (1966).

SYNTHESIS OF 1-(2-DEOXY-β-D-Ribofuranosyl)-5,6-
DIHYDRO-5-METHYL-s-TRIAZIN-2,4(1H,3H)Dione

H. I. SKULNICK

The Upjohn Company

In May of this past year, a report (1) from our laboratories described the isolation of a novel nucleoside antibiotic (<u>1</u>) from
the culture *Streptomyces platensis*. The nucleoside was found
to be active against a variety of gram-negative and gram-positive
bacteria (especially gram-negative) and was active against a
number of DNA viruses (especially the Herpes virus). This nucleoside was active orally as well as subcutaneously and intraven-
ously.

The biosynthesis of this nucleoside has been recently described (2,3). The antiviral activity has been described (4)
and the structure determination of this nucleoside as <u>1</u> was pre-

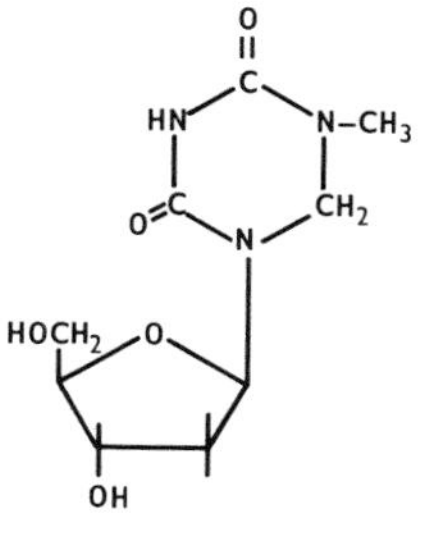

5,6-dihydro-5-azathymidine

<u>1</u>

sented at the recent IUPAC meeting in New Zealand (B. Bannister).
The nucleoside has limited potential as an antibacterial agent
since resistance developed rather quickly and toxicity was seen
at high doses in both the cat and the dog.

Since only low titers were obtained from fermentation, this
prompted us to attempt a total chemical synthesis. We would now
like to describe the synthesis of this nucleoside via the silyl
ether modification of the Hilbert-Johnson condensation.

The structure of this nucleoside was determined to be 1-(2-deoxy-β-D-ribofuranosyl) 5-methyl-5,6-dihydro-*S*-triazine-2,4(1H,3H)-dione,(5,6,-dihydro-5-azathymidine, DHAdT). As one would expect from the 5,6-dihydro structure, this nucleoside is not stable under acidic conditions but is stable under neutral and basic conditions.

The effect of DHAdT on pseudorabies, vaccina, HSV-2, and HSV-1 virus, as determined by H. Renis of our laboratories, is shown in Fig. 1.

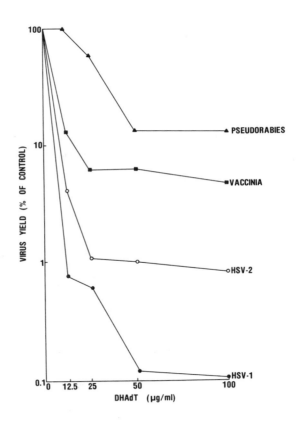

Figure 1. Effect of DHAdT on virus yields. H. E. Renis, B. A. Court, and E. E. Edison, ASM Meeting, Atlantic City, N.J., May 2-7, 1976.

Synthesis of the base portion of DHAdT, 5-methyl-5,6-dihydro-*S*-triazine-2,4-(1H,3H)-dione has been reported (5) via the ring closure of *N*-methylbiuret (2̲) with ethylformate followed by reduction of the resulting *N*-methyl-*S*-triazine (3̲) with either

Raney nickel (W-28) or 5% Rh/C to afford **4**. The biuret (**2**) can be synthesized from ethyl allophonate and methylamine *or* methyl isocyanate and urea. The *N*-methyl triazine (**3**)is recovered as an ethanol or water adduct (across the 5,6 double bond), which can be reduced to afford **4** directly or freed from its adduct by azeotroping from benzene with *p*-TSA added prior to reduction.

A. Piskala and J. Gut, Collection Czechoslov.
Chem. Commun. **26**, 2519 (1961)

Silylation of the dihydro-*S*-triazinone base (**4**) with hexamethyldisilizane gave the *bis*-silyl compound (**5**) as an oil in almost quantitative yield. This oil slowly hydrolyzed to afford the crystalline, stable, monosilyl dihydrotriazinone (**6**). The monosilyl base (**6**) can also be synthesized directly with *bis*-trimethylsilyltrifluoroacetamide in pyridine at 25C. Assignment of this structure was made on the basis of the the PMR spectra shown in Figs. 2,3, and 4. The PMR spectra of 5-methyl-5,6-di-hydro-*S*-triazine-2,4-(1H,3H)-dione (**4**) are shown in Fig. 2. Irradiation of the N_3-H δ 9.3 does not affect the 5-methylene at δ

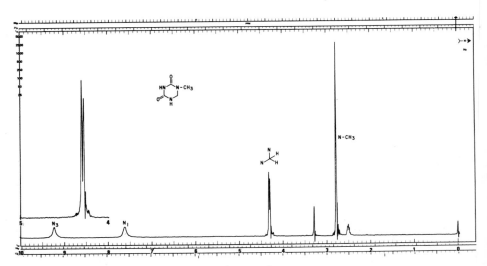

Figure 2.

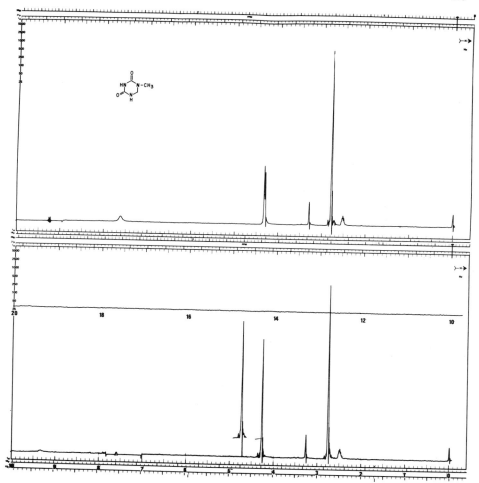

Figure 3.

4.3 as shown in Fig. 3. Irradiation of the N_1-H sharpens the methylene peak to a singlet. We therefore looked at the 5-CH_2 group in the mono-silyl-*S*-triazine compound (6) to determine its tautomeric structure. The 5-CH_2 in the silyl base was a singlet as shown in Fig. 4 and thus this compound was assigned the 2-*O*-silyloxy structure as shown.

Condensation of the mono *or* disilylated dihydrotriazine compound with 3,5-di-*O*-toluoyl-D-ribofuranosyl (7) chloride in an acetonitrile/ethylene dichloride mixture using $SnCl_4$ as the Lewis acid catalyst gave a 1:1 mixture of the α and β anomers of the

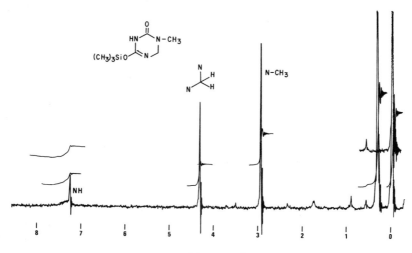

Figure 4.

3'5'-di-O-toluate esters of DHAdT. The β anomer (8) can be crystallized directly from the reaction mixture, making a difficult
separation unnecessary. The overall nucleoside yield of the reaction is 70%, from which a 25-30% yield of the β-nucleoside
(8) can be directly crystallized out. The condensation reaction
was run under a variety of conditions with the optimum conditions
for preparing large quantities of the dihydro-S-triazine nucleoside being those shown. That is, use of SnCL₄ as Lewis acid catalyst, CH₃CN/ethylene dichloride solvent mixture- at an initial
temperature of -25°C followed by warming to +25°C, and allowing
the reaction to proceed at 25°C for a period of approximately 6
hr.
    Table I outlines the types of conditions attempted. To sum-
marize, on a 1-mm scale, maximum yields were obtained using the
polar solvents CH₃CN or CH₃NO₂, at -25°C, using a 2:1 ratio of
silyl triazine base to sugar, an excess of catalyst to triazine
base, and allowing the reaction to proceed over a period of 18
hr. Yields of the β-nucleoside (8) are based on a disk bioassay
vs k.p.
    During scaleup, however, problems were encountered that
when solved gave us an insight into the possible reaction mechanism. It should be noted that in the SnCl₄-catalyzed condensation of bis-silyloxy-5-ethyluracil with 3,5-di-o-toluoyl-2-D-ri-
bofuranosyl halide, Vorbrüggen did not find any variation in the
α/β ratio based on reaction conditions or scale up.
    When the silyl dihydro-S-triazine condensation was scaled
up, the ratio of α/β greatly increased (to 15:1 from 1:1), and

TABLE I.   Reactions Based on 1-mm Base

Ratio mm sugar:mm base	Catalyst mm $SnCl_4$	Solvent	$t^{\circ}C$	Time (hr)	Ratio α β	DHAdT Yield (%)
1:1	0.7	$C_2H_4Cl_2$	5	18	---	0
1:2	0.7	$C_2H_4Cl_2$	5	18	4:1	2.85
2:1	0.7	$C_2H_4Cl_2$	5	18	---	<0.1
1:1	0.7	$C_2H_4Cl_2$	-25	18	---	0
1:2	0.7	$C_2H_4Cl_2$	-25	18	4:1	2.14
2:1		$C_2H_4Cl_2$	-25	18	---	0
1:1	0.7	CH_3CN	5	18	---	0
1:2	0.7	CH_3CN	5	18	1:1	9.1
1:1	0.7	CH_3CN	-25	18	1:1	8.9
1:2	0.7	CH_3CN	-25	18	1:1	19.0
2:1	0.7	CH_3CN	-25	18	1:1	12.0
1:1	0.35	CH_3CN	-25	18	1:1	12.8
1:2	0.35	CH_3CN	-25	18	1:1	8.8
2:1	0.35	CH_3CN	-25	18	1:1	12.2
1:1	1.3	CH_3CN	-25	18	1:1	14.4
1:2	1.3	CH_3CN	-25	18	1:1	24.0
2:1	1.3	CH_3CN	-25	18	1:1	14.9
1:2	0.9	$CH_3N)_2$	-25	5 min.	1:1	29.0
1:2	0.9	CH_3NO_2	-25	18	---	<5.0

the necessary reaction time was also increased (to 7-10 days from 18 hr). It was found that by solubilizing the sugar halide in a minimum amount of ethylene dichloride before addition, and by warming the reaction to +25°C after the initial cooling to -25°C, we could obtain α/β ratios of 1:1 with reaction times of 6 hr or less regardless of the reaction scale. In addition, the α anomer was initially formed (at -25°C) followed by the slow formation of the β anomer (at +25°C).

A possible explanation for the stability of the reaction upon warming (the reaction could be warmed to 40-50°C without any noticeable decomposition of sugar), the high reactivity in polar solvents, and the initial formation of α anomer (at low temperatures) followed by formation of β anomer (at higher temperatures) can be found in two previously published reports.

The first (6) outlines the formation of stable Sn-N π complexes formed from the interaction of the Sn catalyst on the silylated base: in this case either pyrimidines or *as*-triazines (10). We would expect that this stable complex could form at the N_1 position in the 5,6-dihydro-*S*-triazine derivative.

V. Niedballa and H. Vorbrüggen, J. O. C., 41, 2084 (1976)

Dissociation of this complex would be greatest in the polar solvents (CH_3CN or CH_3NO_2) and at higher temperatures. The base would then react with the sugar cation to give the protected nucleoside.

The stable tin-dihydro-*S*-triazine complex could then react with the carbonium ion formed via transannular displacement of

the ditoluoyl sugar halide. The α anomer (that which would react with the 5'-1' carbonium ion 14) is the kinetic product, and
the β-anomer (that which would react with the 3'-1' carbonium
ion 15) the thermodynamic product of the reaction. There is
little evidence to suggest that the chloro sugar is reacting directly with the base via an SN₂ displacement since the initial
product formed is the α nucleoside as opposed to the β, which is
the one expected from an SN₂ attack on the α-chloro sugar. This
simplified series of equilibria was originally proposed by Sorm
and co-workers but outlined here in a paper by Bardos (7).

Kotick, Szantay and Bardos, J.O.C., 34, 3806 (1969)

When the conditions favorable for good β-nucleoside formation were applied to the *bis*-silyl ether of thymine, a greater
than 70% yield of a mixture of α+β nucleosides was obtained.
The proportions of β:α ranged from 4:1 to 5:1.

β:α (5:1 @ -25°)
(4:1 @ -25° to +25°)
nucleoside yield (based on sugar) : 71-75%

De-protection with sodium methoxide in methanol gave the unprotected nucleosides in overall yields of better than 60%. The β nucleoside (18) was formed in greater than 25% overall isolated yields.

The PMR spectra, in D_2O, of the α and β anomers of 1-(2-deoxy-D-ribofuranosyl)-5-methyl-5,6-dihydro-S-triazine-2,4(1H, 3H)dione (19 and 18, respectively) with the characteristic triplet for the l'H at δ 6.2 for the β anomer and the quartet for the α anomer at δ 6.2 are shown in Fig. 5.

The condensation reaction can also be carried out with mercuric bromide as the catalyst in the presence of molecular sieves with acetonitrile being the solvent of choice. Yields are considerably lower than the $SnCl_4$-catalyzed condensations. When molecular sieves were omitted, only the α anomer could be isolated. These results are similar to those reported by Szabolcs in a recent paper on the condensation of 5-alkyluracils with protected 2-deoxyribofuranosyl chlorides in the presence of mercuric bromide and molecular sieves.

The effect of DHAdT on HSV-1 is shown in Fig. 6. The drug was prepared in normal saline and given subcutaneously four times (intravenous innoculation) in mice each day for five days. The percent mortality is plotted against the days after innoculation. The effect of DHAdT is obviously dose dependent, with the greatest number of survivors and largest increase in life span coming at the highest doses. With the 400-mg/kg dose, 65% of the mice survived, 45% survived with a dose of 200 mg/kg, and 35% with the 100 mg/kg dose, with 95% of the nontreated mice dying. One of the problems with the nucleoside was its short half-life in the blood. An attempt was made to increase the length of time that the minimum effective dose of the nucleoside was present in the blood. The use of esters of certain nucleosides as depot forms is well documented. Various 5' and 3' esters of DHAdT were prepared and tested as depot forms of the antibiotic.

The free nucleoside (18) was reacted with the appropriate acid chloride in pyridine at 25°C to give the desired 5' ester directly. Due to the acid lability of the dihydronucleoside the 3' esters were prepared via the 5'-*tert*-butyldimethylsilyl ether as described previously for pyrimidine deoxynucleosides by Ogilvie.

Treatment of DHAdT with TBDMS-chloride-imidazole in dimethylformamide gave the 5'-silyl ether (21) in > 90% yield. Acetylation with the appropriate acid chloride to afford 23 was followed by de-silylation with tetra-*n*-butyl-ammonium fluoride/THF to give the free 3' ester (22).

The antiviral activities of some representative 5' esters *in vitro* and *in vivo* vs HSV-1 are shown in Fig. 7. The virus yield is plotted against the drug concentration in μ*M*. In the upper right-hand corner is a table comparing the *in vivo* activity of various 5' esters of DHAdT vs DHAdT. The compounds are given at a rate of 0.4 m*M*/kg, three times a day for five days. The M.S.T. and percent S for each compound is given. The table shows that some of the esters increase the mean survival times and the percent survivors over the parent compound (DHAdT). The 3' esters, under identical conditions, did not show any significant activity.

When attempts were made to condense the *bis*-silyl-dihydro-*S*-triazine base (5) with tribenzoyl ribofuranosyl bromide or acetate (24) some interesting observations were made. The

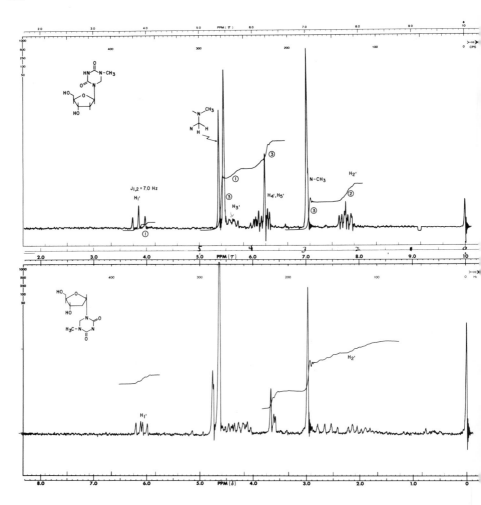

Figure 5

major product from this reaction was the N_3 nucleoside (25).
In addition, depending upon the reaction conditions, three other
nucleosides could also be isolated. Using 2,3,5-tri-O-benzoyl-
D-ribofuranosyl 1-o-acetate (24), N_1-acetyl-N_3-(tribenzoyl-β-D-
ribofuranosyl)-5,6-dihydro-5-methyl-S-triazine-2,4-(1H,3H)-dione
(28) was formed. Using 2,3,5-tri-O-benzoyl-D-ribofuranosyl bromide, the N_3-S-triazine nucleoside (26), plus the methanol adduct
(27) was formed. Deprotection of 25 and 28 gave the 3-(β-D-ribofuranosyl)-5-methyl-5,6-dihydro-S-triazine nucleoside derivative.
The formation of a N_3 nucleoside was not observed when the
2-deoxy sugar derivative was used and the reason for a total lack
of N_1 nucleoside formation in the above cases is unclear.

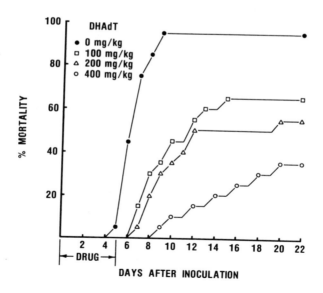

Figure 6. *Effect of DHAdT on HSV-1 (IV Inoculation) in mice DHAdT given SC Q.I.D. x 5 days. H. E. Renis, B. A. Court and E. E. Edison, ASM Meeting, Atlantic City, N.J., May 2-7, 1976.*

The formation of 28 was clearly due to an impurity of CH_3OH in the sugar. When 1-*o*-acetyl-2,3,5-tri-O-benzoyl-D-ribofuranose acetate, dried at 50°C, was used as precursor for the sugar halide no 28 could be detected.

When the *bis*-trimethylsilyl base (5) was condensed with 2, 3,5-tri-O-acetyl-D-ribofuranosyl bromide (29) in ethylene dichloride using $SnCl_4$ as catalyst, both the N_1 and N_3 nucleosides (30 and 31) were formed. When tetraacetyl-D-ribofuranose was used, a fair amount of the N_1 acetyl base (32) was also formed. In CH_3CN, only 32 could be detected and was isolated in very high yields (>75%). Both the N_1 and N_3 ribosides (30 and 31) lacked activity against a variety of RNA and DNA viruses.

In conclusion, I would like to express my appreciation to Dr. Brian Bannister, Dr. Tom Brodasky, Dr. Lester Dolak, Mr. R. D. Birkenmeyer, and Mr. B. Czuk for their collaborative assistance and Mr. Paul Meulman and Mr. Terrance Scahill of our NMR Department for their assistance in structure determination. In addition I would like to thank Dr. H. Renis of our Virology Department for his excellent work on the antiviral properties of the antibiotic and its esters.

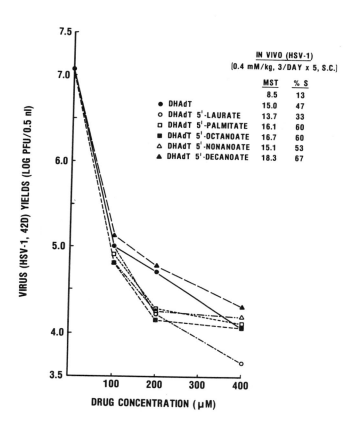

Figure 7. Antiviral activity of 5'-esters of DHAdT Herpes-virus (HSV-1) titers were determined 24 hrs after inoculation.

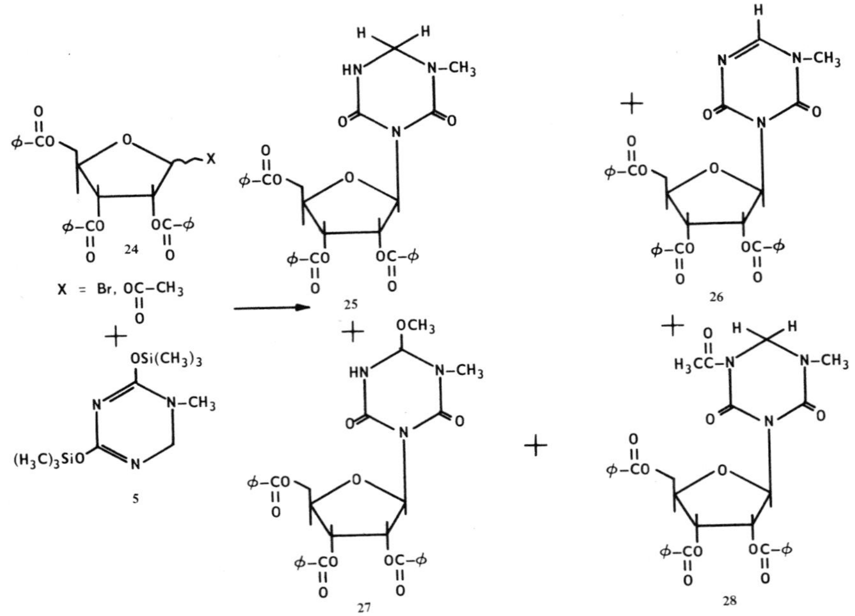
24
X = Br, OC–CH3
5
25
26
27
28

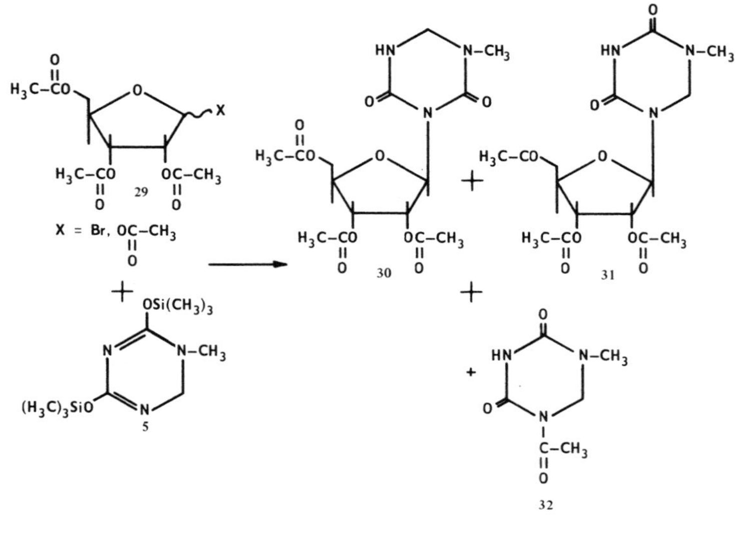
29
X = Br, OC–CH3
5
30
31
32

REFERENCES

1. Bannister and DeBoer, Report to the American Society of Microbiologists, 1976.

2. Slechta and Ciadella, Seminar, Department of Biochemistry, School of Medicine, Temple University, 1976, unpublished.

3. Slechta and Cindella, Proceedings of the 10th International Congress of Biochemistry, Germany, 1976.

4. H. Renis, Report to the American Society of Microbiologists, 1976.

5. A. Piskala and J. Gut, *Coll. Czech. Chem. Commun. 26,* 2519 (1961).

6. U. Nieballa and H. Vorbrüggen, *J. Org. Chem. 41,* 2084 (1976).

7. Kotick, Szntay, and Bardos, *J. Org. Chem. 34,* 3806 (1969).

AN APPROACH TO THE SYNTHESIS OF 1,
9-DIRIBOSYL PURINE

BERNARD RAYNER, CLAUDE TAPIERO, AND JEAN-LOUIS IMBACH

Université des Sciences et Techniques du Languedoc
Laboratoire de Chimie Bio-Organique,
Montpellier Cédex, France

ABSTRACT

The aim of this work was to synthetise 6-imino 1,9-diribosyl
purines and related compounds as part of phosphoribosyl ATP, a
key intermediate in the histidine biosynthetic pathway. Since
no general method has yet been described for introducing a ribofuranosyl group at position 1 of the purine ring, we accomplished
a synthesis of such compounds using 4-cyano 5-aminoimidazole as
our starting material. In all cases we obtained the corresponding purines as an anomeric mixture. These anomeric nucleosides
were separated by chromatography and the data of all new nucleoside derivatives are given.

The primary aim of this work was to obtain 1,9-diribofuranosyl
adenine and related derivatives. This nucleoside is of considerable interest since the nucleotide derivative (the phosphoribosyl ATP) has been proposed as one of the key compounds of histidine biosynthesis. (1)
    The 1-ribofuranosyladenine, per se, had not been synthesized,
therefore, we were very interested in developing a general approach for the introduction of a ribose residue at position 1 of
the purine ring. In fact, only two 1-ribofuranosylpurines were
described in the literature (2,3). 1-β-D-Ribofuranosyl-2-oxo-
purine (2) had been prepared by Fox et al. according to Scheme 1.
    The second 1-ribofuranosylpurine described (3) in the lit-
erature was obtained by Montgomery et al. using the approach

illustrated in Scheme II.   Therefore, we initiated research in
this area, as there was no general synthetic approach available
for the preparation of 1-ribofuranosylpurines.

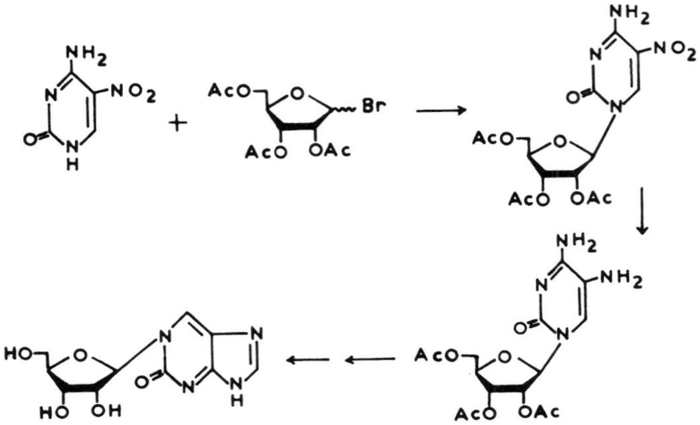

Scheme I

The introduction of an alkyl group specifically at position
1 of a purine ring is a well-known problem. One general approach
has been to transform a 5-amino 4-cyanoimidazole into the corresponding 5-substituted amidine, which is then easily ring
closed to afford a 1-substituted adenine (4).   Therefore, we de-
cided to follow the same approach and selected 5-amino-4-cyano-
1-β-D-ribofuranosyl-imidazole (1) as our starting material
(Scheme III).

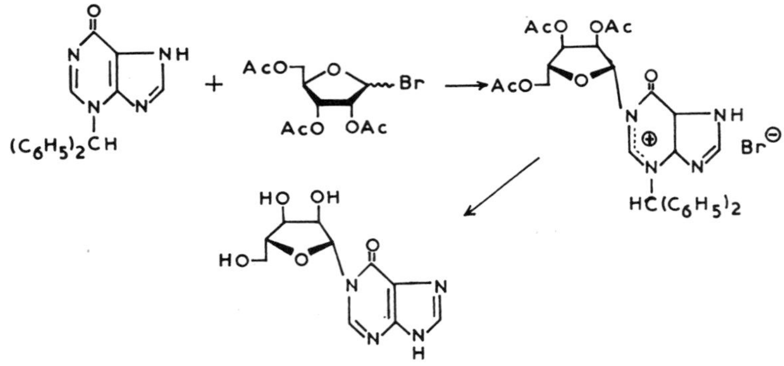

Scheme II

Treatment of this starting material (1) with ethylorthofor-
mate furnished the corresponding iminoether derivative (5) (2).
However, the reaction of 2 with 2,3-O-isopropylidene-D-ribofur-
anosylamine (6) (3) (in CH$_3$CN or benzene) did not give the desired amidine (4), presumably due to extensive decomposition of
the ribosylamine. However, the desired amidine (4) could be obtained by reacting compound 1 with the N-ribosyl formiminoether
derivative (5), which had been synthesized (6) by Shaw *et al.*

Using an excess of the iminoether derivative (4) and pyridine as the solvent, three compounds can be isolated by column
chromatography (the two anomeric 1,9-diribofuranosyladenines
(6) (10%) and the intermediate amidine (4) in a yield of 23%).

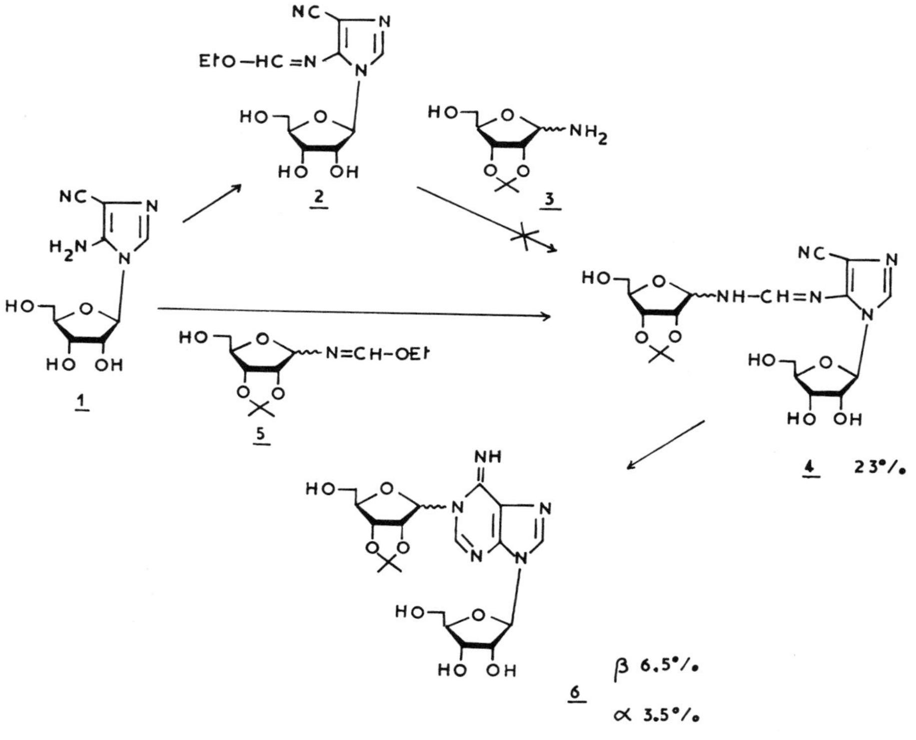

Scheme III

Furthermore, after isolation we have observed a facile ring
closing of the amidine derivative (4) into the corresponding anomeric mixture of 1,9-ribofuranosyladenines (6).

However, this reaction is not as simple as it appears since
the iminoether derivative (5) is not very stable, and in fact the
diribofuranosyladenines are also quite unstable. This will be
discussed in more detail later.

We used the same approach, but started with 5(4)-amino-4(5)-
cyanoimidazole (7) and obtained the same two anomeric 1-D-ribo-
furanosyladenines (8), but with a total yield of 18% for the α-
anomer and 24% for the β-anomer (Scheme IV). For these compounds
the amino form is expected to be the predominant form, however,
no tautomerism studies have yet been performed.

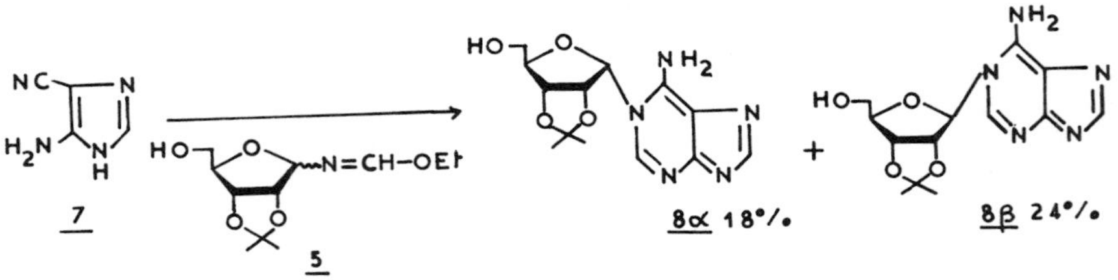

7   5   8α 18%   8β 24%

Scheme IV

We have also tried to obtain the diribofuranosyladenines
by a direct substitution with a peracylated chlorosugar on 1-
D-ribofuranosyladenine, which had been previously been silylated,
or directly with DMAC as solvent. However, we were unable to
obtain any amount of the desired product.

Therefore, we have established that by using the N-ribosyl-
formiminoether derivative (5), it is possible to introduce a ribose moiety at position 1 of adenine, but considerable care
should be taken in selecting the reaction conditions.

Since the starting sugar can undergo mutarotation under all
these reactions, anomeric mixtures were obtained and the anomeric
configuration of the ribose ring at position 1 of each nucleoside was determined using our Δδ criterion (7) (see Table I).
For the diribosyl derivatives, the anomeric configuration of the
ribose moiety at position 9 is obviously β since there should
have been no anomerization under the reaction conditions employed.

The uv spectra observed for all these compounds are in
agreement with their assigned structures, as based on comparisons
with uv spectral data from the literature (see Table II).

All attempts to remove the isopropylidene group in an acidic medium gave rise to extensive decomposition. In a neutral
medium, we have observed a Dimroth type rearrangement. The N-1
riboside derivative 8β was heated at reflux temperature for 1.
5 hr in aqueous medium to give two anomeric ribosides (9α and
9β), presumably N-6, with a 50% yield in a ratio of 2 to 1.
Starting with the α-N-1 anomer (8α) we observed exactly the same
reaction, but with 3 hr of reflux. To confirm the structure of
these presumed N-6 ribosides, we reacted adenine with D-ribose
and obtained, as described by Sanchez and Orgel (8), a 6-ribosyl-
aminopurine with the ribose ring primarily in the pyranose form.

TABLE I.   100-MHz Proton Magnetic Resonance Spectra
(Solvent: DMSO-d$_6$, TMS as Internal Reference)

Compound	δH_2 and H_8 (ppm)	$\delta H_{1'}$ ($J_{1'-2'}$, Hz) (ppm)	Isopropylidene group δCH_3 (ppm)	$\Delta\delta$ (ppm)
6β	8.26 - 8.36	6.30 (2.4) 5.79 (5.6)	1.51 - 1.28	0.23
6α	7.95 - 8.18	6.62 (3.0) 5.78 (6.0)	1.32 - 1.24	0.08
8α	7.89 - 8.03	6.50 (3.8)	1.23 - 1.20	0.03
8β	8.14 - 8.20	6.29 (3.2)	1.56 - 1.31	0.25
9α	8.22 - 8.32	6.31 (6)[a]	1.54 - 1.36	0.18
9β	8.17 - 8.25	6.27 (9)[a]	1.48 - 1.30	0.18

[a]*This signal appeared as a doublet. After exchange with deuterium, the anomeric proton was observed as broad singlet with the line width at half height indicated in parentheses.*

Treatment of this compound with 2,2-dimethoxypropane under acidic conditions gave us the same anomeric mixture (Scheme V).

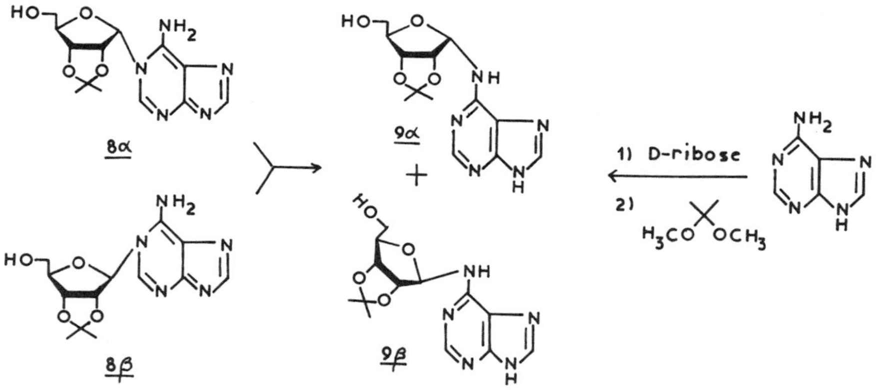

Scheme V

This can be explained because the introduction of a five-membered ring, in this case a dioxolane ring, on the two cis hydroxyl groups of a sugar molecule strongly favors the formation of the furanose form. The synthesis of the 2,3-O-isopropylidene-ribofuranosylamine (6) (3) from the ribopyranosylamine is a good example of this selectivity. Some other examples can be found in recent work (9) by Maffatt *et al.*

TABLE II. uv Data

Compound	Reference	λ_{max}	$(\varepsilon \times 10^{-3})$	λ_{min}	λ_{max}	$(\varepsilon \times 10^{-3})$	λ_{min}	λ_{max}	$(\varepsilon \times 10^{-3})$	λ_{min}
		pH = 1			pH = 7			pH = 11		
6β		260	(12.6)	236.5	260, 264[a]	(11.8)	235	259.5, 265[a], 288[a]	(11.9)	235
6α		259.5	(13.2)	236	259.5, 264[a]	(12.5)	235	259.5, 265[a], 288[a]	(12.7)	235
1-Methyladenosine	14	256.5	(13.6)	231	257, 260-265[a]	(14.6)		257, 260-265[a]	(14.6)	223
1,9-Dimethyladenine	12	259	(10.6)	235				259, 265[a]	(10.4)	235
N6-Methyladenosine	14	261	(16.3)	231	265	(16.3)	229	265	(15.0)	223

Compound	Reference	λ_{max}	$(\varepsilon \times 10^{-3})$	λ_{min}	λ_{max}	$(\varepsilon \times 10^{-3})$	λ_{min}	λ_{max}	$(\varepsilon \times 10^{-3})$	λ_{min}
		pH = 1			pH = 7			pH = 13		
8α		259	(10.3)	228.5	259	(10.1)	230	268.5, 264.5[a]	(12.9)	238
8β		261	(9.5)	230	261	(9.9)	230	270, 266[a]	(13.9)	239
1-Methyladenine	11	257	(11.7)					269	(14.1)	

[a] Shoulder.

TABLE III   uv Data

Com-pound	Refer-ence	pH=1 $\lambda_{max}(\varepsilon x10^{-3})\lambda_{min}$	pH=7 $\lambda_{max}(\varepsilon x10^{-3})\lambda_{min}$	pH=13 $\lambda_{max}(\varepsilon x10^{-3})\lambda_{min}$
$\underline{9\alpha}$		274.5 (19.4)234	264 (19.2)227 270[a]	273 (15.1)238.5 279.5[a]
$\underline{9\beta}$		274.5 (19.1)234	265 (17.6)227.5 271[a]	273 (14.7)238.5 279[a]
N^6-Meth-yladenine	13	267 (16.1)232		273 (15.9)239

[a]Shoulder

The two anomers $\underline{9\alpha}$ and $\underline{9\beta}$ were separated on column chroma-tography, with the less polar compound being the more abundant. The uv spectra of these two compounds and of 6-methylaminopurine in basic medium are directly comparable (see Table III); slight differences are observed in acidic medium presumably because of decomposition. The pmr spectra of these two compounds are in agreement with the proposed structure (see Table I): we observed peaks for two heterocyclic protons, two methyl groups, and a doublet with J=9,5 Hz for the anomeric protons for each compound. After exchange with deuterium, the anomeric protons were observed as broad singlets at $\delta6.31$ and $\delta6.27$ ppm with a linewidth at half height of 9 and 6Hz, respectively.

Although the anomeric signal of the less polar compound was at a lower field than the more polar compound, it was not possible to determine the exact anomeric configuration. Furthermore, our $\Delta\delta$ criterion does not really apply for these products, since they can not be considered as true ribofuranonucleoside. In fact, we have observed a $\Delta\delta$ value of 0.18ppm for these two anomers. There-fore, we have obtained the ^{13}C NMR spectra of these two anomers, and for the less polar product we observed the signals correspond-ing to the C_1 and C_2 atoms of the ribose ring at a higher field than those of the more polar (see Table IV). Therefore, we have assigned the less polar anomer the α configuration (10).

In order to increase the yield of the Dimroth-rearranged product, we heated the β-N-1 riboside ($\underline{8\beta}$) at reflux temperature in a basic medium (1/1 mixture of 1 N NaOH and MeOH). A careful analysis of the crude reaction mixture revealed then 10% of the rearranged product, with a ratio of α/β equal to about 2/1, 17% of starting material, and 50% of α-N-1 riboside derivative $\underline{8\alpha}$. The same results were obtained starting with the α-N-1 anomer.

Concerning the mechanism of this reaction (see Scheme VI), it is a well-known fact that there is an initial nucleophilic attack at position 2 of the purine ring giving intermediate

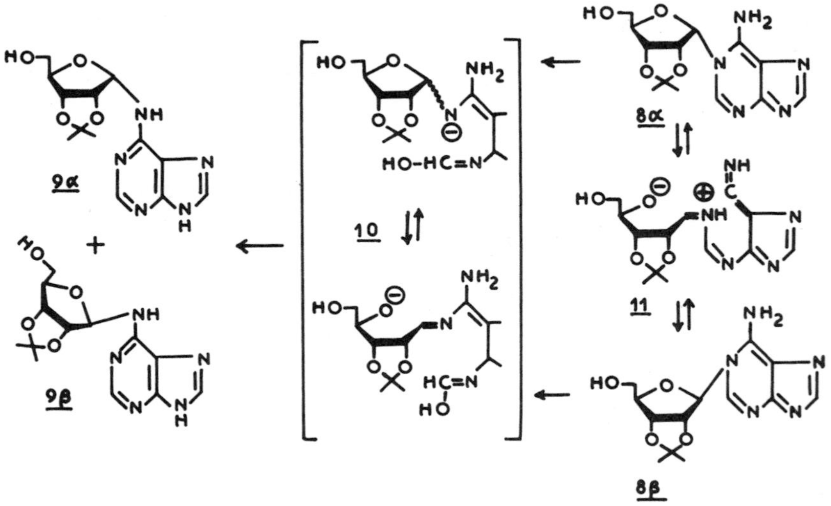

Scheme VI

forms such as <u>10</u>, which can mutarotate. Then <u>10</u> can be rearrange
to Dimroth's N-6 ribofuranosyl compounds <u>9α</u> and <u>9β</u>.

The formation of <u>8α</u> from <u>8β</u> can be rationalized on the basis
of the formation of an intermediate such as <u>11</u> (15) which could
be related to the basicity expected for such 1-substituted
purines.

TABLE IV. ^{13}C NMR Spectra in DMSO-d_6 (δ in ppm, TMS as Internal
            Reference).

Compound	Ribose ring					Isopropylidene group		
	$C_{1'}$	$C_{2'}$	$C_{3'}$	$C_{4'}$	$C_{5'}$	CH_3	CH_3	C
<u>9α</u>	81.69	81.39 (2C)		79.10	62.13	26.04	24.58	111.78
<u>9β</u>	87.20	84.95	84.71	81.83	62.03	26.77	25.02	111.83

Furthermore, during these equilibrating conditions, we
have observed that the cis-1,2 products (i.e., α anomers) are
predominant, which shows that they are thermodynamically more
stable than the β anomers. These observations are in agreement
with some recent results reported (9) by Moffatt *et al.* for some
C-glycosides in the O-isopropylidene ribofuranose serie.

Therefore, we have established that the N-ribosylformimino-
ether derivative is of general use to introduce a ribose moiety
at position one of the purine ring.

REFERENCES

1. B. N. Ames, R. G. Martin, and B. J. Garry, *J. Biol. Chem.* *236*, 2019 (1961).
2. J. J. Fox and D. Van Praag, *J. Org. Chem. 26*, 526 (1961).
3. J. A. Montgomery and H. J. Thomas, *J. Heterocycl. Chem. 5*, 741 (1968).
4. E. C. Taylor and P. K. Loeffler, *J. Am. Chem. Soc. 82*, 3147 (1960).
5. K. Suzuki, T. Meguro, and I. Kumashiro, Japanese patent 7011, 701 (Cl.16E 611.2), 27 April 1970, Applied for, 16 March 1966; *Chem. Abstr. 73*, 45795 m (1970).
6. N. J. Cusack, B. J. Hildick, D. H. Robinson, P. W. Rugg, and G. Shaw, *J. Chem. Soc., Perkin Trans. 1* 1720 (1973).
7. B. Rayner, C. Tapiero, and J.-L Imbach, *Carbohydr. Res. 47*, 195 (1976).
8. W. D. Fuller, R. A. Sanchez, and L. E. Orgel, *J. Mol. Biol. 67*, 25 (1972).
9. H. Ohrui, G. M. Jones, J. G. Moffat, M. L. Maddox, A. T. Christensen, and S. K. Byram, *J. Am. Chem. Soc. 97*, 4602 (1975).
10. H. Sugiyama, N. Yamaora, B. Shimizu, Y. Ishido, and S. Seto, *Bull. Chem. Soc. Japan 47*, 1815 (1974).
11. J. A. Montgomery and H. J. Thomas. *J. Org. Chem. 30*, 3235 (1965); J. W. Jones and R. K. Robins, *J. Am. Chem. Soc. 85*, 193 (1963).
12. A. D. Broom, L. B. Townsend, J. W. Jones, and R. K. Robins, *Biochemistry 3*, 494 (1964).
13. G. B. Elion, E. Burgi, and G. H. Hitchings, *J. Am. Chem. Soc. 74*, 411 (1952).
14. J. W. Jones and R. K. Robins, *J. Am. Chem. Soc. 85*, 193 (1963).
15. This mecanism is tentatively proposed; the equilibrium $\underline{8\alpha} \rightleftarrows \underline{8\beta}$ could be also visualized through $\underline{10}$.

AN INITIAL STUDY ON THE SELECTIVITY AND
STEREOSPECIFICITY OF THE RIBOSYLATION REACTION

JEAN-LOUIS BARASCUT AND JEAN-LOUIS IMBACH

Universite' des Sciences et Techniques du Languedoc,
Laboratoire de Chimie Bio-Organique
Place E. Bataillon, 34060 Montpellier Cédex, France

The problem of creating a glycosyl bond (CN) between a suitably
substituted ribofuranosyl sugar and a heterocyclic base has
been studied by numerous research groups using various techniques. However, this general reaction, which can be called the
"ribosylation reaction," is generally lacking in selectivity and
stereospecificity, which confers to this approach some very
obvious limitations. Among the various techniques used for this
ribosylation reaction using a ribofuranose sugar with a participating group in 2' position, we would like to comment success-
ively on the two methods, which are now the most commonly used.
(1) The acid-catalyzed fusion reaction and (2) the activation
of the base with a trimethylsilyl (TMS) group and the subsequent
reaction with a suitable ribofuranosyl derivative.
   Concerning the fusion reaction, one can find in the literature (1,2) a lot of contradictory data, which prompted us to
reexamine this reaction. To initiate our studies in this area
we selected a very simple model: the reaction between indazole
(1) (as there are only two sites for ribosylation) and 1,2,3,
5 - tetra-O-acetyl-β-D-ribofuranose (2β). An additional advantage of using this reaction as a model was that the four possible
products (3 - 6) were known (3) (see Scheme I).
   Using PMR spectroscopy as an analytical tool, we systemati-
cally studied the influence of the nature of the catalyst, the
percentage of the catalyst, the temperature of the fusion, the
time of the reaction, and the sterochemistry of the starting
sugar.

We first elected to use toluene sulfonic acid, on the basis
of previous studies (4), as the catalyst. The curves presented
in Scheme II represent the influence of the amount of catalyst
on the product distribution *with all other factors remaining
constant.*

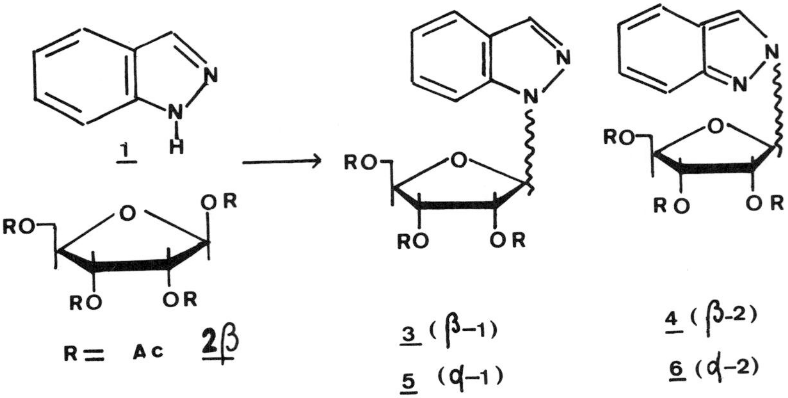

Scheme I

Curve A represents the total yield in nucleoside products
and one can see that with 4% (or more) of catalyst we essentially
obtained a quantitative yield of products. By a close examination
of the other curves, i.e., B (percentage of N-1 ribosylation),
E(percentage of N-2 ribosylation), and also the anomeric distribution in each of these series (curves C, D, F, and G), one can
see that *the product distribution is closely related to the amount
of catalyst used.* With between 0.3 and 2% of catalyst with a maximum for 1.4% we obtained mainly N-2 substitution (i.e., the
kinetic products), but with 7% of catalyst the reaction affords
almost entirely the N-1 isomer (the thermodynamic products).
Furthermore, with this β sugar, the steroselectivity of the reaction is about 80%, regardless of the amount of catalyst used.
Similar curves were obtained when we studied the influence of the
other factors such as the time of the reaction and the temperature of the fusion (5).

In all cases, the kinetic N-2 nucleosides are initially
formed and then by increasing the reaction conditions the thermodynamic products are found to be predominant.

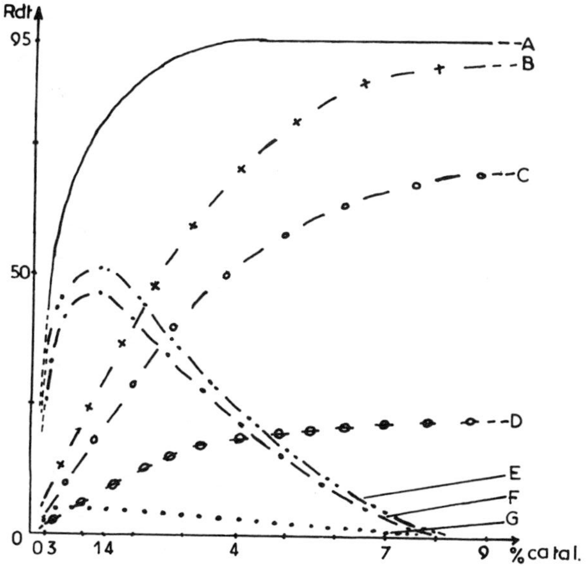

%*p-T.S.	Yield %	5 α-1	3 β-1	6 α-2	4 β-2	% N-1	% N-2
0,3	40	traces	3	5	32	3	37
0,5	52	3	5	5	39	8	44
0,7	61	5	10	5	41	15	46
1	70	6	14	5	45	20	50
1,5	86	10	23	5	48	33	53
2	88	12	30	5	41	42	46
2,5	90	14	36	4	36	50	40
3,5	94	17	47	2	28	64	30
4	95	18	51	2	24	69	26
5	95	20	58	–	17	78	17
6	95	21	64	–	10	85	10
7	95	22	68	–	5	90	5
8	95	22	70	–	3	92	3
9	95	22	71	–	2	93	2

Scheme II. Percentage of p-T.S. in mole/base-Influence of the amount of catalyst. Product distribution. Curve A, yield total; B, N-1 isomers; C, 3 (β-1); D, 5 (α-1); E, N-2 isomers; F, 4 (β-2); G, 6 (α-2).

TABLE I   Influence of the Stereochemistry of the Sugar on Product Distribution

Compounds	Anomer	Total	Yield (%) N-1		N-2		N-1	N-2
			cis-1',2'	trans-1',2'	cis-1',2'	trans-1',2'		
AcO, OAc, OAc, AcO	α:cis-1,2	86	15	30	5	36	45	41
	β:trans-1,2	88	18	31	8	31	49	39
AcO, OAc, OAc, OAc	α:cis-1,2	77	20	25	3	29	45	32
	β:trans-1,2	80	23	30	2	25	53	27
AcO, R, R, OAc (R=OAc)	α:trans-1,2	81	13	40	5	23	53	28
	β:cis-1,2	79	11	38	3	27	49	30
AcO, OAc, AcO	β:cis-1,2	72[a]	13	15	4	40	28	44
	α:trans-1,2							

[a] Yield after isolation by column chromatography.

242

%catal	A	α-1	β-1	α-2	β-2	%N-1	%N-2
0,3	5	-	-	3	2	-	5
0,7	13	3	-	6	4	3	10
1,4	27	3	2	11	11	5	22
2,6	47	8	4	10	25	12	35
3,9	69	16	12	9	32	28	41
5	86	22	23	8	33	45	41
6,3	94	25	32	7	30	57	37
7	95	26	33	6	30	59	36
8,7	95	25	37	5	28	62	33

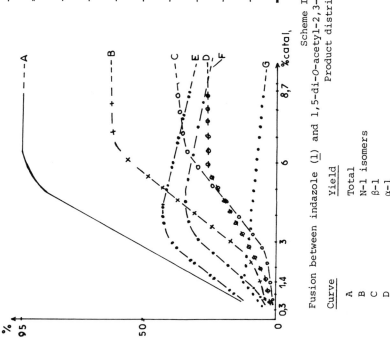

Scheme III

Fusion between indazole (1) and 1,5-di-O-acetyl-2,3-O-isopropylidene β-D-ribofuranose (7β).
Product distribution

Curve	Yield
A	Total
B	N-1 isomers
C	β-1
D	α-1

Curve	Yield
E	N-2 isomers
F	β-2
G	α-2

Furthermore, the sterochemistry of the starting sugar does not appear to influence the course of the acid-catalyzed fusion since we obtained the same curves as shown in Scheme II when we used (5) 1,2,3,5 tetra-O-acetyl-α-D-ribofuranose (2α) (α vs β). With other tetra-O-acetyl-D aldofuranoses the same remarks can be made (Table I). This result can be easily explained by an equilibration of the starting sugar and/or of the final products of the reaction since we have already shown (4) that both mechanisms can occur.

We have also studied "nonparticipating" sugars such as 1,5-di-O-acetyl 2,3-O-isopropylidene-β-D-ribofuranose (7β), where we obtained the same kinds of curves (Scheme III). However, 7β appears to be slightly less reactive than 2β. We also obtained a better overall yield of α nucleosides. Therefore, it appears that the acid-catalyzed fusion reaction is very sensitive to the experimental conditions (which explains the profusion of seemingly contradictory data in the literature).

If we carefully choose the reaction conditions it is possible to obtain preferentially the kinetic isomer, using "mild" fusion conditions (6), or the thermodynamic isomer using "strong" fusion conditions. In all cases, the βanomers are predominant. Therefore, we have now shown with this example that we can confer some selectivity to the acid-catalyzed fusion reaction.

However, the fusion reaction is sometimes unsuitable for the preparation of certain nucleosides (i.e., in the case of high melting points or too acidic aglycones) so other approaches are often used.

The activation of an aglycone with a T.M.S. group is now of general use in nucleoside chemistry. The silylated heterocycle can react directly with a halosugar or with an acetoxy with sugar in presence of $SnCl_4$ (7). Both of these approaches are often performed with CH_3 CN as the solvent.

We will consider now, as a model, the ribosylation of four different triazolopyrimidinones (8,10) (Scheme IV). We will also study comparatively, the results we have obtained using these various techniques.

The sites of glycosylation for all nucleoside products were established unambiguously by uv (5,8,9) or ^{13}C nmr spectroscopy (11), and their anomeric configuration was determined with the aid of our Δδ criterion (12,13,14). In the s-triazolo (1,5-a) pyrimid-7-one (11), the observed results are presented in Scheme V, and the following comments can be made.

(1) the acid-catalyzed fusion using 1% of catalyst, gives an anomeric mixture of the two isomers substituted, respectively, on the six-and the five-membered ring. But with 7% of catalyst, only an anomeric mixture of the isomer substituted on the six-membered ring was obtained.

According to our previous results on the fusion reaction, the isomer we obtained with 7% of the catalyst should be the thermodynamic product. This can be easily confirmed by simple equilibration experiments starting with each lf the β nucleosides (12 and 13) (Table II).

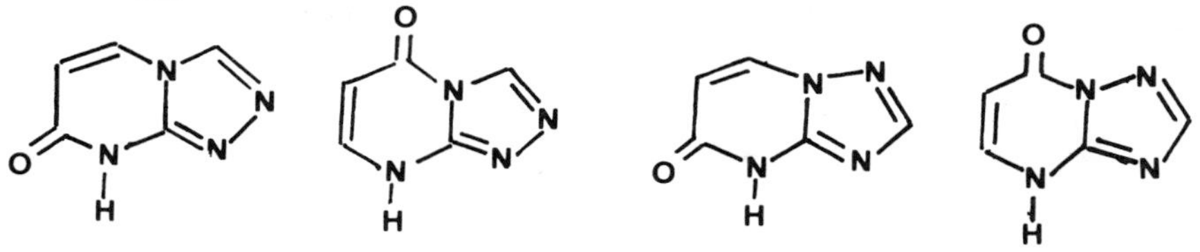

s-Triazolo(1,3-a)pyrimidinone   *s*-Triazolo(1,5-a)pyrimidinone.

Scheme IV

TABLE II   Transposition of 12 and 13 Under Fusion Reaction Conditions (180°C, 15 or 30 min; Catalyst:I_2, 7%)

Starting nucleoside	Time (min)	Products obtained %			
		12	14	13	15
12	15	80	20		
13	15	59	7	22	12
13	30	67	33		

(2) By the reaction of the T.M.S. heterocycle with 2,3,5 tri-*O*-acetyl-D-ribofuranosyl chloride (16) at room temperature using CH_3CN as solvent, we obtained a mixture of the thermodynamic and kinetic nucleosides, but only the β anomers.

(3) The reaction of the acetoxy sugar (2β) with the catalyst $SnCl_4$ at room temperature in CH_3CN emphasizes that the overall yield of the reaction is very sensitive to the percentage of Lewis acid. Furthermore, it should be noted that this reaction gives mainly the thermodynamic β anomer (12) the percentage of which increases with the amount of catalyst used (Table III).

TABLE III

$SnCl_4$ (eq.)	12 (%)	13 (%)	Total yield (%)
0.35	83	17	30
1	84	16	55
1.5	91	9	100

In every case, only β nucleosides are obtained at room temperature, but with refluxing conditions the yield of the reaction increases and α nucleosides were also observed (see Scheme
V).

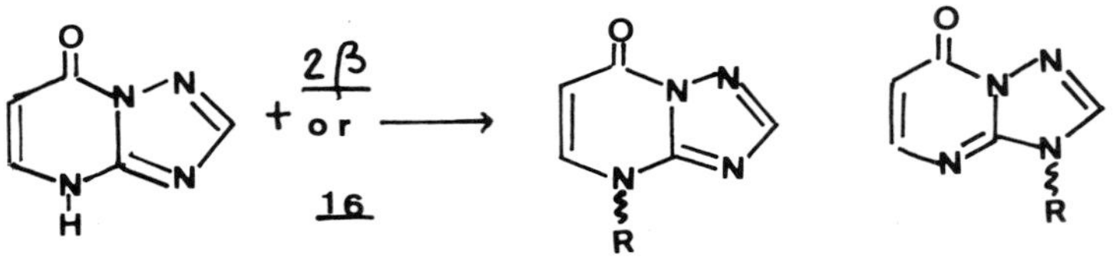

Scheme V

If we now consider the isomeric s-triazolo(4,3-a) pyrimid-
5-one (9) the same observations can be made (Table IV).

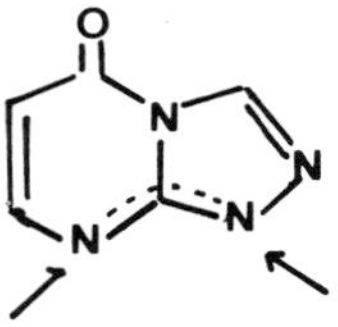

Thermodynamic   Kinetic

TABLE IV

		Conditions		Kinetic (%)	Thermodynamic (%)	Total yield (%)
Fusion	2β	180°C	1% I_2	35		35
		200 C	7% I_2	2	35	37
T.M.S. derivative of 9	16	Room		22		22
	2β	temp- era- ture	1 eq. $SnCl_4$	11	57	68
	2β	CH_3CN	1.5 eq. $SnCl_4$	9	91	100

[a]*During this fusion reaction we observed a Dimroth transposition of the thermodynamic nucleoside (15).*

(2) Two fusion reactions using "mild" and "strong" conditions allowed us to determine the relative stability of the two isomers. As in the previous series, the thermodynamic nucleosides have the ribosyl moiety residing on the pyrimidine ring and the kinetic nucleosides have the ribosyl moiety on the five-membered ring.

(2) In this case, the chlorosugar 16 gives *only* the β kinetic nucleoside.

(3) We used $SnCl_4$ and acetoxy sugar at room temperature to obtain mainly the β thermodynamic product. In the same series, but with the ketogroups in a α position to the N atom in the pyrimidine ring, the same kind of remarks may be made (Scheme VI): (a) condensation of the chlorosugar and the silylated aglycone gives the kinetic nucleosides (16) and (b) acid-catalyzed fusion with "strong" conditions gives mainly the thermodynamic nucleoside. It appears to us that it is then possible to make some remarks concerning the selectivity and the stereospecificity of the glycosylation reactions. We have studied this series as follows: (a) at room temperature in CH_3CN (17), starting with a silylated aglycone (only β nucleosides are generally obtained); (b) kinetic nucleosides are mainly obtained at room temperature with a halosugar or by a fusion using "mild" conditions (they can be obtained alone or as a mixture with the thermodynamic isomer, their ratio being a function of the energy difference between the two forms). (c) thermodynamic nucleosides are best obtained at room temperature with an excess of $SnCl_4$ (1.5 eq) (17); and (d) generally (16), we can observe α anomers if the glycosylation is performed with a catalyst and at high temperature.

On the basis of the above, we would like to propose an initial attempt to rationalize nucleoside synthesis. According to the desired site of glycosylation we propose (a) to determine the structure and the relative stability of the various isomers with the acid-catalyzed fusion reaction using "mild" and "strong" conditions and (b) to apply these results to choose, with a silylated base, the technique to be used in order to direct the glycosylation (the β configuration being optimum at room temperature). It was of considerable interest to see if our approach could be applied to other heterocycles and we should now like to comment on a recent report in the literature from another laboratory.

In the imidazole series, Cook and co-workers (18) wanted to introduce a ribosyl moiety with the β configuration on the N atom adjacent to the CH_2CN group of 17. They tried all the approaches presented on the Scheme VII and found that the best results were obtained using 1.44 eq. of $SnCl_4$ and an acetoxy sugar.

By perusal of their data, keeping our rationalizing approach in mind, one can predict that their fusion reactions, which was

Scheme VII

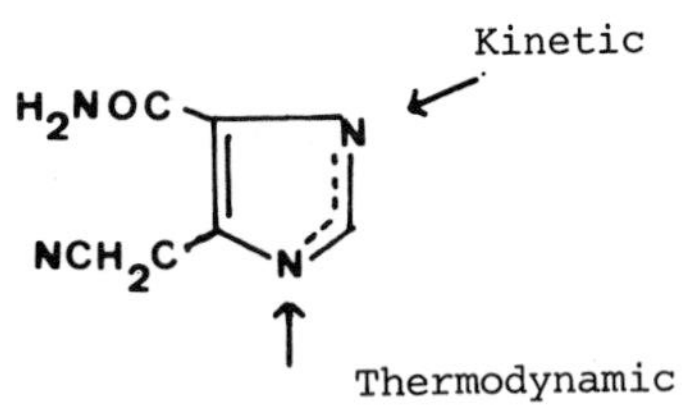

$\underline{17}$

				Kinetic		Thermodynamic	
				α	β	α	β
Fusion	2β	155°C	0.58% catalyst	12	47	4, 7	19
T.M.S. derivative	Bromo sugar	Room temperature	CH₃CN		48		9
of 17	2β	Room temperature	0.36 eq. SnCl₄ 0.72 eq. SnCl₄ 1.44 eq. SnCl₄	CH₃CN	10 34.5		20 29.5 100

performed under "mild" conditions, will give mainly the kinetic nucleosides.

Since they obtained a preferentially substitution on the N atom adjacent to the carboxamido group, this shows that it was the kinetic site of substitution. Since they wished to obtain the other isomer (thermodynamic) with β configuration, using our proposed criterion the reaction conditions would have been evident (room temperature in CH_3CN to have β configuration and silylated base and acetoxy sugar with 1.5 eq. of $SnCl_4$ to have the best yield for the thermodynamic isomer), and that is in fact what they observed (Scheme VII).

The rationalizing approach we have proposed here must obviously be used with care and we are now studying its scope and limitations using various models. However, we hope that this initial attempt, will be of some utility for the nucleoside chemist.

REFERENCES

1. H. Iwamura, M. Miyakado, and T. Hashizume, *Carbohyd. Res.* *27*, 149 (1973).

2. K. Imai, A. Nohara, and M. Honjo, *Chem. Pharmacol. Bull.* *14*, 1377 (1966).

3. B. L. Kam and J. L. Imbach, *J. Carbohyd. Nucleosides Nucleotides 1*, 287 (1974).

4. B. L. Kam, Thèse Doctorat es Sciences Montpellier, September 1976.

5. J. L. Barascut, B. L. Kam, and J. L. Imbach, *Bull. Soc. Chim.* 1983 (1976).

6. "mild" conditions: low amount of catalyst; low temperature, i.e., 155°C; short time, i.e., 10 min. *"strong" conditions:* about 7% of catalyst, high temperature, time around 20 min.

7. H. Vorbruggen and U. Niedballa *Angew. Chem. Int. Ed. 9*, 461 (1970); *J. Org. Chem. 39*, 3654, 3661, 3668 (1974). *J. Org. Chem. 41*, 2084 (1976).

8. J. L. Barascut and J. L. Imbach, *Bull. Soc. Chim.* 2651 (1975).

9. J. L. Barascut, C. Ollier de Marichard, and J. L. Imbach, *J. Carbohyd. Nucleosides Nucleotides 3*, 281 (1977).

10. Studies on the ribosylation of some triazolo (1,5-a)pyrimidinones have been also previously reported: see M. W. Winkley, G. F. Judd, and R. K. Robins, *J. Heterocycl. Chem. 8*, 237 (1971); G. R. Revankar, R. K. Robins, and R. L. Tolman, *J. Org. Chem. 39*, 1256 (1974).

11. P. Dea, G. R. Revankar, R. L. Tolman, R. K. Robins, and
     M. P. Schweitzer, *J. Org. Chem. 39*, 3226 (1974).

12. J. L. Imbach, J. L. Barascut, B. L. Kam, B. Rayner, C.
     Tamby, and C. Tapiero, *J. Heterocycl. Chem. 10*, 1069
     (1973).

13. J. L. Imbach, J. L. Barascut, B. L. Kam, and C. Tapiero,
     Tetrahedron Lett. 2, 129 (1974).

14. J. L. Imbach, *Ann NY Acad. Sci. 255*, 177 (1975).

15. C. Ollier de Marichard, Thèse de spécialité, Montpellier,
     (1976).

16. In the case of 10 the kinetics isomers α and β were obser-
     ved at room temperature.

17. Stannic chloride promotes the formation of acyloxonium ions
     of such acetylated ribofuranose and also of catalyst-
     heterocycle complex as recently shown by Vorbruggen. Fur-
     ther, the utilization of polar solvent such as CH_3CN,
     decreases the stability of the complex between $SnCl_4$
     and the aglycone. Thermodynamic nucleosides are thus
     preferentially obtained (N-1 substituted for the uracil
     series).

18. P. D. Cook, R. J. Rousseau, A. M. Mian, P. Dea, R. B.
     Meyer Jr., and R. K. Robins, *J. Am. Chem. Soc. 98*,
     1492 (1976).

ON THE MECHANISM OF NUCLEOSIDE SYNTHESIS

HELMUT F. VORBRÜGGEN, ULRICH NIEDBALLA,
KONRAD KROLIKIEWICZ, BÄRBEL BENNUA

Research Laboratories of Schering A.G.
Berlin/Bergkamen, Germany and

GERHARD HÖFLE

Department of Organic Chemistry, Technical University
Berlin 12, Germany

ABSTRACT

The mechanism of the Friedel-Crafts-catalyzed silyl Hilbert-Johnson reaction is discussed with special emphasis on the interaction between silylated uracils and $SnCl_4$ or $(CH_3)_3SiSO_3CF_3$. The combination of chemical and ^{13}C-NMR data permits the conclusion that only the free silylated bases give the desired N_1 nucleosides, whereas the N_1-o complexes lead slowly to the unwanted N_3 nucleosides. Apparently both, the silylated N_1 as well as the N_3 nucleosides, can react further to afford the N_1,N_3 bisribosides.

For a number of years we have studied Friedel-Crafts-catalyzed nucleoside synthesis starting from silylated bases and peracylated sugars (1-7). And it is only this reaction and our current ideas about its mechanism that we want to discuss here.
During nucleoside synthesis two major processes occur: (a) the reaction of the sugar moiety with the catalyst to give the electrophilic sugar cation, and (b) the interaction of the catalyst with the silylated base.
In the first process (a) (Scheme I), $SnCl_4$, or the new

$(CH_3)_3SiSO_3CF_3$ $((CH_3)_3SiClO_4)$, reagent converts the stable, crystalline peracylated sugars, the most convenient and commercially available starting materials, into the rather stable cyclic 1, 2-acyloxonium ions with formation of either $SnCl_4OAc^-$ or CF_3 SO_3^- as counterions. With $(CH_3)_3SiSO_3CF_3$, the l-acetate is transformed into the silylester of acetic acid. Since the $SnCl_4$ OAc^-anion is bulkier than $CF_3SO_3^-$anion it can be anticipated that the reactions of the corresponding reactive 1,2-acyloxonium ions as intimate ion pairs should reflect this difference in bulk.

The same stable cation is obtained by treatment of the rather unstable acylated 1-halosugars with $AgClO_4$ in benzene, whereupon the silver halide precipitates (8-10).

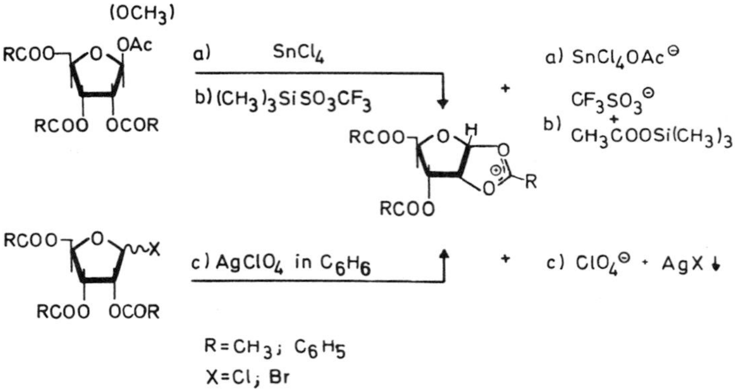

R = CH_3; C_6H_5
X = Cl; Br

Scheme I

Under these thermodynamically controlled and at least partially reversible conditions the silylated base (Scheme II) can *only* attack the stable sugar cation from the top (the β side) to give exclusively the β-nucleoside (11,12). During this pro-

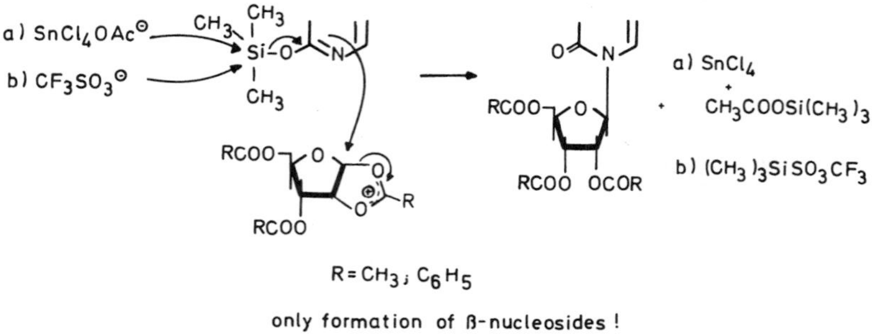

R = CH_3; C_6H_5

only formation of β-nucleosides!

Scheme II

cess the $SnCl_4OAc^-$ anion reacts with the silyl group at the base to give $SnCl_4$ and the trimethylsilylester of acetic acid, whereas the triflate anion gives again the starting trimethyl-silyltriflate (7).

For these two processes, formation of the sugar cation and subsequent (or synchronous) nucleoside formation, no fur-ther reagent is necessary or desirable (13).

As yet only one exception to the exclusive formation of β-nucleosides in the presence of a 2α-acyloxygroup in the sugar moiety has been described (14). Brink and Jordan (Scheme III) isolated up to 25% of the α-nucleoside using only about 0.15 eq. of $SnCl_4$. Apparently, the basic acetamide group in the sugar

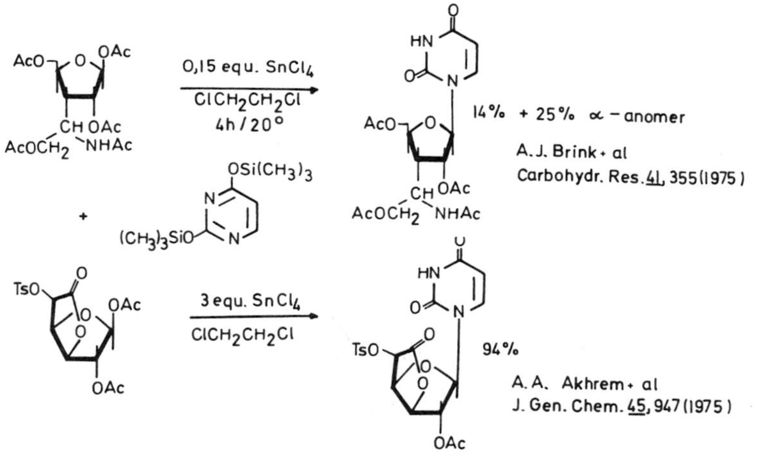

Scheme III

moiety binds one equivalent of $SnCl_4$ to form the rather stable amide-$SnCl_4$ complex, which seems to impede the formation of the stable 1,2-acyloxonium ion. Whether or not the 5-acetoxy group leads to the formation of a cyclic β-1,5-acyloxonium ion and thus to α-nucleoside formation deserves further studies. However, re-peating the reaction with 2 eq. of $SnCl_4$ might give the "normal" 1,2-acyloxonium ion exclusively and consequently only the β-nucleoside.

Since $SnCl_4$ complexes with polar groups, e.g., γ-lactons, Akhrem *et al.* (15) had to employ 3 eq. of $SnCl_4$ to obtain a 94% yield of the pure β-nucleoside (16,17). Unfortunately we have to refrain from discussing other sugar moieties *without* a 2α-acyl-oxy group. Thus, let us close this first part with one final example.

Since the replacement of oxygen by sulfur leads to a de-creased stability of the acylated sugar cation, the correspond-ing 4-thio-sugar (Scheme IV) gives, in the presence of 1.4 eq.

of SnCl$_4$, 65% of the desired β-nucleoside as well as a 3% yield
of the α-nucleoside (18). The rate of formation and the stability of the corresponding 1,2-acyloxonium ion might be increased
by using the stronger Lewis acid (TiCl$_4$) instead of SnCl$_4$, thus
giving only the desired β-nucleoside.

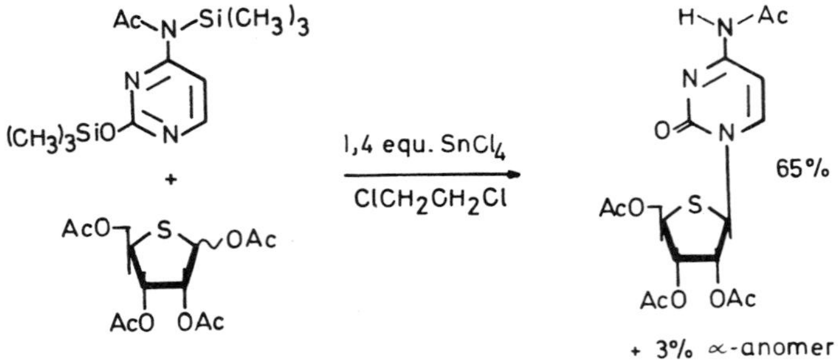

N. Ototani + R.L. Whistler, J. Med. Chem. **17**, 535 (1974)

Scheme IV

The second topic, the interaction of Friedel-Crafts catalysts with silylated bases, has up to now been practically neglected. The starting point for our investigation was the surprising result that silylated 5-nitro-uracil (Scheme V) reacted
very rapidly in the presence of only 0.2 eq. of SnCl$_4$ to give
the benzoylated 5-nitro-uridine in nearly quantitative yield.

This induced us to study the influence of 5-substituents in
silylated uracils on nucleoside formation (Scheme VI). As can
be expected, with increasing basicity more and more stable and
less reactive donor-acceptor complexes between the silylated
bases and the Friedel-Crafts catalysts are formed. Thus, even-
tually one equivalent of Friedel-Crafts catalyst is inactivated
or "neutralized." In that case the formation of the sugar cation
and nucleoside synthesis is only effected if more than one equivalent of catalyst is employed.

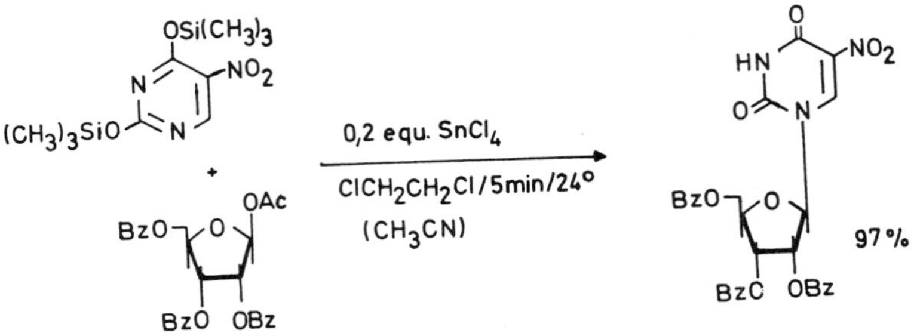

Scheme V

Compared to 1,2-dichloroethane, more polar solvents like acetonitrile lead to diminished stability of these complexes. Furthermore, the nature of the catalyst is important (2).

B) Interaction of catalysts with silylated bases

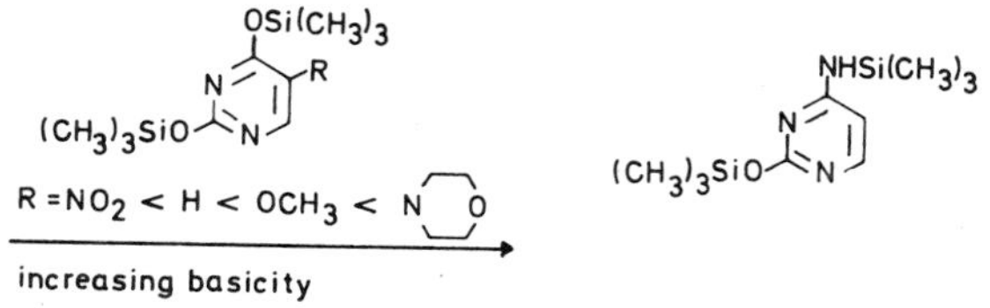

R = NO$_2$ < H < OCH$_3$ < N   O

increasing basicity

Formation of stable Donor – Acceptor Complexes of basic pyrimidines with Lewis – acids in unpolar solvents like ClCH$_2$CH$_2$Cl.
Diminished stability of these complexes in polar CH$_3$CN.
Nature of catalysts is important.

Scheme VI

To get an idea about the varying basicities of the silylated bases, we have compiled some pK$_a$ values of O-methyl pyrimidines and O-methyl pyridines (Scheme VII) from the literature (19), since O-methyl groups are chemically very similar to O-silyl groups.

pK$_a$ –Values of Bases

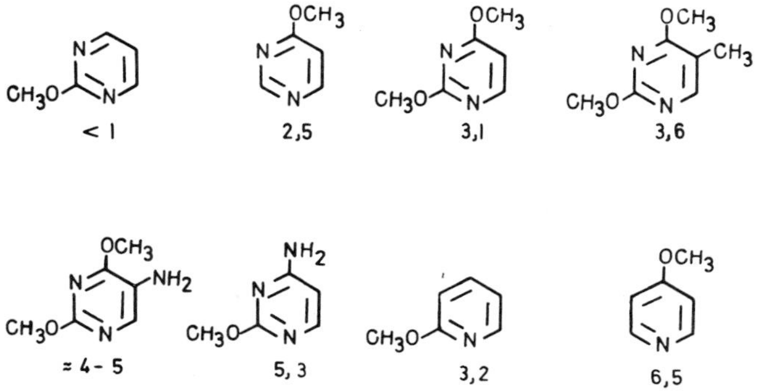

Ar – O – CH$_3$ ≈ Ar – O – Si(CH$_3$)$_3$

Scheme VII

It can be readily seen that the basicity of 2,4-dimethoxy pyrimidine (20), the model for silylated uracil, is increased dramatically by introduction of electron-donating 5-substituents. However, the estimated value for 5-amino-2,4-dimethoxypyrimidine, pK_a 4--5, is still lower than the value for 2-methoxy-4-amino pyrimidine (pK_a = 5.3), the model for silylated cytosine (21). The last values for 2-methoxy-pyridine (pK_a = 3.2) and 4-methoxy-pyridine (6.5) demonstrate why silylated 2-pyridone readily gives nucleosides, whereas the much more basic silylated 4-pyridone reacts only under forcing conditions to give the corresponding nucleoside (4).

Besides the structure and basicity of the silylated base the nature of the catalyst (Scheme VIII) is also important. We have

Nature of Catalysts

$$TiCl_4 > SnCl_4$$
$$- - - - - -$$

$$(CH_3)_3SiSO_3CF_3 > (CH_3)_3SiClO_4$$

$$\delta\,^{29}Si-NMR \qquad 44,6 \qquad\qquad 43,4$$

$$\left[(CH_3)_3SiSO_3F \quad ; \quad (CH_3)_3SiBF_4\right]$$

$$(CH_3)_3SiCl + CF_3SO_3H \xrightarrow[5\,h]{\Delta} (CH_3)_3SiSO_3CF_3 + HCl\uparrow$$

$$bp\ 133°$$

Scheme VIII

tried most of the Friedel-Crafts catalysts, but have usually preferred $SnCl_4$ because it is a weaker Lewis acid than $TiCl_4$ and gives generally homogeneous reactions.

From our new catalysts, we prefer $(CH_3)_3SiO_3CF_3$ to $(CH_3)_3SiCl_4$ because it is a slightly stronger Lewis acid (as measured by its downfield ^{29}Si *nmr* shift (22 and because it is thermally completely stable, whereas *pure* $(CH_3)_3SiClO_4$ explodes above 60°C (23).

We have just tested $(CH_3)_3SiSO_3F$, which is readily available by the reaction of $(CH_3)_3SiF$ with SO_3 (24), but our first experiments showed that $(CH_3)_3SiSO_3F$ gives rise to complicated side reactions during nucleoside synthesis. Since $(CH_3)_3SiSO_3CF_3$ (b.p. 133°C) can be readily prepared and has just become commercially available (25), it should be generally preferred to $(CH_3)_3SiClO_4$. However, if $(CH_3)_3SiSO_3CF_3$ or trifluouomethane sulfonic acid cannot be obtained, $(CH_3)_3SiClO_4$ can be easily prepared in solution by the reaction of $(CH_3)_3SiCl$ with a solution of the anhydrous $AgClO_4$ in benzene followed by filtration of AgCl (23).

When we changed from the electron-attracting 5-nitro substituent in silylated uracil to the electron-donating 5-methoxy group (Scheme IX), we observed a dramatically lowered reactivity.

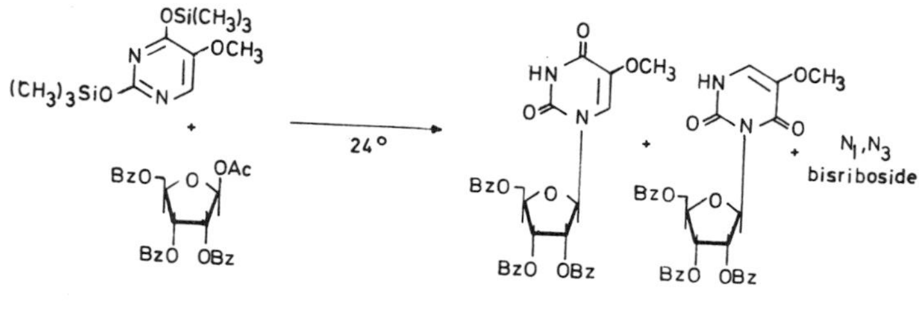

ClCH$_2$CH$_2$Cl (2,5h) 1,4 equ. SnCl$_4$		53%	27%	13%
CH$_3$CN (12h) "		90%	3%	traces
ClCH$_2$CH$_2$Cl (4h) 1,1 equ.(CH$_3$)$_3$SiSO$_3$CF$_3$	90%	—		traces

Scheme IX

Due to strong complex formation with the silylated base, one equivalent of SnCl$_4$ was "inactivated" and only an excess of SnCl$_4$ gave the reactive sugar cation and thus nucleoside formation. In 1,2-dichloroethane a 53% yield of the desired N_1-nucleoside was obtained (6). However, in the polar acetonitrile, in which the base -- Lewis acid complex is dissociated, a more than 90% yield of the thermodynamically favored N_1-nucleoside was isolated.

In contrast to SnCl$_4$, $(CH_3)_3SiSO_3CF_3$ gave, even in 1,2-dichloroethane, a 90% yield at the N_1 product (26).

Increasing the basicity by using a 5-morpholino substituent (Scheme X) instead of a 5-methoxy substituent with SnCl$_4$ in acetonitrile afforded only a 53% yield of the N_1-nucleoside (6).

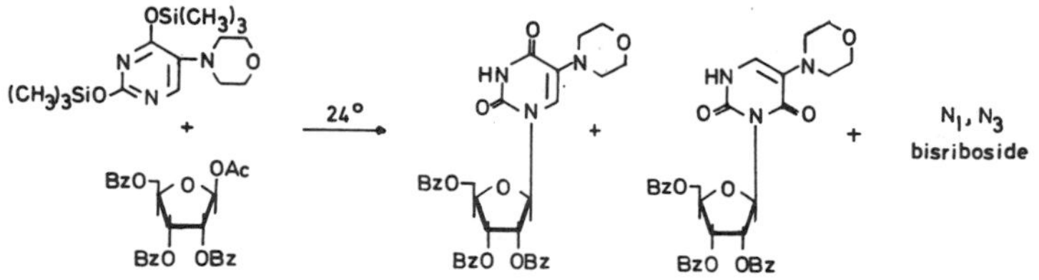

ClCH$_2$CH$_2$Cl (2h) 1,4 equ. SnCl$_4$		39%	18%	42%
CH$_3$CN (1h) "		53%	32%	12%
ClCH$_2$CH$_2$Cl (10h) 1,1 equ.(CH$_3$)$_3$SiSO$_3$CF$_3$	95%	—		traces

Scheme X

Again, replacement of $SnCl_4$ by $(CH_3)_3SiSO_3CF_3$ gave the N_1-nucleo-side in 95% yield (26).

Due to the difference in acid strength between $SnCl_4$ and $(CH_3)_3SiSO_3CF_3$, but possibly also due to the already discussed different size in bulk of the reactive sugar salts, silylated 6-methyl uracil (Scheme XI) gave (in acetonitrile) the desired N_1-riboside in 70% yield using $(CH_3)_3SiSO_3CF_3$.

On keeping the reaction mixture in acetonitrile over night or longer at 24°C, the N_3-nucleoside rearranges (Scheme XII) mainly to the N_1-nucleoside, probably via the depicted mecha-nism (26).

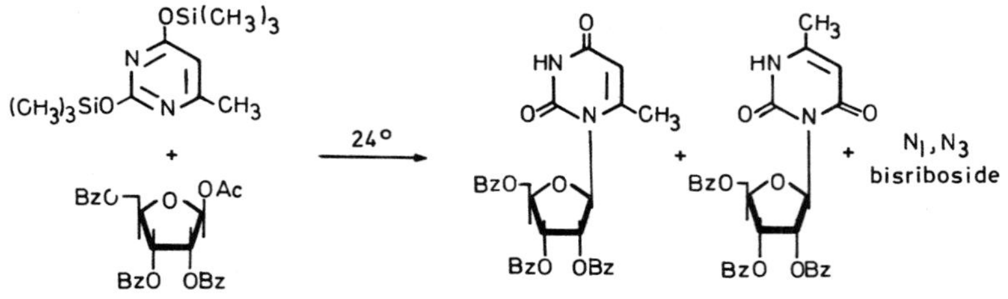

$ClCH_2CH_2Cl$	1,1 equ. $SnCl_4$	13%	68%	
CH_3CN	"	41%	52%	3%
$ClCH_2CH_2Cl$	1,1 equ. $(CH_3)_3SiSO_3CF_3$	47%	18%	12%
CH_3CN	"	71%	2%	11%

Scheme XI

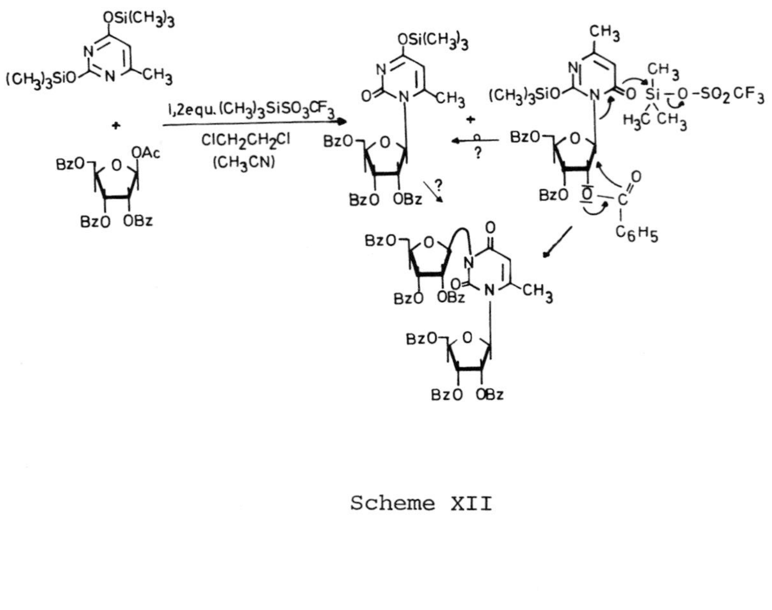

Scheme XII

Addition of a second methyl group in 5-position of the sil-ylated uracil (Scheme XIII) gave quite a dramatic rise in the yield of the desired N_1-nucleoside with $SnCl_4$ in acetonitrile and with $(CH_3)_3SiSO_3CF_3$ in 1,2-dichloroethane (26). We attribute this improvement in the yield of the desired N_1-nucleoside mainly to sterical factors since the 5-methyl group pushes the bulky 4-trimethylsilyloxy group over towards the N_3-nitrogen, impeding the attack at N_3 (6). Finally, the introduction of the rather

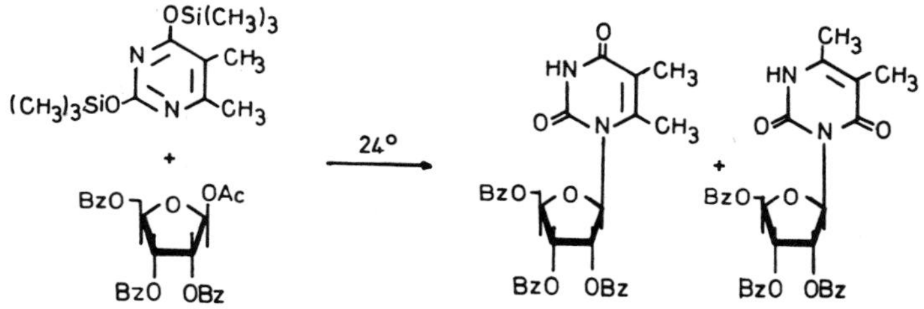

ClCH$_2$CH$_2$Cl	l equ. SnCl$_4$	10%	60%
CH$_3$CN	"	66%	17%
ClCH$_2$CH$_2$Cl	1,1 equ.(CH$_3$)$_3$Si SO$_3$CF$_3$	82%	9%

Scheme XIII

bulky 5-nitro group (Scheme XIV) into silylated 6-methyluracil increases this sterical factor and dramatically reduces the ba-sicity of the silyl compound. Thus, with $SnCl_4$ in 1,2-dichlor-oethane a 84% yield of the desired N_1-nucleoside was obtained (6).

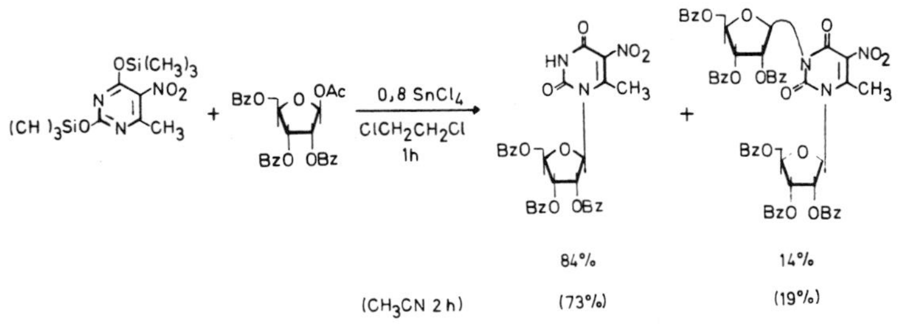

| | | 84% | 14% |
| (CH$_3$CN 2 h) | | (73%) | (19%) |

Scheme XIV

R. K. Robins (Scheme XV) recently reported similar observa-tions with substituted silylated imidazole,-methyl carboxylates obtaining only a high yield of the desired imidazole nucleoside, when more than one equivalent of $SnCl_4$ was used. Apparently, only a silylated base fully complexed with $SnCl_4$ gives the de-sired imidazole nucleoside (27).

In a recently published synthesis of bredinin, a new nucleo-side antibiotic, similar $SnCl_4$-complexes are probably responsible for the low yield obtained (28).

Concluding the experimental examples in nucleoside synthesis let us briefly mention a recent Japanese paper (Scheme XVI),
in which an efficient transglycosidylation is described in the
presence of $(CH_3)_3SiClO_4$ (29). The mechanism of this reaction

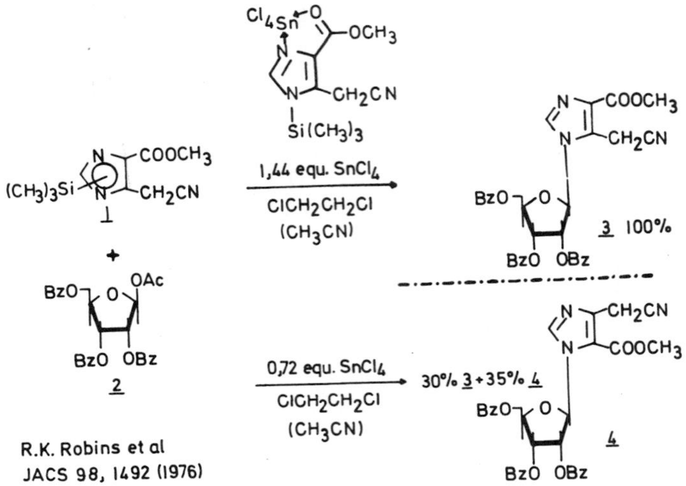

Scheme XV

is probably the same as the rearrangement of the N_3-nucleoside
of 6-methyluracil described above (compare Scheme XII).

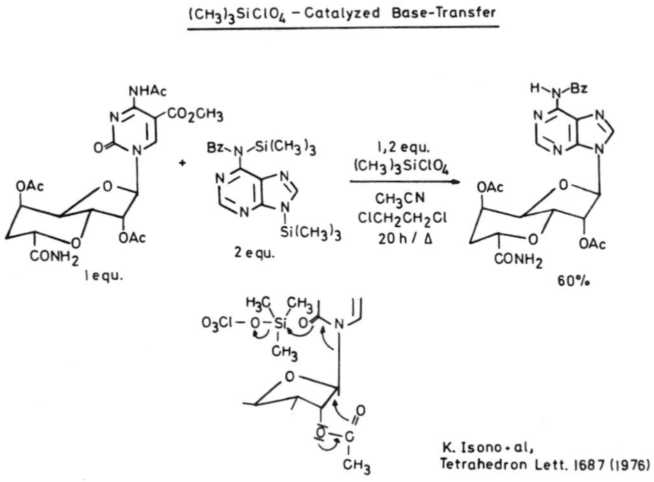

$(CH_3)_3SiClO_4$ – Catalyzed Base-Transfer

Scheme XVI

What is the nature of these complexes between silylated bases and the Friedel-Crafts catalysts? To elucidate the structure of these complexes, we have measured the ^{13}C NMR spectra in $CDCl_3$ (30). As depicted in Scheme XVII, the 2- and 6-carbon atoms adjacent to the nitrogen in pyridine show a pronounced upfield shift on protonization. The same effect is observed for the 6-complexes (31) of the nitrogen in silylated 2-pyridone with either $(CH_3)_3SiSO_3CF_3$ or $SnCl_4$: the 2- and 6-carbons are again shifted upfield. As is readily seen $SnCl_4$, as the stronger Lewis acid, leads to more pronounced upfield shifts of the 2- and 6-carbon atoms.

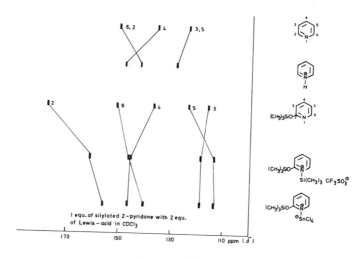

Scheme XVII

Looking at one detailed example, Scheme XVIII depicts the 6-complexes between silylated 5-methoxyuracil and increasing amounts of $(CH_3)_3SiSO_3CF_3$.

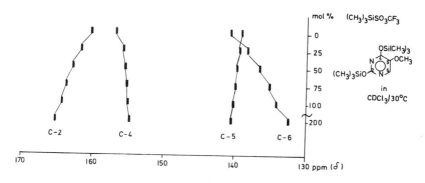

Scheme XVIII

Again, the upfield shift of the C-6 signal is striking, whereas the C-4 signal is only slightly shifted upfield. Therefore, the contribution of the N_3-σ complex (Scheme XIX) can only be of minor importance in the equilibrium.

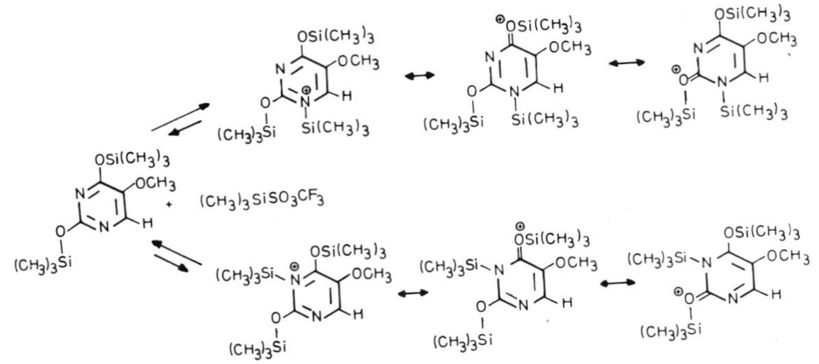

Scheme XIX

Compared to silylated 5-methoxyuracil, very similar ^{13}C NMR results were obtained for other silylated 5- or 6-, and 5,6-disubstituted uracils with either $SnCl_4$ or $(CH_3)_3SiSO_3CF_3$, permitting the same conclusions about the major complex formation at N_1 (30).

What happens in nucleoside synthesis? The answer seems to be quite simple as again exemplified by the reaction of silylated 5-methoxyuracil in the presence of $SnCl_4$ (Scheme XX):

(1) Only the free silylated base seems to react with the 1,2-acyloxonium ion to give the desired N_1-nucleoside.

(2) However, the N_1-6 complex reacts slowly with the sugar cation to give the undesired silylated N_3-Nucleoside (32).

(3) Both the silylated N_3- and perhaps also the silylated N_1-nucleoside can react further with the sugar cation to afford the N_1,N_3-*bis*-riboside. These reactions are presently being investigated.

As already emphasized, the ratio of the free and complexed silylated base is dependent on the solvent. In the more polar solvent, acetonitrile, less σ-complex, and consequently more free base is present than in 1,2-dichloroethane, leading to more of the desired N_1-nucleoside.

The corresponding reaction with $(CH_3)_3SiSO_3CF_3$ is very similar (Scheme XXI). But since $(CH_3)_3SiSO_3CF_3$ is the weaker Lewis acid forming less complexed base, more free base is present and thus more N_1-nucleoside is obtained.

Furthermore, $(CH_3)_3SiSO_3CF_3$ and $(CH_3)_3SiClO_4$ seem to form the N_1-, the N_3-, as well as the N_1,N_3-bis-ribosides in *reversible* reactions (compare Schemes XII and XVI).

Our present results already permit a few simple conclusions.

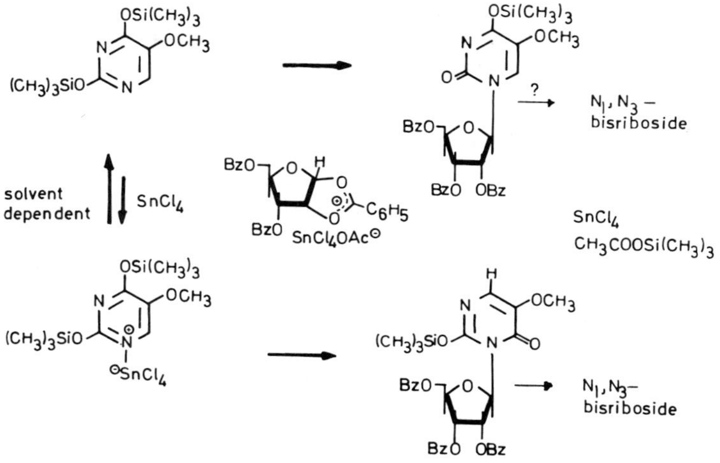

Scheme XX

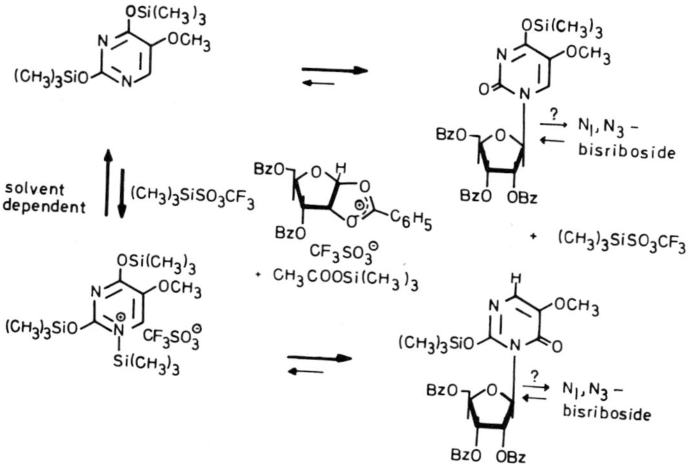

Scheme XXI

For silylated cytosines, which can only react at N_1, and weakly basic silylated uracils, the cheap $SnCl_4$ will usually give high yields of the desired N_1-nucleosides if the proper amount of catalyst is employed. However, with more basic or sterically hindered silylated uracils as well as for transglycosidylations $(CH_3)_3SiSO_3CF_3$ or $(CH_3)_3SiClO_4$ should be preferred in 1,2-dichloroethane or acetonitrile.

Since a number of problems are still open, this account should be considered as just a progress report. But we intend to conclude our studies on the synthesis of pyrimidine and purine nucleosides in the near future and hope that our results will make nucleoside synthesis more rational and thus easier to reproduce.

REFERENCES

1. Niedballa, U., and Vorbrüggen, H. *J. Org. Chem. 39,* 3654 (1974).
2. Niedballa, U., and Vorbrüggen, H. 1974. *J. Org. Chem. 39,* 3660 (1974).
3. Niedballa, U. and Vorbrüggen, H. 1974. *J. Org. Chem. 39,* 3664 (1974).
4. Niedballa, U. and Vorbrüggen, H. 1974. *J. Org. Chem. 39,* 3668 (1974).
5. Niedballa, U., and Vorbrüggen, H. 1974. *J. Org. Chem. 39,* 3672 (1974).
6. Niedballa, U., Vorbrüggen, H. 1976. *J. Org. Chem. 41,* 2084 (1976).
7. Vorbrüggen, H. and Krolikiewicz, K. *Angew. Chem. Inst. Edit. 14,* 421 (1975).
8. Birkhofer, L., Ritter, A. and Kühlthau, H. P. *Chem. Ber. 97,* 934 (1964).
9. Vorbrüggen, H. and Strehlke, P. *Chem. Ber. 106,* 3039 (1973).
10. Igarashi, K., Honma, T., and Irisawa, J. *Carbohyd. Res. 15,* 329 (1970).
11. Baker, B. R., Joseph, J. P., Schaub, R. E., and Williams, J. H. *J. Org. Chem. 19,* 1786 (1954).
12. Fox, J. J., and Wempen, I. *Adv. Carbohydr. Chem. 14,* 283 (1959).
13. Prisbe, E. J., Smejkal, J., Verheyden, J. P. H., and Moffatt, J. G. *J. Org. Chem. 41,* 1836 (1976).
14. Brink, A. J., and Jordan, A. *Carbohydr. Res. 41,* 355 (1975).
15. Akhrem, A. A., Timoshchuk, V. A., and Mikhailopulo, I. A. *J. Gen. Chem. 45,* 947 (1975).
16. For complexes of peracylated sugars with Lewis acids like

TiCl$_4$ compare: Csürös, Z., Deák G., Holly S., Török-Kalmár, A., and Zára-Kaczián, E. *Acta Chim. Acad. Sci. Hung.* *62*, 95 (1969).

17. Csürös, Z., Deák, G., Gyurkovics, I., Haraszthy-Papp, M. and Zára-Kacznián, E. *Acta Chim. Acad. Sci. Hung.* *67*, 93 (1971).
18. Ototani, N., and Whistler, R. L. *J. Med. Chem.* *17*, 535 (1974).
19. Perrin, D. D. *Dissociation Constants of Organic Bases in Aqueous Solution,* Butterworths, London, 1965, pp. 143-228.
20. Wong, J. L., and Fuchs, D. S. *J. Org. Chem.* *35*, 3786 (1970).
21. Silylated N$_4$-acetyl-cytosine is certainly less basic than cytosine. But since silylated cytosine can only react at N$_1$, the N$_4$-acetylation is normally not necessary.
22. Marsmann, H. C., and Horn, H. G. *A. Naturforsch.* *27 b*, 1448 (1972).
23. Wannagat, U., and Liehr, W. *Angew. Chem.* *69*, 783 (1957).
24. Schmidbaur, H. *Chem. Ber.* *98*, 83 (1965). We thank Prof. Schmidbaur for the gift of $(CH_3)_3SiSO_3F$.
25. Fluka No. 91 741.
26. Vorbrüggen, H., Bennua, B. and Krolikiewicz, K. unpublished.
27. Cook, P. D., Rousseau, R. J., Mian, A. M., Dea P., Meyer, R. B., Jr., and Robins, R. K. *J. Am. Chem. Soc.* *98*, 1492 (1976).
28. Hayashi, M., Hirano, T., Yaso, M., Mizuno, K. and Ueda, T. *Chem. Pharm. Bull.* *23*, 245 (1975).
29. Azuma, T., and Isono, K. *Tetrahedron Lett.* 1687 (1976).
30. With $(CH_3)_4Si$ as internal standard. The full details of our ^{13}C NMR studies with silylated bases (including purines) will be published elsewhere.
31. For a recent review on the literature and nomenclature of Π and σ complexes compare: Perkampus, H.-H. *Wechselwirkung von π-Elektronensystemen mit Metallhalogeniden,* Springer-Verlag, Berlin, 1973.
32. This reaction with the sugar cation probably occurs via the dissociated form of the complex, in which the trimethylsilyl group stays close to the N$_1$-nitrogen.

SYNTHESIS AND CHEMISTRY OF CERTAIN AZOLE NUCLEOSIDES

JOSEPH T. WITKOWSKI AND ROLAND K. ROBINS

ICN Pharmaceuticals, Inc.,
Nucleic Acid Research Institute,
Irvine, California

ABSTRACT

The synthesis and properties of a number of nucleosides, 1,2,4-triazoles, 1,2,3-triazoles, pyrazoles, and tetrazoles have been investigated. Synthetic methods, which are the most suitable for glycosylation of various azole heterocycles, are discussed. Methods utilized include the acid-catalyzed fusion procedures, treatment of the appropriate trimethylsilyl derivatives with blocked glycosyl halides, and others. Glycosylation of these azole heterocycles generally results in mixtures of isomeric nucleoside products. The ratio of the isomers formed depends on the substituents on the heterocycle and the method of glycosylation. A number of nucleosides structurally related to the antiviral nucleoside 1-β-D-ribofuranosyl-1,2,4-triazole-3-carboxamide (ribavirin) are discussed.

Azole nucleosides encompass a number of both naturally occurring (imidazole, pyrazole) and synthetic (1,2,3-triazole, 1,2,4-triazole, tetrazole) nucleosides. Certain azole nucleosides of both the naturally occurring and synthetic groups have shown significant biological activity. Among these are the C-nucleoside antibiotic pyrazomycin (pyrazofurin, 1) (1) and the synthetic nucleoside 1-β-D-ribofuranosyl-1,2,4-triazole-3-carboxamide (ribavirin, 2) (2,3). Interest in structure-activity relationships has prompted the investigation of nucleosides structurally related to these bioactive compounds.

Biological data, including antiviral activity (4-6), enzymatic studies (7), and investigations, on the mechanism of action (8,9) of most of the synthetic nucleosides and nucleotides described here have been reported elsewhere. This report is primarily concerned with the synthesis and chemistry of 1,2,4-triazole nucleosides and nucleosides of certain 1,2,3-triazoles, pyrazoles, and tetrazoles. Factors that determine the distribution of products obtained by various glycosylation methods are discussed.

| 1 | 2 |
| Pyrazomycin | Ribavirin |

Structures 1 & 2

1,2,4-TRIAZOLE NUCLEOSIDES

The most successful methods for the synthesis of 1,2,4-triazole nucleosides have been the acid-catalyzed fusion procedure (10) and glycosylation of the trimethylsilyl (TMS) derivatives of these heterocycles with suitably blocked glycosyl halides (2). Both methods generally give good total yields of nucleoside products, but the isomeric distribution varies greatly with the substituents on the heterocycle. In most cases, with a monosubstituted or disubstituted 1,2,4-triazole, two isomeric products are formed. These are the 1-glycosyl-3-substituted-1,2,4-triazole and the 1-glycosyl-5-substituted-1,2,4-triazole, which are identified by spectral properties (11) or by chemical conversions to products of known structure. The results obtained with these glycosylation procedures as applied to a number of 1,2,4-triazole are described below.

Methyl 1,2,4-triazole-3-carboxylate (*3*, R=COOCH$_3$) provides an essentially quantitative yield of nucleoside products on treatment of the TMS derivative of this heterocycle with an acyl-block ribofuranosyl halide in acetonitrile at room temperature (2). Th two blocked nucleosides are readily separated by crystallization or by chromatography. The ratio of isomers obtained by this procedure, is about 1:1 (Table I). With the acid-catalyzed fusion

TABLE I. Formation of 1,2,4-Triazole Nucleosides from Methyl 1,2,4-Triazole-3-Carboxylate and 3-Cyano-1,2,4-Triazole

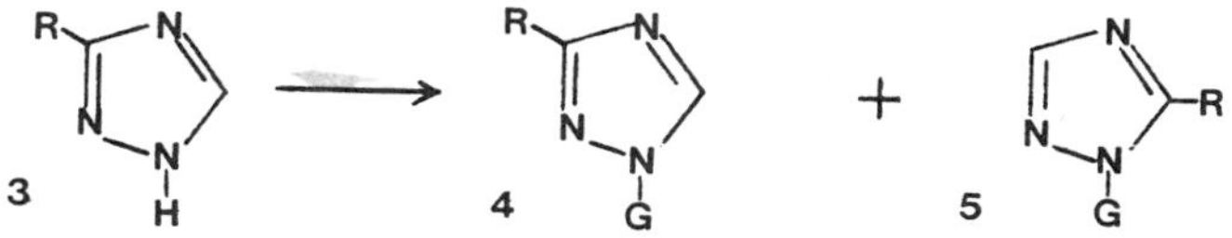

G= 2,3,5-Tri-O-acyl-β-D-ribofuranosyl

Structures 3-5

R	TMS-glycosyl halide method		Acid-catalyzed fusion	
	%4	%5	%4	%5
-COOCH$_3$	51	46	78	7
-CN	80	6	80	<10

fusion procedure, using methyl 1,2,4-triazole-3-carboxylate, an acyl-blocked ribofuranose and *bis* (p-nitrophenyl)phosphate as catalyst, the total yield of nucleoside products is somewhat lower (85%) than with the above route using a glycosyl halide. However, the isolated yield of isomer 4, which leads to 1-β-ribofuranosyl-1,2,4-triazole-3-carboxamide, is substantially higher (78%) than that obtained with the first procedure. This may be due in part to the higher temperature (165°C) of the fusion reaction. In support of this, we have observed isomerization of the 1,5-disubstituted isomer (6) to the 1,3-isomer (7) under conditions of the acid-catalyzed fusion procedure (12).

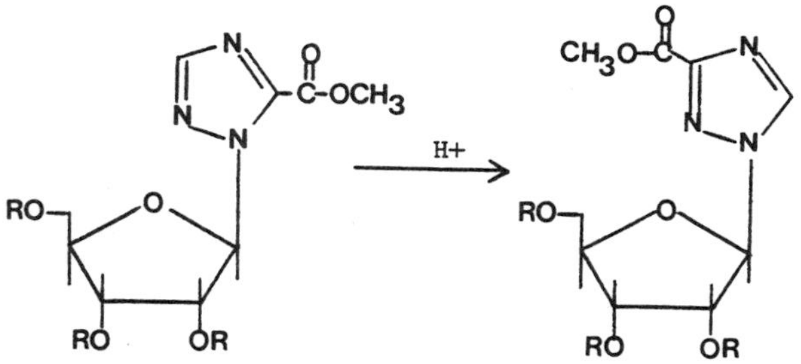

R = acyl

Structures 6 & 7

Inspection of a molecular model of 6 shows that the glycosyl
moiety at N-1 and the carbomethoxy group at the 5-position of
the triazole are crowded, and this steric factor must contribute
to the predominant formation of isomer 7 in the acid-catalyzed
fusion procedure. It is of interest to note that the carboxylic
acid corresponding to isomer 6, 1-β-D-ribofuranosyl-1,2,4-tria-
zole-5-carboxylic acid (8) decarboxylates at room temperature,
again demonstrating that 6 is sterically unfavorable. In contrast, 1-β-D-ribofuranosyl-1,2,4-triazole (10) is the product
formed (2).

Structures 8-10

With 3-cyano-1,2,4-triazole (3, R=CN), two isomeric products
are again obtained with both the acid-catalyzed fusion procedure
and the TMS-glycosyl halide method (13). However, with this triazole both methods give essentially the same ratio of products
(Table I), with only a low yield of the 1,5-isomer (5, R = CN)
Purines with electron withdrawing substituents give similar results in the fusion procedure (10). With 3-cyano-1,2,4-triazole
the yield of 4 (R = CN) obtained by this procedure was increased
to 80% by addition of *bis* (*p*-nitrophenyl)-phosphate as an acid
catalyst.

Factors that influence the ratio of products formed by the
TMS-glycosyl halide method are not clear. The actual structures
of the intermediate TMS derivatives are seldom established. In
those cases where mixtures of isomeric nucleoside products are
formed, it is uncertain whether the TMS intermediates were also
mixtures of isomeric derivatives. The questions of how the struc
tures of the nucleosides formed are related to the TMS derivative
and the actual mechanism of this reaction are interesting areas
for investigation.

The nucleosides obtained from both methyl 1,2,4-triazole-
3-carboxylate and 3-cyano-1,2,4-triazole are versatile intermediates for the synthesis of nucleosides with various substituents
on the triazole ring (Scheme I). The 1,2,4-triazole-3-carboxamid

nucleoside (2) is readily prepared from both the carbomethoxy (7) and cyano (12) 1,2,4-triazoles. In addition, the carboxylic acid (11, R_1 = OH), thiocarboxamide (13, X = S), carboxamidine (13, X = NH·HCl), and related derivatives were obtained (2, 13).

The structures of these nucleosides have been established by ^{1}H NMR and ^{13}C NMR spectral studies (11), and in the case of 1-β-D-ribofuranosyl-1,2,4-triazole-3-carboxamide (2) by x-ray crystallography (14).

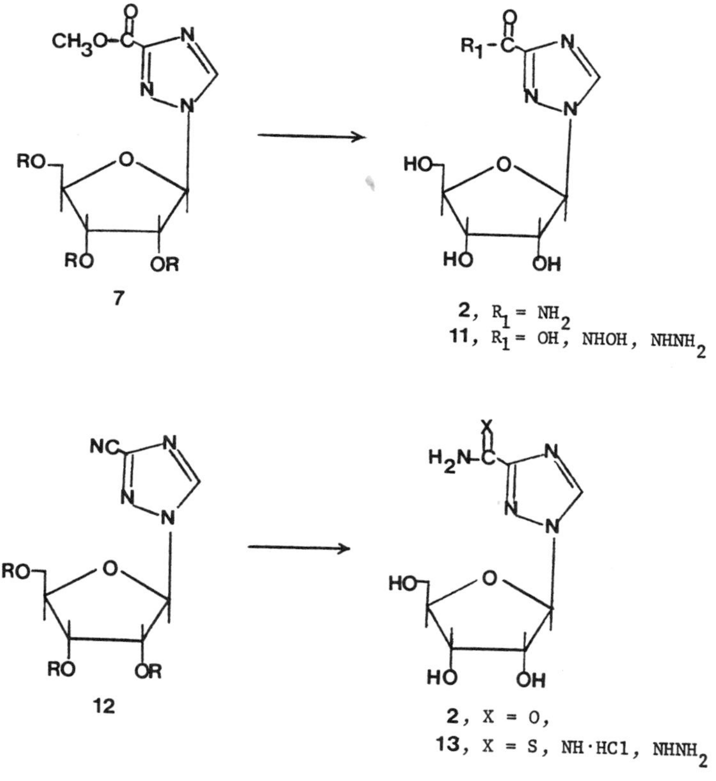

7

2, R_1 = NH$_2$
11, R_1 = OH, NHOH, NHNH$_2$

12

2, X = O,
13, X = S, NH·HCl, NHNH$_2$

Scheme 1

We have also investigated 1,2,4-triazole-3-carboxamide nucleosides, substituted at the 5-position of the heterocycle (15). Electrophilic substitution has not been successful with these azole nucleosides and the acid-catalyzed fusion procedure with disubstituted 1,2,4-triazoles was utilized to obtain the 5-substituted products. The distribution of products formed by this procedure varied greatly with the substituents on the heterocycle. Also, the total yield of nucleoside products ranged from low to greater than 80%. This appears to depend on the pK$_a$ of the heterocycle; those triazoles with strongly electron-with-

drawing substituents give the most satisfactory results with
this procedure. Similar effects have been observed in the synthesis of purine nucleosides by the acid-catalyzed fusion procedure (10). Steric factors also appear to affect the ratio of
isomers formed (Table II).

TABLE II. Formation of 3,5-Disubstituted-1,2,4-Triazole Nucleosides by the Acid-Catalyzed Fusion Procedure

G = 2,3,5-Tri-O-acyl-β-D-ribofuranosyl

R	%15	%16
H	78	7
CH$_3$	25	21
Cl	36	50
NO$_2$	Not isolated	77

Methyl 3-methyl-1,2,4-triazole 5-carboxylate (14, R = CH$_3$)
with a pK$_a$ of 8.46 (16) gives a low total yield (46%) of nucleosides with a 1:1 ratio of isomers 15 and 16 (R = CH$_3$). In contrast, methyl 1,2,4-triazole-3-cargoxylate (14, R = H), with a
pK$_a$ of 7.70 (16) and no steric hindrance at N-1, gives an 85%
yield of nucleoside products in a 10:1 ratio. The acid-catalyzed
fusion procedure with methyl 3-chloro-1,2,4-triazole-5-carboxylat
(14, R - Cl) with a pK$_a$ of 5.20 (16) gave an 86% total yield of
nucleoside products. In this case the major isomer is 16 (R = Cl
which results from ribosylation at the site adjacent to the carbomethoxy group. The only product isolated from the 3-nitro-1,
2,4-triazole 14 (R = NO$_2$) was 16 (R = NO$_2$). With this triazole
(14, R = NO$_2$), which has a pK$_a$ of 3.55 (16), the fusion procedure
again proceeded readily without an acid catalyst. The 5-chloro-
1,2,4-triazole nucleoside (17) was used to obtain nucleosides
(18) substituted at the 5-position by nucleophilic displacement.
The isomeric nucleoside, 3-chloro-1-β-D-ribofuranosyl-1,2,4-tria-
zole-5-carboxamide, remained unchenged under similar conditions.
The synthesis of nucleosides of halogen derivatives of 1,
2,4-triazole has also been investigated (17). The results of

the acid-catalyzed fusion procedure with 3-fluoro-, 3-chloro-, 3-bromo-, and 3-iodo-1,2,4-triazole are summarized in Table III.

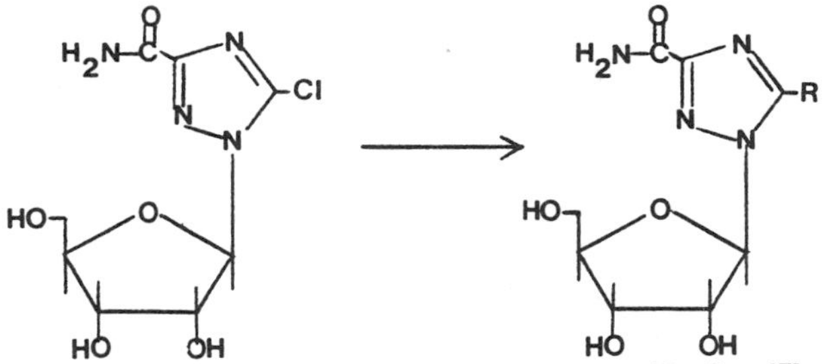

17 → 18, R = NH$_2$, SH

Structures 17 & 18

TABLE III. Nucleosides of Halogen Derivatives of 1,2,4-Triazole

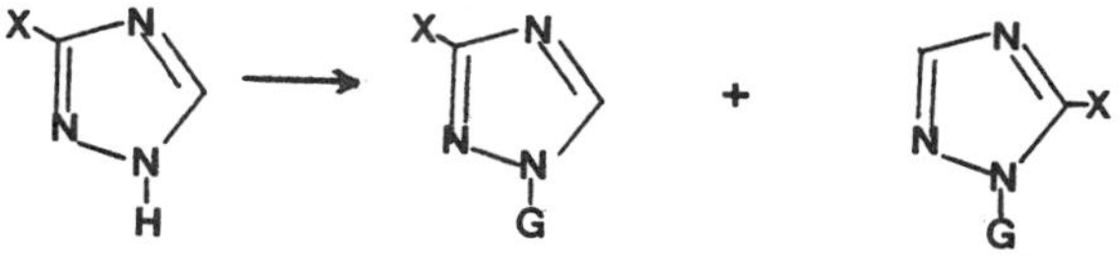

19 → 20 + 21

G = 2,3,5-Tri-O-acyl-β-D-ribofuranosyl

X	%20	%21
F	47	Not isolated
Cl	54	21
Br	48	15
I	44	Not isolated

Only one isomer has been isolated from 3-fluoro-1,2,4-triazole and 3-iodo-1,2,4-triazole. In the case of the iodo derivative, steric hindrance at the nitrogen adjacent to the iodo group may account for the formation of only the more sterically favorable isomer. With 3-fluoro-1,2,4-triazole, the failure to isolate the 5-fluoro-1,2,4-triazole nucleoside (21, X = F) may be due to hydrolysis of the fluoro group, which would be expected to be much more readily displaced in this isomer than in the 3-fluoro-1,2,4-triazole nucleoside (21, X = F). This 3-fluoro isomer

(21, X = F) survived deacylation under the usual conditions to give the deblocked nucleoside, 3-fluoro-1-β-D-ribofuranosyl-1,2,4-triazole.

Displacement of the halogen at the 5-position again proceeds readily, in contrast to the isomeric 3-halogen-1,2,4-triazole nucleosides, in which the halogen is resistant to nucleophilic substitution. Thus, treatment of 5-chloro-1-β-D-ribofuranosyl-1,2,4-triazole (22) with ammonia provided 5-amino-1-β-D-ribofuranosyl-1,2,4-triazole (23).

Another triazole nucleoside containing a reactive halogen at the 5-position is 5-bromo-3-nitro-1-(2,3,5-tri-O-acetyl-β-D-ribofuranosyl)-1,2,4-triazole (24) (18). Treatment of 24 with liquid ammonia provided 5-amino-3-nitro-1-β-D-ribofuranosyl-1,2,4-triazole (25) which was reduced to give 3,5-diamino-1-β-D-ribofuranosyl-1,2,4-triazole (26) (12). This compound (26) is a ribonucleoside of guanozole (3,5-diamino-1,2,4-triazole), a heterocycle that exhibits anticancer activity (19, 20).

Two isomeric ribonucleosides of 3-amino-1,2,4-triazole (27) have been described. Reduction of 3-nitro-1-β-D-ribofuranosyl-1,2,4-triazole provided (18) the corresponding 3-amino-1,2,4-triazole nucleoside (28). A second isomer, 5-amino-1-β-D-ribofuranosyl-1,2,4-triazole (23), was obtained by treatment of 5-chloro-1-β-D-ribofuranosyl-1,2,4-triazole with ammonia, as de-

scribed earlier (17). Ribosylation of the TMS derivative of 3-
amino-1,2,4-triazole with 1-O-acetyl-2,3,5-tri-O-benzoyl-β-D-
ribofuranose in the presence of a Lewis acid has been found to
give a mixture of three isomeric nucleosides (21). Two of these
were identified after deacylation as 3-amino-1-β-D-ribofuranosyl-
1,2,4-triazole (28) and 5-amino-1-β-D-ribofuranosyl-1,2,4-tria-
zole (23) by comparison with authentic samples of these nucleosides. The third isomer was established as 3-amino-4-β-D-ribo-
furanosyl-1,2,4-triazole (29) by comparison of spectral data with
3-amino-4-methyl-1,2,4-triazole (22). This is one of the few
cases in which ribosylation at the 4-position of the triazole
ring has been observed.

R = β-D-ribofuranosyl

Structures 27-29

A number of nucleosides of 1,2,4-triazole-3-carboxamide with
modified glycosyl moieties have been synthesized. These nucleosides and the synthetic methods used are listed in Table IV.

By the appropriate use of blocking groups (participating
acyl or nonparticipating benzyl) both the α- and β-anomers of the
ribo- and *arabino*-nucleosides have been obtained. In the case
of the 2'-deoxynucleosides, it was found convenient to use 1-O-
acetyl-3,5-di-O-p-toluoyl-2-deoxy-D-*erythro*-pentofuranose in the
acid-catalyzed fusion procedure. This sugar contains both the
1-O-acetyl group, which is known to give good results in the
fusion procedure, and the p-toluoyl protecting groups, which facilitate the chromatographic separation of the anomers formed in
this procedure. One additional modified triazole nucleoside, 1-
β-D-psicofuranosyl-1,2,4-triazole-3-carboxamide, has recently
been reported (28).

Two methods have been employed to obtain 5'-deoxy and 5'-
substituted nucleosides in this series of triazoles (Scheme II).
The first is the acid-catalyzed fusion procedure with 3-cyano-
1,2,4-triazole and the 1,2,3-tri-O-acyl-D-ribofuranose (30) modified at the 5-position. The 3-cyano-1,2,4-triazole nucleosides
(31) thus obtained were utilized (12) to prepare the 5'-fluoro-
5'-deoxy nucleoside (32) and the 5'-deoxy-1,2,4-triazole-3-thio-
carboxamide nucleoside (33). Alternatively, use of the Rydon
reagent (29) on a blocked ribonucleoside (34) gave the 5'-iodo-
5'-deoxy nucleoside (35, R = I), which was converted to 5'-deoxy
and 5'-amino-5'-deoxy nucleosides (35) (12).

TABLE IV. Methods for Synthesis of Glycosyl-1,2,4-Triazole-3-Carboxamides

Glycosyl	Method[a]	Reference
β-D-Ribofuranosyl	A, B, C	2, 13
α-D-Ribofuranosyl	D	23
β-D-Arabinofuranosyl	D	23
α-D-Arabinofuranosyl	A	23
β-D-Xylofuranosyl	C	23
2-Deoxy-β-D-erythro- pentofuranosyl	A, B	23
2-Deoxy-α-D-erythro- pentofuranosyl	A, B	23
3-Deoxy-β-D-erythro- pentofuranosyl	C	23
α-L-Lyxopyranosyl	B	23
α-L-Arabinopyranosyl	C	23
4'-Thio-β-D-ribofuranosyl	B	24
2'-O-Methyl-β-D-ribofur- anosyl	E	25
3'-O-Methyl-β-D-ribofur- anosyl	E	25
β-D-Erythrofuranosyl	B	27

[a]Methods: (A) acid-catalyzed fusion procedure with methyl 1,2,4-triazole-3-carboxylate; (B) acid-catalyzed fusion procedure with 3-cyano-1,2,4-triazole; (C) glycosylation of the TMS derivative of methyl 1,2,4-triazole-3-carboxylate with an O-acyl glycosyl halide; (D) glycosylation of the TMS derivative of methyl 1,2,4-triazole-3-carboxylate with an O-benzyl glycosyl halide; (E) diazomethane and stannous chloride (26).

The 5'-phosphates of certain of these triazole nucleosides are of interest in studying the biochemical mechanism of action of these compounds. The use of phosphoryl chloride in a trialkyl phosphate (30) with the unprotected nucleoside gave (8) the 5'-phosphate (36) of 1-β-D-ribofuranosyl-1,2,4-triazole-3-carboxamide. To investigate structure-activity correlations with related synthetic nucleotides, 1-β-D-arabinofuranosyl-1,2,4-triazole-3-carboxamide 5'-phosphate (37), 1-(2-deoxy-β-D-erythro-pentofuranosyl)-1,2,4-triazole-3-carboxamide 5'-phosphate (38), 1-β-ribofuranosyl-1,2,4-triazole-3-thiocarboxamide 5'-phosphate (39), 1-β-D-ribofuranosyl-1,2,4-triazole-3-carboxamidine 5'-phosphate (40), and 1-β-D-ribofuranosyl-1,2,4-triazole-3-carboxylic acid 5'-phosphate (41) were prepared (31). In addition, the 3',5'-cyclic phosphate (42) and 2',3'-cyclic phosphate (43) of 1-β-D-ribofuranosyl-1,2,4-triazole-3-carboxamide were synthesized by conventional methods (32).

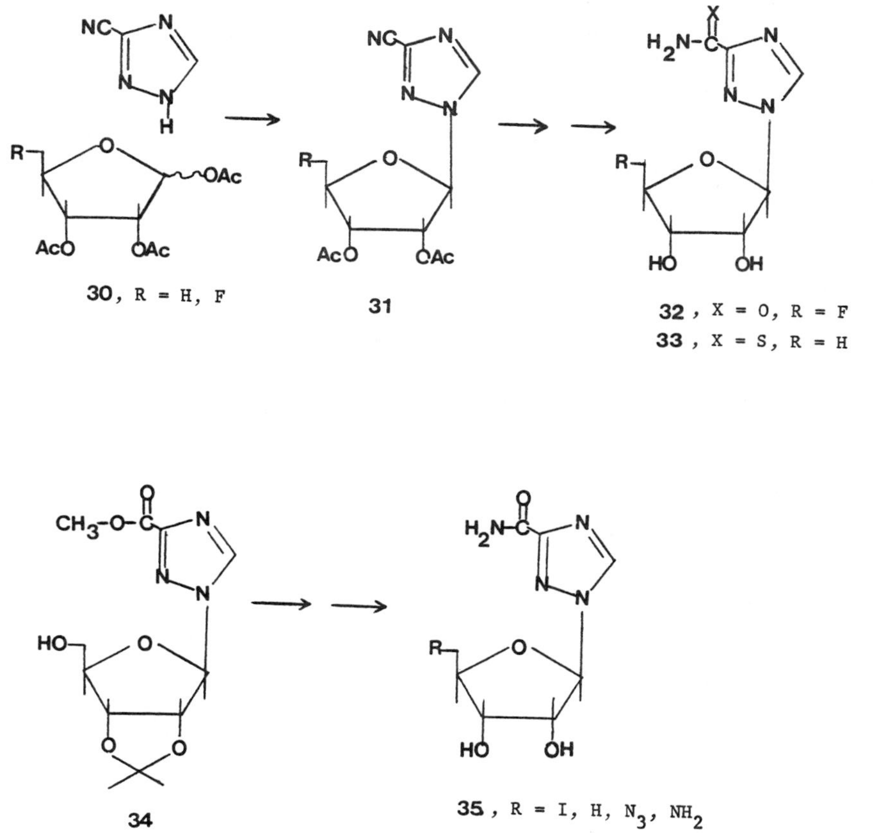

30, R = H, F

31

32 , X = O, R = F
33 , X = S, R = H

34

35 , R = I, H, N$_3$, NH$_2$

Scheme II

Penetration of intact nucleotides into the cell has been an area of active investigation (33). One class of nucleotide analogs that would not be subject to enzymatic cleavage is the class of 5'-methylene derivatives, and these analogs have been investigated in the purine series (34, 35). The 5'-methylene analog (44) of 1-β-D-ribofuranosyl-1,2,4-triazole-3-carboxamide 5'-phosphate was synthesized (36) using the 5'-aldehyde and the appropriate Wittig reagent. Some related nucleosides (45 and 46), with an ionizable group (carboxylic acid) and a hydrogen-bonding substituent (carboxamide) at the site normally occupied by the phosphate group, were similarly prepared. The enzymatic activity of these compounds compared to the 5'-phosphate has been reported (36).

36, X = O, R$_1$ = H, R$_2$ = OH   39, X = S, R$_1$ = H, R$_2$ = OH

37, X = O, R$_1$ = OH, R$_2$ = H   40, X = NH, R$_1$ = H, R$_2$ = OH

38, X = O, R$_1$ = R$_2$ = H

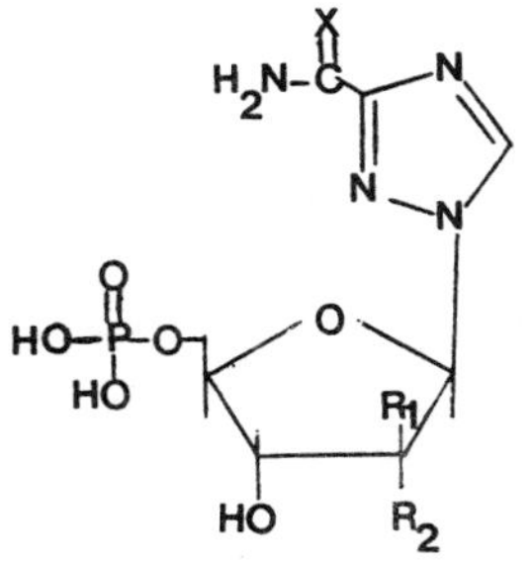

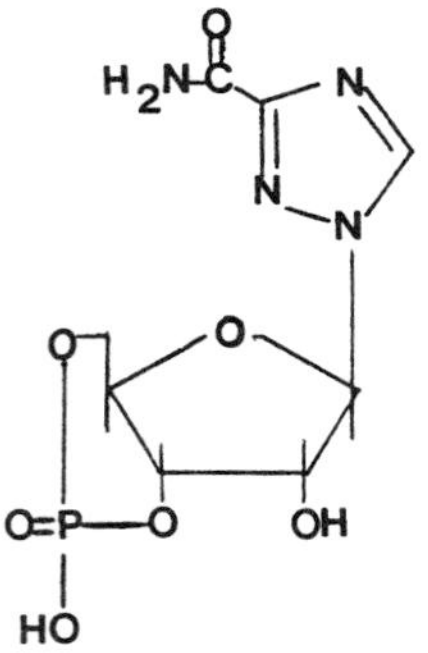

42 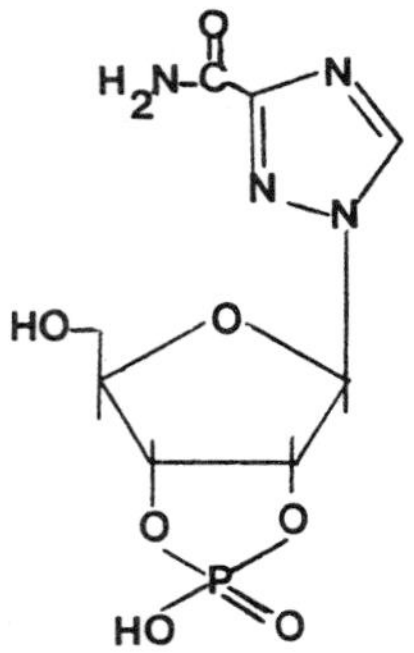   43

Structures 36-43

Certain 5'-modified nucleosides in the purine series have been investigated. These include the 5'-O-sulfamate (37, 38), which is related to the antibiotic nucleocidin, the 5'-O-nitrate (39) and the 5'-O-carbamoyl (40,41) derivatives. Use of the appropriate reagents with the 2',3'-O-isopropylidene nucleoside (47) followed by deblocking provided the corresponding 1,2,4-triazole-nucleosides (48) followed by deblocking provided the corresponding 1,2,4-triazole-nucleosides (48) (12).

1,2,3-TRIAZOLE NUCLEOSIDES

A number of 1,2,3-triazole nucleosides have been prepared by cycloaddition reactions of glycosyl azides with substituted acetylenes and other reagents (42-51). This method provides 1-glycosyl-1,2,3-triazoles substituted at the 4- and/or 5-positions.
It was of interest to investigate the 1,2,3-triazole-4-car-boxamide ribonucleosides isoteric with 1-β-D-ribofuranosyl-1,2,4-triazole-3-carboxamide (ribavirin, 2). These nucleosides (49 and 50) are structurally very similar to 2 but contain hydrogen-bonding sites at different positions on the azole ring.
The synthesis of 1-β-D-ribofuranosyl-1,2,3-triazole-4-car-boxamide (49) via cycloaddition of 2,3,5-tri-O-benzoyl-β-D-ribo-furanosyl azide with methyl propiolate has been reported (47). Ribosylation of the appropriate triazole base with a blocked

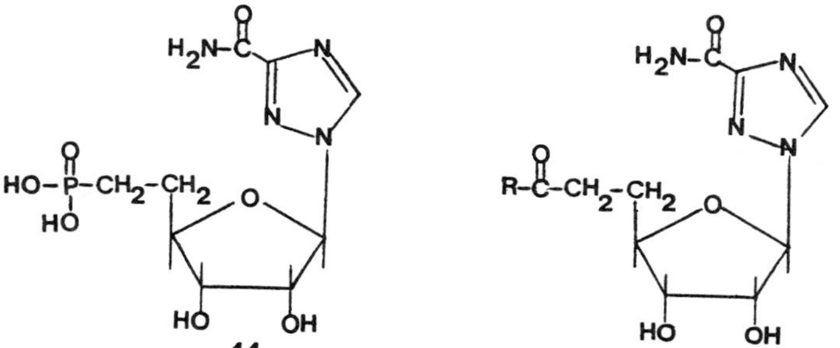

Structures 44-46

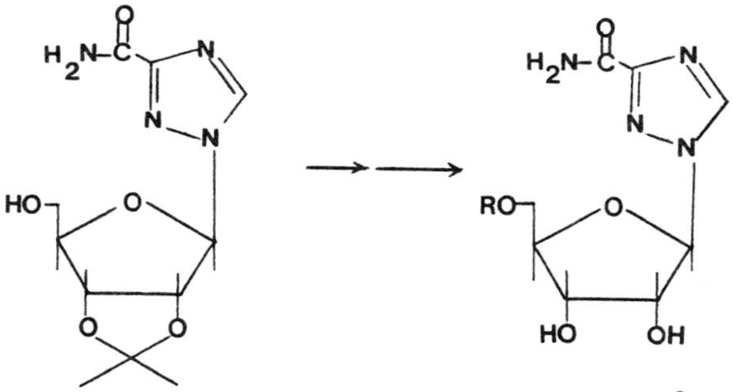

Structures 47 & 48

glycosyl halide in the presence of mercuric cyanide provides another route for the synthesis of 49 (52). A novel ring contraction of 5-diazouridine resulting in the formation of the 1,2,3-triazole nucleoside 49 has also been described (53).

As a route to 2-glycosyl-1,2,3-triazoles, we utilized (54) the acid-catalyzed fusion procedure with 4-substituted-1,2,3-triazoles (51). This method is successful with 4-cyano-1,2,3-triazole, methyl 1,2,3-triazole-4-carboxylate, 4-nitro-1,2,3-triazole, and unsubstituted 1,2,3-triazole. Both the 1-glycosyl-4-substituted-1,2,3-triazoles (52) and the 2-glycosyl-4-substituted-1,2,3-triazoles (53) were obtained by this procedure. The yields of isomers formed in each case are given in Table V. In the case of methyl 1,2,3-triazole-4-carboxylate, a small amount of the 1-glycosyl-5-substituted-1,2,3-triazole was also isolated.

The 1,2,3-triazole-4-carboxamide nucleosides (49 and 50) were
obtained by treatment of 52 and 53 (R = COOCH$_3$) with methanolic
ammonia.

Structures 49 & 50

TABLE V. Formation of 1,2,3-Triazole Nucleosides by Acid-Catalyzed Fusion Procedure

G = 2,3,5-Tri-*O*-acyl-β-D-ribofuranosyl-

R	%47	%48
H	53	10
-CN	39	39
-COOCH$_3$	30	57
-NO$_2$	24	58

A 1,2,3-triazole nucleoside, which is structurally similar
to the C-nucleoside pyrazomycin (1) and also the naturally occurring imidazole nucleoside, bredinin (54) (55), is 5-hydroxy-1-
β-D-ribofuranosyl-1,2,3-triazole-4-carboxamide (55). This triazole nucleoside (55) was synthesized (56) by the two routes shown
in Scheme III. Ribosylation of the TMS-derivative of 4-hydroxy-
1,2,3-triazole-5-carboxamide (56) with 1-*O*-acetyl-2,3,5-tri-*O*-
benzoyl-β-D-ribofuranose in the presence of a Lewis acid provided
after deacylation, the nucleoside 55. The second route to 55 was
by cycloaddition of the glycosyl azide 57 with ethyl malonamate.

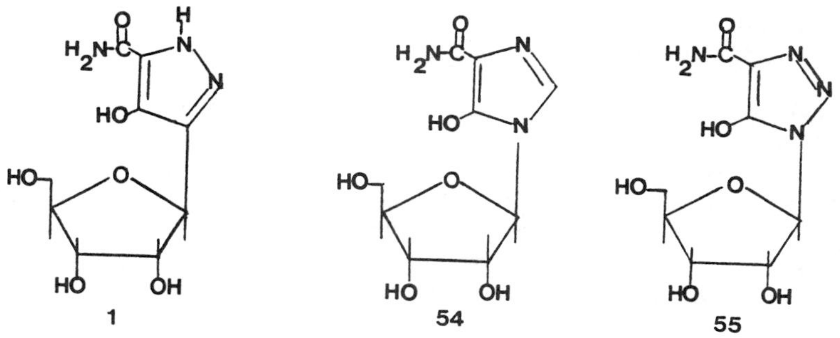

Structures 54 & 55

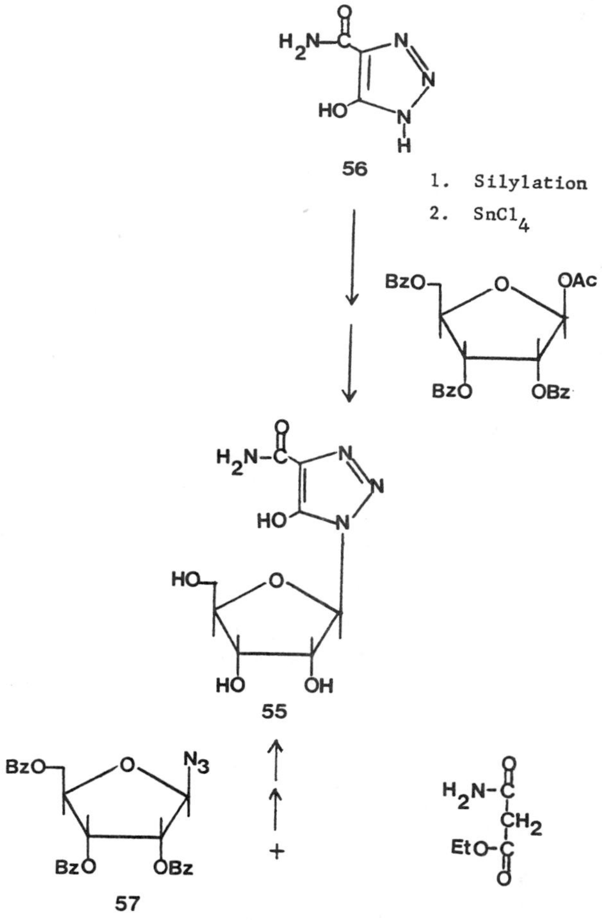

Scheme III

PYRAZOLE NUCLEOSIDES

The syntheses of several pyrazole nucleosides (57-61) have been described. We (62), and others (59,60), were interested in pyrazole nucleosides structurally related to 1-β-D-ribofuranosyl-1,2,4-triazole-3-carboxamide (ribavirin, 2). These nucleosides, 1-β-D-ribofuranosylpyrazole-3-carboxamide (58) and 1-β-D-ribofuranosylpyrazole-4-carboxamide (59) were readily prepared by the acid-catalyzed fusion procedure from the corresponding pyrazole esters (59, 60, 62). An interesting study using the TMS

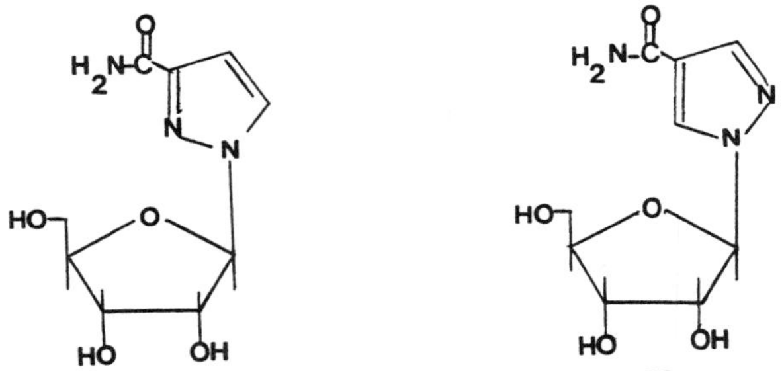

Structures 58 & 59

derivative of ethyl pyrazole-3-carboxylate to obtain related pyrazole nucleosides has been reported (63).

TETRAZOLE NUCLEOSIDES

Tetrazoles are an obvious extension of the investigation of diazole and triazole nucleosides. The ribonucleoside (60) of tetrazole-5-carboxamide was obtained from the corresponding ethyl ester via the acid-catalyzed fusion procedure (62). A similar synthesis of tetrazole ribonucleosides has recently been reported (64).

ACKNOWLEDGMENTS

We thank Marie Therese Campbell and O. P. Crews for assistance in the preparation of chemical intermediates used in this work.

Structure 60

REFERENCES

1. Gutowski, G. E., Sweeney, M. J., DeLong, D. C., Hamill, R. L., Gerzon, K., and Dyke, R. W., *Ann. N.Y. Acad. Sci.* 255, 544 (1975) and references therein.
2. Witkowski, J. T., Robins, R. K., Sidwell, R. W., and Simon, L. N., *J. Med. Chem.* 15, 1150 (1972).
3. "Ribavirin" is the name approved by the US Adopted Names Council for 1-β-D-ribofuranosyl-1,2,4-triazole-3-carboxamide; Virazole is the ICN Pharmaceuticals, Inc. trademark for this compound.
4. Sidwell, R. W., Huffman, J. H., Khare, G. P., Allen, L. B., Witkowski, J. T., and Robins, R. K., *Science* 177, 705 (1972).
5. Sidwell, R. W., Simon, L. N., Witkowski, J. T., Robins, R. K., *Progr. Chemotherap.* 2, 889 (1974).
6. Witkowski, J. T., Robins, R. K., Sidwell, R. W., and Simon, L. N., Abstracts, 167th National Meeting of the American Chemical Society, Los Angeles, California, 1974, No. MEDI 39.
7. Streeter, D. G., Simon, L. N., Robins, R. K., and Miller, J. P., *Biochemistry* 13, 4543 (1974).
8. Streeter, D. G., Witkowski, J. T., Khare, G. P., Sidwell, R. W., Bauer, R. J., Robins, R. K., and Simon, L. N., *Proc. Natl. Acad. Sci. USA* 70, 1174 (1973).
9. Scholtissek, C., *Arch. Virol.* 50, 349 (1976).
10. Goodman, L., in Basic Principles in Nucleic Acid Chemistry, Vol. 1 (P.O.P. Ts'o, ed.), Academic Press, New York, 1974, p. 102 and references therein.

11. Kreishman, G. P., Witkowski, J. T., Robins, R. K., and
    Schweizer, M. P., *J. Am. Chem. Soc. 94*, 5894 (1972)..
12. Witkowski, J. T., and Robins, R. K., unpublished results.
13. Witkowski, J. T., Robins, R. K., Khare, G. P., and Sidwell,
    R. W., *J. Med. Chem. 16*, 935 (1973).
14. Prusiner, P., and Sundaralingam, M., *Nature New Biology
    244*, 116 (1973).
15. Naik, S. R., Witkowski, J. T., and Robins, R. K., *J. Heter-
    ocycl. Chem. 11*, 57 (1974).
16. Bagal, L. I., Pevzner, M. S., and Lopyrev, U. A., *Khim.
    Geterotsikl. Soedin, Sb. 1: Azotsoderzhaschchie Getero-
    tsikly 70*, 77876t (1969).
17. Fuertes, M., Naik, S. R., Robins, R. K., and Witkowski,
    J. T., Abstracts, Fifth International Congress of Hetero-
    cyclic Chimistry, Ljubljana, Yugoslavia, 1975, p. 348.
18. Witkowski, J. T., and Robins, R. K., *J. Org. Chem. 35*, 2635
    (1970).
19. Brockman, R. W., Shaddix, S., Laster, W. T., Jr., and Schabe.
    F. M., Jr., *Cancer Res. 30*, 2358 (1970).
20. Hahn, M. A., and Adamson, R. H., *J. Natl. Cancer Inst. 48*,
    783 (1972).
21. Fuertes, M., Robins, R. K., and Witkowski, J. T., *J. Carbo-
    hydr., Nucleosides Nucleotides 3*, 169 (1976).
22. Barascut, J. L., Claramont, R. M., and Elguero, J., *Bull.
    Soc. Chim. Fr.* 1849 (1973).
23. Witkowski, J. T., Fuertes, M., Cook, P. D., and Robins, R.
    K., *J. Carbohydr. Nucleosides Nucleotides 2*, 1 (1975).
24. Pickering, M. V., Witkowski, J. T., and Robins, R. K., *J.
    Med. Chem. 19*, 841 (1976).
25. Naik, S. R., Witkowski, J. T., Robins, R. K., unpublished
    results.
26. Robins, M. J., Naik, S. R., and Lee, A.S.K., *J. Org. Chem.
    39*, 1891 (1974).
27. Szekeres, G. L., Robins, R. K., Stout, M. G., and Witkowski,
    J. T., unpublished results.
28. Prisbe, E. J., Smejkal, J., Verheyden, J.P.H., and Moffatt,
    J. G., *J. Org. Chem. 41*, 1836 (1976).
29. Verheyden, J.P.H., and Moffatt, J. G., *J. Org. Chem. 35*,
    2319 (1970).
30. Yoshikawa, M., Kato, T., and Takenishi, T., *Tetrahedron
    Lett.* 5065 (1967).
31. Witkowski, J. T., Fuertes, M., Cook, P. D., and Robins, R.
    K., unpublished results.
32. Shuman, D. A., Meyer, Jr., R. B., Witkowski, J. T., and
    Robins, R. K., unpublished results.
33. Cohen, S. S., and Plunkett, W., *Ann. NY Acad. Sci. 255*,
    269 (1975).

34. Jones, G. H., and Moffatt, J. G., *J. Am. Chem. Soc.* 90, 5337 (1968).

35. Hampton, A., Sasaki, T., and Paul, B., *J. Am. Chem. Soc.* 95, 4404 (1973).

36. Fuertes, M., Witkowski, J, T., Streeter, D. G., and Robins, R. K., *J. Med. Chem.* 17, 642 (1974).

37. Shuman, D. A., Robins, M. J., and Robins, R. K., *J. Am. Chem. Soc.* 92, 3434 (1970).

38. Jenkins, I. A., Verheyden, J.P.H., and Moffatt, J. G., *J. Am. Chem. Soc.* 98, 3346 (1976).

39. Lichtenthaler, F. W., and Muller, H. J., *Synthesis* 199 (1974).

40. Baker, B. R., Tanna, P. M., and Jackson, G.D.F., *J. Pharmacol. Sci.* 54, 987 (1965).

41. Fleming, W. C., Lee, W. W., and Henry, D. W., *J. Med. Chem.* 16, 570 (1973).

42. Michael, F., and Baum, G., *Chem. Ber.* 90, 1595 (1957).

43. Baddiley, J., Buchanan, J. G., and Osborne, G. O., *J. Chem. Soc.* 1651 (1958).

44. Baddiley, J., Buchanan, J. G., and Osborne, G. O., *J. Chem. Soc.* 3606 (1958).

45. Garcia-Muñoz, G., Iglesias, J., Lora-Tamzao, M., and Madroñero, R., *J. Heterocycl. Chem.* 5, 699 (1968).

46. Garcia-Lopez, M. T., Garcia-Muñoz, G., Iglesias, J., Madroñero, R., and Rico, M., *J. Heterocycl. Chem.* 6, 639 (1969).

47. Alonso, G., Garcia-Lopez, M. T., Garcia-Muñoz, G., Madroñero, R., Rico, M., *J. Heterocycl. Chem.* 7, 1269 (1970).

48. El Khadem, H., Horton, D., and Mershreki, M. H., *Carbohyd. Res.* 16, 409 (1971).

49. Harmon, R. E., Earl, R. A., and Gupta, S. K., *J. Org. Chem.* 36, 2553 (1971).

50. Harmon, R. E., Earl, R. A., and Gupta, S. K., *Chem. Commun.* 296 (1971).

51. Hutzenlaub, W., Tolman, R. L., and Robins, R. K., *J. Med. Chem.* 15, 879 (1972).

52. Makabe, O., Fukatsu, S., and Umezawa, S., *Bull. Chem. Soc. Japan* 45, 2577 (1972).

53. Thurber, T. C., and Townsend, L. B., *J. Am. Chem. Soc.* 95, 3081 (1973).

54. Lehmkuhl, F. A., Witkowski, J. T., and Robins, R. K., *J. Heterocycl. Chem.* 9, 1195 (1972).

55. Mizuno, K., Tsujino, M., Takada, M., Hayashi, M., Atsumi, K., Asano, K., and Matsuda, T., *J. Antibiotics* 27, 775 (1974).

56. Lehmkuhl, F. A., Witkowski, J. T., Khare, G. P., Robins, R. K., and Sidwell, R. W., Abstracts, 165th National Meeting of the American Chemical Society, Dallas, Texas, 1973, No. MEDI 56.

57. Tanaka, H., Hayashi, T., and Nakayama, K., *Agr. Biol. Chem.*
     37, 1731 (1973).
58. Barascut, J. L., Tamby, C., and Imbach, J. L., *J. Carbohydr.*
     Nucleosides Nucleotides 1, 77 (1974).
59. Korbukh, I. A., Preobrazhenskaya, M. N., and Judina, O. N.,
     J. Carbohydr. Nucleosides Nucleotides 1, 363 (1974).
60. Makabe, O., Nakamura, M., and Umezawa, S., *Bull. Chem. Soc.*
     Japan 48, 3210 (1975).
61. Montgomery, J. A., and Thomas, H. J., *J. Med. Chem. 15*,
     182 (1972).
62. Lehmkuhl, F. A., Witkowski, J. T., and Robins, R. K., un-
     published results.
63. Preobrazhenskaya, M. N., Korbukh, I. A., and Blanco, F. F.,
     J. Carbohydr. Nucleosides Nucleotides 2, 73 (1975).
64. Poonian, M. S. Nowoswiat, E. F., Blount, J. G., Williams,
     T. H., Pitcher, R. G., and Kramer, M. J., *J. Med. Chem.*
     19, 286 (1976).

THE SYNTHESIS OF HETEROCYCLIC ANALOGS OF PURINE
NUCLEOSIDES AND NUCLEOTIDES
CONTAINING A BRIDGEHEAD NITROGEN ATOM

GANAPATHI R. REVANKAR AND ROLAND K. ROBINS

ICN Pharmaceuticals, Inc.,
Nucleic Acid Research Institute,
Irvine, California

ABSTRACT

The synthesis of several nucleosides and nucleotide derivatives
containing a bridgehead nitrogen atom is described. Glycosylation of the trimethylsilyl derivative of the appropriate aglycon
with either 2,3,5-tri-*O*-acetyl-β-D-ribofuranosyl bromide in acetonitrile or tetra-*O*-acetyl-β-D-ribofuranose in the presence of
a Lewis-acid catalyst in dichloroethane has been investigated.
Stereospecific and regioselective methods have been devised to
obtain the inosine and guanosine analogs of the 1,2,4-triazolo
(1,5-*a*)pyrimidine, imidazo(1,2-*c*)pyrimidine, and imidazo(1,2-*a*)
pyrimidine ring systems. The synthesis of certain *N*-glycosides
of pyrazolo(1,5-*a*)pyrimidines and pyrazolo(1,5-*a*)-1,3,5-triazines
has also been accomplished. The site of glycosylation of these
unnatural nucleosides has been determined unequivocally by a combination of ^{1}H NMR and ^{13}C NMR spectroscopic methods. The anomeric configurations have been established by using ^{1}H NMR of the
2',3'-*O*-isopropylidene derivatives, as well as periodate oxidation,
followed by a reduction with sodium borohydride. None of these
compounds exhibited significant antiviral or antimicrobial activity *in vivo*.

In recent years, the naturally occurring purine and pyrimidine nucleosides and nucleotides as well as a host of synthetic
analogs have exhibited a wide variety of biological activity.
The biological activity of these nucleoside analogs is largely
due to their structural similarities to the natural enzyme substrates. It was therefore of considerable interest to synthesize

287

structural analogs that would have the potential either to emu-
late or antagonize the functions of the naturally occurring
purine nucleosides and nucleotides.

   In recent years many such nucleosides have been described,
which resemble at first glance the purine nucleosides, adenosine,
guanosine, or inosine, but which actually differ in some minor
aspect. The number of these so-called "counterfeits" that could
be prepared and identified employing alternative heterocyclic
systems is limited because of the arduous task of the organic
chemist to assign unequivocally the site of glycosylation and anomeric configuration. In many cases the stereospecific and regioselective syntheses of such nucleosides are very difficult and
require tedious separation and characterization of a mixture of
isomers and anomers.

   It was the primary goal of our investigation to prepare some
inosine and guanosine analogs of the purine skeleton containing
a bridgehead nitrogen atom. These heterocyclic systems are also
of particular interest since the corresponding nucleosides lack
an NH function at position 1 of the purine ring, therefore, hydrogen bonding of the Watson-Crick type would not be possible.

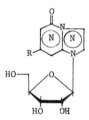

Purine nucleoside   Purine nucleoside analogs with
R=H=Inosine   a bridgehead nitrogen atom
R=NH$_2$=Guanosine

   We have recently described (1,2) the first chemical synthesis
of a purine nucleoside analog containing a bridgehead nitrogen
atom. The heterocyclic moiety, 1,2,4-triazolo(1,5-a)pyrimidine,
can be pictured as purine in which N-1 and C-5 are interchanged.
The synthesis of the inosine analog is summarized in Scheme I.
The coupling of the trimethylsilyl derivative (1) of 1,2,4-tri-
azolo(1,5-a)pyrimidin-7-one (3) with 2,3,5-tri-O-benzoyl-D-ribo-
furanosyl bromide (2) in acetonitrile led to the formation of two
blocked nucleosides. The separation of these nucleosides was
achieved by column chromatography on alumina and they were subsequently assigned the isomeric structures 3 and 4. Removal of the
blocking groups from the syrupy 3 and 4 with methanolic ammonia
at room temperature gave the crystalline nucleosides, 3-(β-D-ribofuranosyl)-1,2,4-triazolo(1,5-a)pyrimidin-7-one (5) and the N-4
isomer (6) in overall yields from 1 of 45 and 20%, respectively.

The site of ribosylation was originally assigned by ultra-violet absorption studies comparing the uv-spectra data of the nucleosides with the appropriate model methyl compounds. A comparison of the ultraviolet spectra (pH 1,7, and 11) of 3-methyl-1,2,4-triazolo (1,5-a)pyrimidin-7-one (1) with 5 and 4-methyl-1,2,4-triazolo(1,5-a)pyrimidin-7-one (1) with 6, were found to be superimposable and established the site of ribosylation. Additional support was obtained by using ^{13}C NMR spectroscopy. We have recently documented (4) the use of ^{13}C chemical shifts for a determination of the glycosylation site in nucleosides of fused nitrogen heterocycles. The reported assignments were made on the basis of α and β substitution shifts (an upfield shift for the α carbon and downfield shifts for the β and γ carbon atoms)

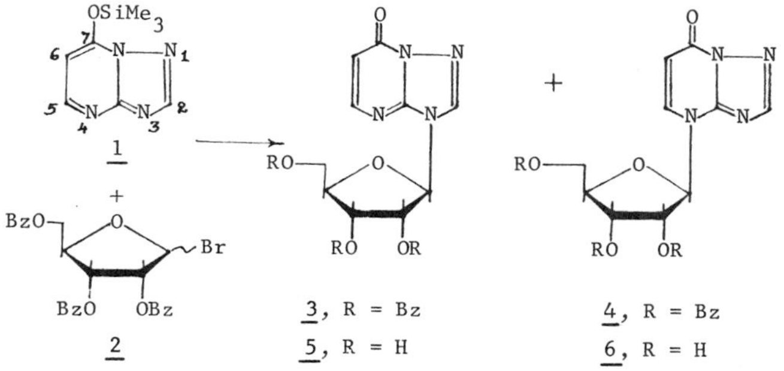

Scheme I

observed when the ribofuranosyl derivatives were compared with the corresponding ionized heterocyclic base. The ^{13}C chemical shifts obtained for the heterocyclic anion (A) and its two ribo-furanosides (5 and 6) are summarized in Table I.

By comparing the ^{13}C chemical shifts of the base anion (A), 5 and 6, we note the upfield shifts of 11.7 and 8.7 ppm for C_2 and C_{3a} (α carbons to N-3) and the essentially unchanged C_5 resonance, which confirms the site of glycosylation in 5 as N-3. Likewise, the upfield shift of 15.2 ppm for both C_{3a} and C_5, and the slight change in C_2 resonance, confirms the site of glycosylation in 6 as N-4. The possibility of N-1 ribosylation can be safely eliminated from the positive β shift of 3.3 and 4.7 ppm observed for C_7 for nucleosides 5 and 6, respectively. The N-1 ribosylation could also be ruled out since the alkylation at this position would be expected to produce a bathochromic shift of about 10 nm due to the favored conjugation and similarity of N-7 substitution noted in the purine series, which we did not observe.

The synthesis of the guanosine analog was considered next (2), and the route envisaged is illustrated by Scheme II. 5-Chloro-1,2,4-triazolo(1,5-a)pyrimidin-7-one (5) was selected as our starting material, since the halogen was predicted to deactivate its neighboring nitrogen in glycosylation reaction, thereby

TABLE I   Comparison of [13]C Chemical Shifts for the Anion of 1,2,
4-Triazolo(1,5-a) pyrimidin-7-one and its Nucleosides 5 and 6.

| Compound | Chemical shift, δppm^a | | | | |
	C_2	C_5	C_6	C_7	$C_{b}3a$
A	152.5	154.1	97.2	160.1	158.3
5	140.8	153.3	104.0	156.8	149.6
6	151.6	138.9	100.2	155.4	143.1
ΔδA-5	11.7	0.8	-6.8	3.3	8.7
ΔδA-6	0.9	15.2	-3.0	4.7	15.2

^{a}Chemical shifts are measured from DMSO-d_6, converted to
TMS scale using the relationship $\delta_{TMS} = \delta_{DMSO} +39.5$ ppm.

producing the requisite N-glycosyl derivative. Condensation of
the trimethylsilyl derivative of 5-chloro-1,2,4-triazolo(1,5-a)-
pyrimidin-7-one (7) with 2,3,5-tri-O-acetyl-D-ribofuranosyl bromide (9) in acetonitrile at room temperature furnished only one
isolable blocked nucleoside (10).   Deacetylation of 10 with methanolic ammonia furnished 5-chloro-3-(β-D-ribofuranosyl)-1,2,4-
triazolo(1,5-a)pyrimidin-7-one (18), which was shown by elemental
analysis to have retained the 5-chloro group.   Dehalogenation of
18 with palladium on carbon gave 3-(β-D-ribofuranosyl)-1,2,4-tri-
azolo (1,5-a)pyrimidin-7-one (12), identical in all respects with
5, thereby confirming the directive effect of the 5-chloro group
to give exclusively the N-3 glycosyl derivative.   Nucleophilic
substitution of the 5-chloro group of 18 gave some interesting nu
cleosides (13 to 17), including the guanosine analog, 5-amino-3-
(β-D-ribofuranosyl)-1,2,4-triazolo(1,5-a)pyrimidin-7-one (13).
Synthetic procedures, complete characterization of products, and
the PMR spectral trends of these nucleosides have been reported (
     A similar glycosylation of the O-trimethylsilyl-5-methyl-
1,2,4-triazolo(1,5-a)pyrimidin-7-one (8) gave the blocked nucleoside 11, which on subsequent deacetylation with methanolic ammonia furnished 5-methyl-3-(β-D-ribofuranosyl)-1,2,4-triazolo(1,5-a)
pyrimidin-7-one (19).
     Although the anomeric configuration of 18 could be tenta-
tively assigned as β on the basis of several empirical rules
(6,7), a more rigorous proof (Scheme III) was in order for this
unusual heterocyclic nucleoside series.   Isopropylidenation of

the 5-chloro nucleoside (18) with 2,2-dimethoxypropane and acetone in the presence of 70% perchloric acid gave nucleoside (20).

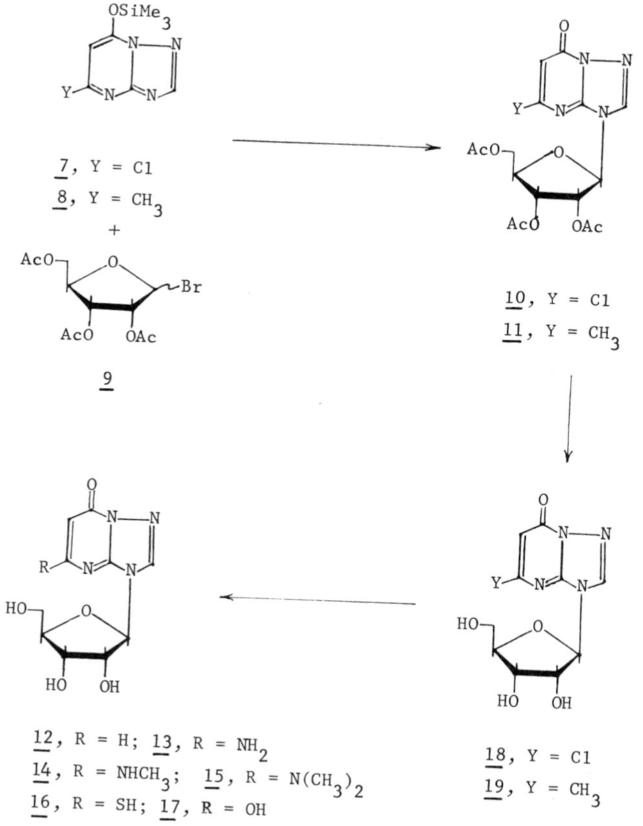

Scheme II

The ^{1}H NMR spectrum of 20 revealed a difference between the
chemical shift of the two methyl signals of the isopropylidene
group of 0.20 ppm, a difference characteristic for the β configuration (8,9). Further treatment of 20 with p-toluenesulfonyl
chloride in pyridine furnished the 5'-O-p-toluenesulfonyl-2',3'-
O-isopropylidene derivative (21). Treatment of 21 with DMSO or
acetonylacetone at 100-110°C for 2-4 hr did not produce the anticipated cyclonucleoside (23). This observation would indicate
that 18 has either the α-configuration (10) or that the N-4 (because of the adjacent electron-withdrawing chloro group) is not
nucleophilic enough to displace the 5'-tosylate. However, dehalogenation of 21 with palladium on carbon gave crystalline 22,
which when heated at 100°C for 4 hr in dimethyl sulfoxide effec-

ted the cyclonucleoside formation (24), thereby establishing the
anomeric configuration of 18, hence 12-17, as β. The identity
of 24 as the cyclonucleoside was confirmed by the presence of an
ionic sulfonate absorption in the infrared spectrum at 1200 cm^{-1}
and the drastic decrease in chromatographic mobility in nonpolar
solvents of 24 as compared to 22.

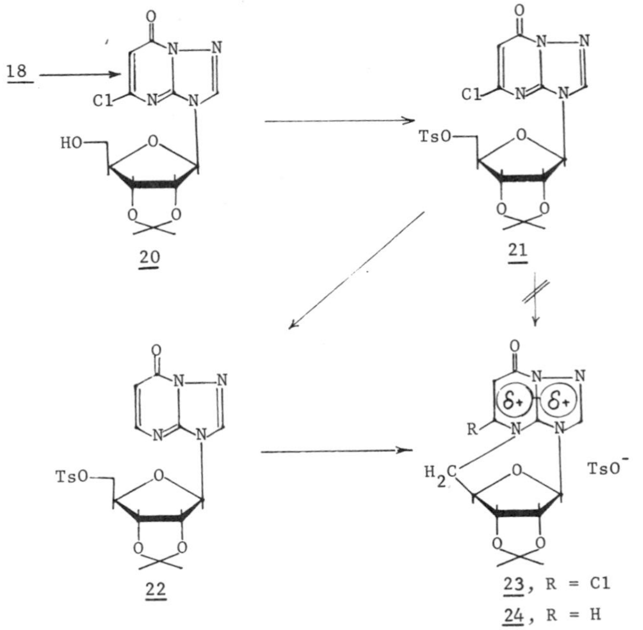

Scheme III

        Since the ability of the halogen atom to influence the site
of glycosylation had been established, several other halogen-sub-
stituted N-bridgehead heterocycles were considered for glycosylation studies to determine the regioselectivity. The imidazo
(1,2-c)pyrimidine ring system, which may be regarded as 3-deaza-
purine with a bridgehead nitrogen atom, was next elected for the
glycosylation studies (Scheme IV). The logical starting material
7-chloroimidazo(1,2-c)pyrimidin-5(6H)-one (11) (26) was prepared
in excellent yield by the ring annulation of 4-amino-6-chloropy-
rimidin-2-one (12) (25) with bromoacetaldehyde diethyl acetal in
aqueous media. Condensation of the O-trimethylsilyl derivative
of 26 (27) with 9 in acetonitrile at room temperature gave a
93% yield of an anomeric mixture of 7-chloro-1(2,3,5-tri-O-acetyl-
D-ribofuranosyl) imidazo(1,2-c)pyrimidin-5-one (31), which resis-
ted all attempts to separate the pure anomers by silicic acid
column chromatography. However, greater resolution of the anomers
was obtained by delaying the separation until after the deacetylation reaction. Deacetylation of 31 with methanolic ammonia

gave 7-chloro-1-(ß-D-ribofuranosyl)imidazo(1,2-c)pyrimidin-5-one (**30**) and the α-anomer **29**. These nucleosides were readily separated by a combination of fractional crystallization and column chromatography; the anomeric ratio being approximately 1:1. The formation of **29** and **30** established that the 7-chloro group was still present and also confirmed the apparent directive effect of the halogen group.

In an effort to improve the yield of the ß-anomer (**30**), we also examined the Friedel-Crafts catalyzed glycosylation procedure (13). Treatment of the trimethylsilyl derivative (**27**) in 1,2-dichloroethane with a poly-O-acetyl sugar in the presence of stannic chloride gave a 64% yield of 7-chloro-1-(2,3,5-tri-O-acetyl-ß-D-ribofuranosyl)imidazo(1,2-c)pyrimidin-5-one, which on subsequent deacetylation with methanolic ammonia furnished **30**.

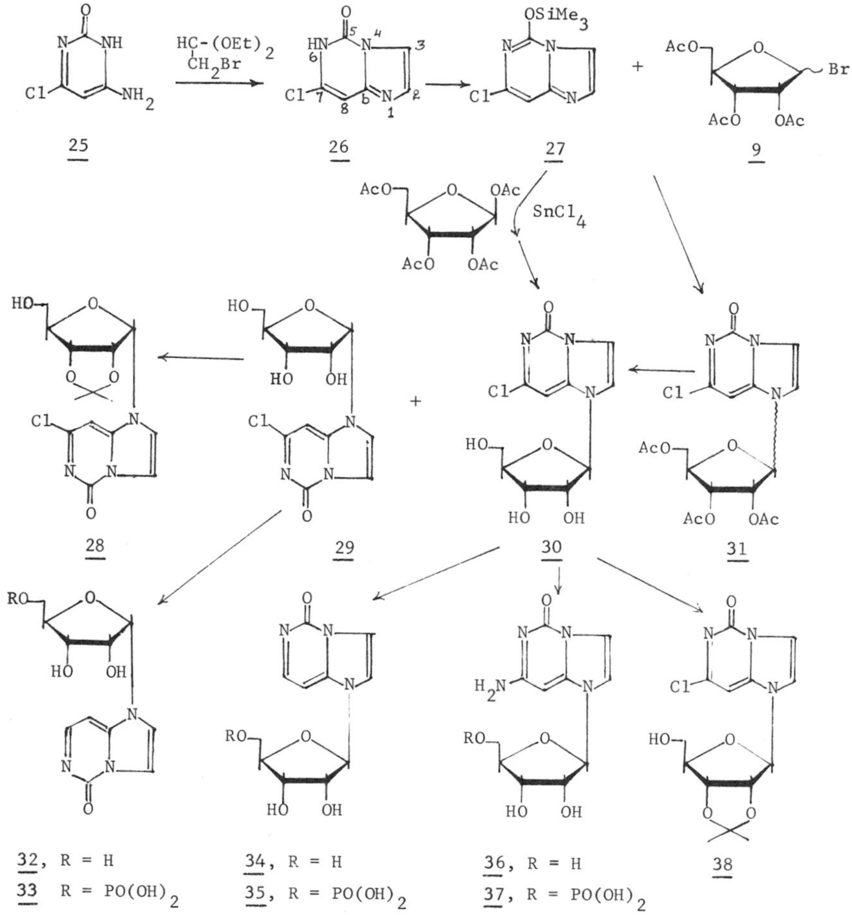

Scheme IV

The anomeric configurations of 29 and 30 were ascertained
by preparing the corresponding 2',3'-O-isopropylidene derivatives
(28 and 38). The ^{1}H NMR spectrum of 38 revealed the difference
between the chemical shift of the two methyl signals of the isopropylidene group to be 0.23 ppm, a difference characteristic of
the β configuration. Similarly, compound 28 revealed a chemical
shift difference of almost 0.0 ppm, indicating the α-configuration. The site of glycosylation was established (11) as N-1,
with the use of ^{13}C NMR, by observing the α,β, and γ carbon
shifts when compared to the heterocyclic base anion. It was
also substantiated by a comparison of the reported ultraviolet
absorption spectra of 6-methyl- (14) or 6-ribosyl- (15) imidazo
(1,2-c) pyrimidin-5-one with that of 32 or 34. These data were
dissimilar, and conclusively proved that the ribosylation has
not taken place at N-6.

Catalytic dehalogenation of 30 and 29 with palladium on carbon gave 1-(β-D-ribofuranosyl)imidazo(1,2-c)pyrimidin-5-one (34)
and the corresponding α-anomer (32) in high yields, respectively.
Treatment of 30 with methanolic ammonia at elevated temperature
and pressure furnished a good yield of 7-amino-1-(β-D-ribofuranosyl)imidazo(1,2-c)pyrimidin-5-one (36). Phosphorylation of unprotected 34 with phosphorus oxychloride in trimethyl phosphate
provided 1-(β-D-ribofuranosyl)imidazo(1,2-c)pyrimidin-5-one 5'-
monophosphate (35). Similar phosphorylation of 32 gave 1-(α-D-
ribofuranosyl)imidazo(1,2-c)pyrimidin-5-one 5'-monophosphate
(33). Application of this procedure to 36 provided 7-amino-1-
(β-D-ribofuranosyl)imadazo(1,2-c)pyrimidin-5-one 5'-monophosphate
(37).

In view of the interesting biological properties of the nucleosides derived from β-D-arabinofuranose (16), it was of considerable interest to synthesize the inosine and guanosine analogs of the β-D-arabinofuranosylimidazo(1,2-c)pyrimidines (17).
The synthesis of these analogs was accomplished (Scheme V) by a
direct glycosylation of the trimethylsilyl derivative 27 with
2,3,5-tri-O-benzyl-α-D-arabinofuranosyl chloride (39) (18) in
benzene. After silicic acid column chromatography a 92% yield
of crystalline 7-chloro-1-(2,3,5-tri-O-benzyl-β-D-arabinofuranosy
imidazo(1,2-c)pyrimidin-5-one (40) was obtained. A careful investigation of the mother liquor furnished chromatographic evidence for the presence of another nucleoside, which was not isolated, and is presumably the anomeric 7-chloro-1-(2,3,5-tri-O-
benzyl-α-D-arabinofuranosyl)imidazo(1,2-c)pyrimidin-5-one. This
nucleoside occurs in a very small amount. Catalytic dehalogenation of 40 with palladium on carbon gave 41 in excellent yield.
Hydrogenolysis of 41 over palladium in 2-methoxyethanol furnished
1-(β-D-arabinofuranosyl)imidazo(1,2-c)pyrimidin-5-one (43). Compound 43 can also be directly obtained by a reductive hydrogenolysis of 40. Treatment of 40 with methanolic ammonia at an
elevated temperature and pressure gave a 42% yield of 42, after

silicic acid column chromatography. Subsequent hydrogenolysis
of <u>42</u> furnished 7-amino-1-(β-D-arabinofuranosyl)imidazo(1,2-*c*)
pyrimidin-5-one (<u>44</u>).

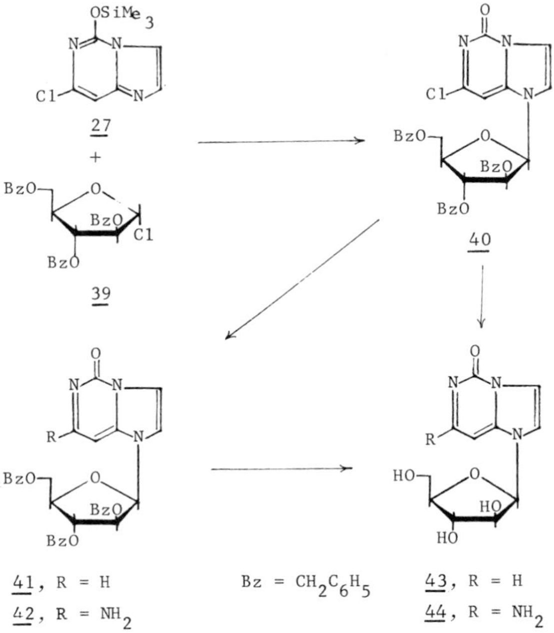

Scheme V

The site of glycosylation for these arabinosides (<u>43</u> and <u>44</u>)
was assigned as N-1 by a direct comparison of their ultraviolet
absorption spectra, which are superimposable, with the uv spectral data of the ribosides of this ring system (<u>34</u> and <u>36</u>), the
structures of which have been unequivocally established. The
anomeric configuration was unequivocally established as β by
subjecting <u>43</u> to periodate oxidation followed by reduction with
sodium borohydride (18).

Glycosylation studies of the imidazo(1,2-*a*)pyrimidine ring
system were undertaken next (Scheme VI). For the aforementioned
reasons 7-chloroimidazo(1,2-*a*)pyrimidin-5-one (<u>48</u>) was selected
as the starting material for the proposed nucleoside synthesis
and was prepared by a three-step reaction from 2-aminoimidazole.
Condensation of 2-aminoimidazolium sulfate (<u>45</u>) with diethyl
malonate in the presence of sodium ethoxide in ethanol produced
5,7-dihydroxyimidazo(1,2-*a*)pyrimidine (<u>46</u>) (<u>19</u>), which on
subsequent chlorination with phosphorus oxychloride gave
5,7-dichloroimidazo(1,2-*a*)pyrimidine (<u>47</u>) (20) in good yield.
Treatment of <u>47</u> with 5% aqueous sodium hydroxide solution gave
the requisite 7-chloroimidazo(1,2-*a*)pyrimidin-5-one (<u>48</u>). Coupling of the trimethylsilyl derivative (<u>54</u>) with the poly-*O*-ace-

tylated sugar bromide (9) in acetonitrile gave a good yield of
syrupy 7-chloro-1-(2,3,5-tri-O-acetyl-β-D-ribofuranosyl)imidazo
(1,2-a)pyrimidin-5-one (52) which on deacetylation with methanolic ammonia at room temperature furnished the free nucleoside
(53). Dehalogenation of 53 with palladium on carbon gave 1-
(β-D-ribofuranosyl)imidazo(1,2-a)pyrimidin-5-one (50). Treatment of 53 with methanolic ammonia at an elevated temperature and
pressure furnished 7-amino-1-(β-D-ribofuranosyl)imidazo(1,2-a)py-
rimidin-5-one (51). The anomeric configuration of these nucleosides was assigned as β on the basis of the observed difference
(0.20 ppm) between the chemical shift of the two methyl signals

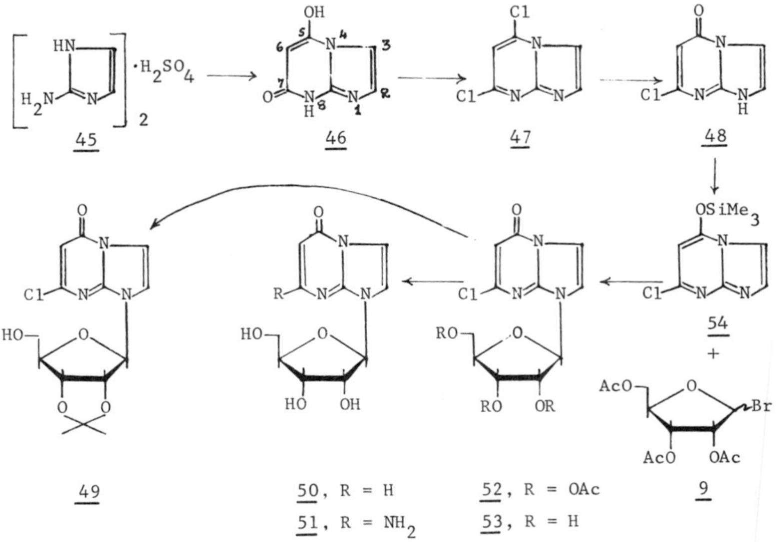

Scheme VI

of the isopropylidene group (8,9) in 49. The site of glycosylation for 53 was determined using ^{13}C NMR and observing the α,β,
and γ carbon shifts. A comparison of ^{13}C chemical shifts between the heterocyclic base anion and 53 revealed upfield shifts
of 9.9 and 4.9 ppm for C_2 and C_{8a} and a downfield shift of -1.5
ppm for C_3 establishing N-1 is the site of glycosylation. The
downfield shift of -1.9 ppm for C_7 also indicates that glycosylation had not occurred at N-8.

After these successful glycosylation studies with polyaza-
indenes to provide some novel inosine and guanosine analogs, we
have extended our investigations to other closely related ring
systems, particularly the pyrazolo(1,5-a)pyrimidines (Scheme VII)
Glycosylation of the trimethylsilyl derivative (54) of 7-hydroxy-
pyrazolo(1,5-a)pyrimidine (21) with 2,3,5-tri-O-acetyl-D-ribosyl
bromide (9) in acetonitrile at room temperature gave a 76% yield

of the blocked nucleoside, which was subsequently deacetylated with methanolic ammonia to furnish 4-(β-D-ribofuranosyl)pyrazolo (1,5-a)pyrimidin-7-one (56). A similar glycosylation of the O-trimethylsilyl derivative (55) of 5-methylpyrazolo(1,5-a)pyrimidin-7-one (22) followed by deacetylation gave 5-methyl-4-(β-D-ribofuranosyl)pyrazolo(1,5-a)pyrimidin-7-one (57).

Evidence that the glycosylation of this ring system had occurred at the N-4 position was obtained by comparing the ^{1}H NMR spectra of 56 and 57 with those of several related heterocycles. It has been observed that *N*-methylation results in an increase in the coupling constant between protons on neighboring carbons (Table II). In the case of N-1 methyl compound, the

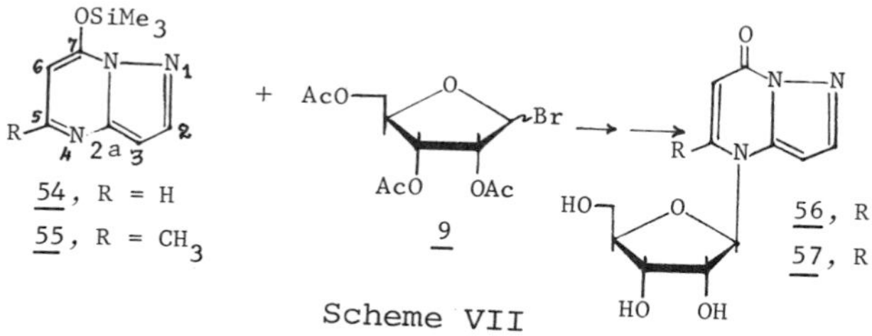

OS iMe$_3$
N—N$_1$
R N $_{2a3}$
54, R = H
55, R = CH$_3$

+ AcO— O —Br
AcO OAc
9

→

O
N—N
R N
HO O
HO OH
56, R = H
57, R = CH$_3$

Scheme VII

H$_{2-3}$ coupling constants are of the order of 3.5 Hz, whereas N-4 methylation results in smaller couplings of 2.0 Hz (21,23). Since ribosylation or methylation is expected to produce similar inductive effects, and the fact that the observed $J_{H_{2-3}}$ in 56 and 57 is only 2.0 Hz leads to the conclusion that N-4 is the site of glycosylation. Additional conclusive support for the ribosylation site by ^{13}C NMR is also available (4).

TABLE II   Some Typical Changes in Proton Coupling Constants as a Result of *N*-Methylation

Compound	J_H
1,5-Dimethylprazolo(1,5-a)pyrimidin-7-one	3.55
1,2-Dimethylpyrazol-3-one	3.50
4,5-Dimethylpyrazolo(1,5-a)pyrimidin-7-one	2.05
4-(β-D-Ribofuranosyl)pyrazolo(1,5-a)pyrimidin-7-one (56)	2.00
5-Methyl-4-(β-D-ribofuranosyl)pyrazolo(1,5-a)pyrimidin-7-one (57)	2.00

Substitution of N for CH at C-6 position in pyrazolo(1,5-a) pyrimidine ring gives an other heterocyclic system: pyrazolo(1, 5-a)-1,3,5-triazine (Scheme VIII). Ribosylation of the trimethylsilyl derivative (58) of 2-methylthiopyrazolo(1,5-a)-1,3,5-triazin-4-one (24) with 2,3,5-tri-O-acetyl-D-ribofuranosyl bromide (9) gave an 83% yield of a nucleoside material, which was subsequently identified as 2-methylthio-3-(2,3,5-tri-O-acetyl-β-D-ribofuranosyl) pyrazolo(1,5-a)-1,3,5-triazin-4-one (59). The structure was assigned on the basis of ultraviolet absorption comparison (24) and ^{1}H NMR chemical shift data of the dethiated product, 3-(2,3,5-tri-O-acetyl-β-D-ribofuranosyl)pyrazolo(1,5-a)-1,3,5-triazin-4-one (60). The J_{H7-8} value of 2.0 Hz for 60 or the deacetylated nucleoside 61 enabled (*vide supra*) the possibility of ribosylation at N-6 to be ruled out. Conclusive evidence for the glycosylation site as N-3 was obtained by ^{13}C NMR. As has been observed in all of the nucleosides prepared for this study, the large positive α shifts observed for C_2 and C_4 (4.6 and 10.5 ppm) for 59 are consistant with the assigned structure. The detailed discussion about the glycosylation site in 59 is available (4).

Preliminary *in vitro* antiviral and antimicrobial evaluation indicated that these new classes of synthetic inosine and guanosine analogs with a bridgehead nitrogen atom are devoid of any significant activity.

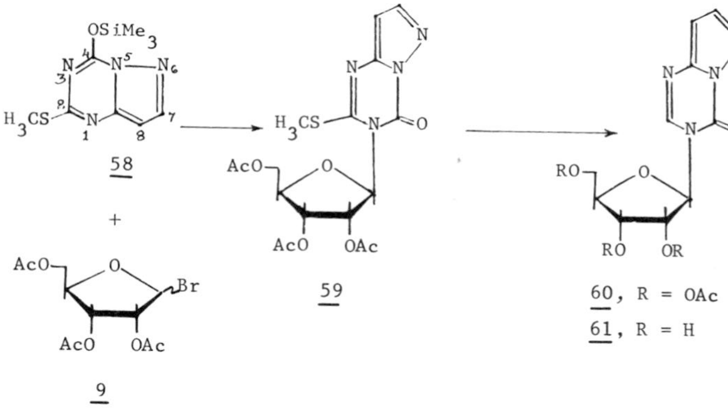

Scheme VIII

ACKNOWLEDGMENTS

    The authors wish to gratefully acknowledge the assistance, counsel, and pleasant association of Drs. Richard L. Tolman, Phoebe Dea, and David G. Bartholomew.

REFERENCES

1. Winkley, M. W., Judd, G. F., and Robins, R. K., *J. Heterocycl. Chem.* *8*, 237 (1971).
2. Revankar, G. R., Robins, R. K., and Tolman, R. L., *J. Org. Chem.* *39*, 1256 (1974).
3. Makisumi, Y., and Kano, H., *Chem. Pharm. Bull.* 7, 907 (1959).
4. Dea, P., Revankar, G. R., Tolman, R. L., Robins, R. K., and Schweizer, M. P., *J. Org. Chem.* *39*, 3226 (1974).
5. Makisumi, Y., *Chem. Pharm. Bull.* 9, 801 (1961).
6. Baker, B. R., *Ciba Found. Symp. Chem. Biol. Purines* 120 (1957).
7. Karplus, M., *J. Chem. Phys.* *30*, 11 (1959).
8. Imbach, J. L., Barascut, J. L., Kam B. L., Rayner B., Tamby C., and Tapiero, C., *J. Heterocycl. Chem.* *10*, 1069 (1973).
9. Imbach, J. L., Barascut, J. L., Kam, B. L., and Tapiero, C., *Tetrahedron Lett.* 129 (1974).
10. Clark, V. M., Todd, A. R., and Zussman, J., *J. Chem. Soc.* 2952 (1951).
11. Bartholomew, D. G., Dea, P., Robins, R. K., and Revankar, G. R., *J. Org. Chem.* *40*, 3708 (1975).
12. Israel, M., Protopapa, H. K., Schlein, H. N., and Modest, E. J., *J. Med. Chem.* 7,5 (1964).
13. Niedballa, U., and Vorbrüggen, H., *Angew. Chem. Int. Ed. Engl.*,*9* 461 (1970).
14. Kochetkov, N. K., Shibaev, V. N., and Kost, A. A., *Tetrahedron Lett.* 1993 (1971).
15. Barrio, J. R., Secrist, J. A., and Leonard, N. J., *Biochem. Biophys. Res. Commun.* *46*, 597 (1972).
16. Revankar, G. R., Huffman, J. H., Allen, L. B., Sidwell, R. W., Robins, R. K., and Tolman, R. L., *J. Med. Chem.* *18*, 721 (1975) and references cited therein.
17. Bartholomew, D. G., Huffman, J. H., Matthews, T. R., Robins, R. K., and Revankar, G. R., *J. Med. Chem.* *19*, 814 (1976).
18. Glaudemans, C. P. J., and Fletcher, H. G., Jr., *J. Org. Chem.* *28*, 3004 (1963).
19. Rao, R. P., Robins, R. K., and O'Brien, D. E., *J. Heterocycl. Chem.* *10*, 1021 (1973).
20. Revankar, G. R. and Robins, R. K., *Ann. N. Y. Acad. Sci.* *255*, 166 (1975).
21. Reimlinger, H., Peiren, M. A., and Merenyi, R., *Chem. Ber.* *103*, 3252 (1972).
22. Makisumi, Y., *Chem. Pharm. Bull.* *10*, 612 (1962).
23. Dorn, H., and Zubek, A., *J. prakt. Chem.* *313*, 969 (1971).
24. Kobe, J., Robins, R. K., and O'Brien, D. E., *J. Heterocycl. Chem.* *11*, 199 (1974).

DETERMINATION OF THE SITE OF GLYCOSYLATION
IN VARIOUS NUCLEOSIDES BY CARBON-13 NMR SPECTROSCOPY

PHOEBE DEA* AND ROLAND K. ROBINS

ICN Pharmaceuticals, Inc.,
Nucleic Acid Research Institute
Irvine, California

ABSTRACT

High-resolution, Fourier transform carbon-13 nuclear magnetic
resonance spectroscopy has been used to study the effect of N-
ribosylation on the chemical shifts of neighboring carbons in
various nucleosides. In heteroaromatic systems, N-ribosylation
has been observed to produce an upfield shift in the carbon-13
NMR signal of the carbon α to the substituted nitrogen and a
downfield shift in the signal of the carbon β to that nitrogen
when the neutral species is compared with the corresponding
anion. The use of these shift parameters provides a convenient
general method for the assignment of ribosylation site in nitrogen heterocycles where several possible sites are available.
The efficiency of this method is extremely advantageous as it
may be used in lieu of the more tedious chemical procedures requiring chemical synthesis.

Determination of the position of attachment of the sugar moiety
to heterocyclic bases in nucleosides where several ribosylation
sites are available is sometimes a most difficult problem. In
cases where related methylated bases of known structures are
available, the site of ribose substitution on the heterocyclic
base can often be located by comparison with the ultraviolet
spectrum of the nucleoside product with that of the appropriate
N-alkyl derivative of the aglycon. A comparison of the extinction
coefficient values and wavelengths of maxima and minima under

*Present address: Chemistry Department, California State
University at Los Angeles, Los Angeles, California 90032.

different pH conditions is usually sufficient to establish the
structure (1); although recently, in the case of 1-β-D-ribofur-
anosyl-s-triazolo(4,5-c)pyridine and 1-methyl-s-triazolo(4,5-c)
pyridine, it has been reported that the empirical model methyl
rule does not apply (2). Ultimate proof of the structure often
involves X-ray analysis of the single crystal or synthesis by
unambiguous means.

With the recent advances in instrumentation in carbon-13
NMR spectroscopy, in particular, the development of pulsed Fourier transform NMR techniques, and the use of wide-band proton
decoupling, there has been an increasing use of carbon-13 NMR
in structural and stereochemical applications (3). The chemical
shifts of carbon-13 nuclei occur over a range of more than 200ppm
saturated carbons appear over a broad range of approximately
45 ppm (δ-2 to 43), olefinic carbons absorb over a span of 65 ppm
in the δ range 100-165, and the sp carbons absorb in a region
intermediate between that of alkanes and alkenes (δ=67-92). Car-
bon-13 NMR spectroscopy is particularly applicable in the structure determination of heterocyclic compounds, where very few protons and in some cases none are present in highly substituted
heterocycles such that the proton NMR spectra yield little information. Several NMR studies in this laboratory (4-8), as
well as others performed elsewhere (2,9-12) have demonstrated
the potential use of carbon-13 NMR spectroscopy as a general unequivocal method for the assignment of glycosylation sites in
heteroaromatic compounds. The assignments were based upon the
reports by Pugmire and Grant (13-15) that when the free pair of
electrons in a nitrogen heterocycle is protonated, an upfield
shift was observed for the carbon α to the protonated nitrogen,
while downfield shifts were observed for the β and γ carbons.
From results of theoretical calculations on these systems, the
observed α-substitution shifts have been explained on the basis
of a decrease in bonding between the nitrogen and the α-carbon,
while β and γ shifts are the result of charge polarization
effects (13). Similar chemical shift changes were observed in
the heterocyclic systems when the free pair electrons on the nitrogen in the anions are protonated (14).

The first use of carbon-13 NMR for assignment of the site
of glycosylation was reported by this laboratory in 3- and 5-
substituted 1-β-D-ribofuranosyl-1,2,4-triazoles (4,16). Several
of these compounds have been shown to exhibit broad antiviral
action in tissue culture and in animal systems (17). As summarized in Table I, the chemical shifts of the base carbons of the
nucleosides were compared to those of the 1,2,4-triazole-3-car-
boxamide anion (1) formed by neutralization of the triazole by
lithium hydroxide in DMSO-d6. In the case of 1-β-D-ribofuranosyl-
1,2,4-triazole-3-carboxamide (2,ribavirin), the α-carbon (C-5)
was shifted downfield by 1.9 ppm as compared to the ionized heterocyclic base. α- and β-substitution shifts of +7.7 and -0.6 ppm

TABLE I.   Carbon-13 Chemical Shifts of 3- and 5-Substituted
1-β-D-Ribofuranosyl-1,2,4-Triazoles

| Compound | Chemical shift, δ, ppm[a] | | |
	C-3	C-5	-C(=O)-
1	156.5	150.9	164.9
2	158.4	146.3	161.8
3	151.5	148.8	160.0
Δδ1-2	-1.9	+4.6	
Δδ1-3	-0.6	+7.7	
R = β-D-Ribofuranosyl			

[a]*Carbon-13 NMR spectra of 20% DMSO-d$_6$ solutions were obtained on a Bruker HX-90 NMR spectrometer operating at 22.62
MHz in the Fourier transform mode at a probe temperature of
30°C. Chemical shifts are measured relative to DMSO-d$_6$, con-
verted to TMS scale using the relationship $\delta_{TMS} = \delta_{DMSO-d_6} + 39.5$ ppm.*

were observed for the 5-substituted isomer (3). The substitution
parameters observed here are found to be comparable to the previously reported α- and β-protonation parameters of +9.04 and
-1.59 ppm, respectively, for the five-membered azines (14) and
the corresponding values of +7.8 and -4.4 ppm reported for the
six-membered azines (13). The assignment of the structure of
ribavirin (2) was later confirmed by single-crystal x-ray studies
by Prusiner and Sundaralingam (18).

We also studied certain imidazole nucleosides, which are
key intermediates for the synthesis of 3-deazaguanosine and 7-
β-D-ribofuranosyl-3-deazaguanine (8). The ribosylation sites
of methyl 5(4)-cyanomethylimidazole-4(5)-carboxylate were assigned using α- and β-methylation shifts reported for N-methyl
imidazole as an approximate model. The carbon-13 chemical
shifts of methyl 5-cyanomethylimidazole-4-carboxylate anion (4)
and its ribosylated derivatives (5,6) are summarized in Table II.

TABLE II.   Carbon-13 Chemical Shifts of 4- and 5-Substituted
Imidazole Anion and Its Nucleosides[a]

| | | Chemical shift | | |
| | | (δ, ppm) | | |
	Compound	C-2	C-4	C-5
4		143.5	124.8	136.4
5		136.8 (136.5)	128.0 (126.8)	130.7 (129.4)
6		139.9 (136.5)	138.4 (138.4)	118.4 (117.8)

R = 2,3,5-tri-O-acetyl-β-D-ribofuranosyl

[a]Values in parentheses are predicted chemical shifts using
α- and β-substitution shifts of +7 and -2 ppm, respectively.

A detailed discussion of the chemical shift assignments on these compounds has been reported (8). The predicted values of theoretical shifts shown in parentheses are computed from the chemical shift of the base anion *4* using a substitution shift parameter of +7 ppm upfield for the α-carbon and a downfield shift of −2 ppm for the β-carbon. These substitution shifts correspond to the chemical shift difference between *N*-methylimidazole and the imidazole anion (15). Reasonable agreement between the predicted and experimental values could be obtained only if the indicated assignment of the two isomers was used.

The use of carbon-13 NMR to assign the site of ribosylation in six-membered heterocyclic system has also been reported (5). In this report, the carbon-13 chemical shifts of the nucleoside of *as*-triazin-3 (4H)-one-1-oxide (7) were first compared to those of the heterocyclic base (*8*). As shown in Table III, the chemical shifts of the C-6 carbon before and after ribosylation are identical, while the C-5 carbon was shifted upfield by 3.8 ppm and the C-3 carbon downfield by 1.6 ppm. Since the C-3 carbon is α to the ribosylated nitrogen whether ribosylation had occurred at N-2 or N-4, the anomaly observed in the shift changes in the C-3 carbon was probably caused by the dynamic proton exchange among the N-H and O-H tautomeric species. The chemical structure of *7* can be established by comparing its carbon-13 chemical shifts with those of the base anion (*9*), where upfield α-shifts of 11.5 and 13.6 ppm were observed for the C-3 and C-5 carbons and a downfield β-shift of 2.0 ppm was observed for C-6.

Although the existence of tautomeric structures can complicate the analyses, Thurber *et al.* (9) were able to assign the structures for the methyl and ribosylated derivatives of N-substituted 1,2,3-triazole-4-carboxamide with the exclusive use of carbon-13 chemical shifts. Their results indicated that N-substitution (methyl or ribofuranosyl) of the neutral species produces shift changes qualitatively similar to the α- and β-protonation parameters. In another study, structural assignments for the isomeric N-1 and N-2 ribofuranosyl derivatives of tetrazoles were aided using this method (11, 12).

The applications for the assignment of glycosylation sites have been extended to fused-ring heteroaromatic compounds. Several nucleosides of heterocyclic fused-ring series have been assigned, namely: 7-amino-*v*-triazolo(4,5-*d*)pyrimidine, pyrazolo (1,5-*a*)pyrimidin-7-one, *s*-triazolo(1,5-*a*)pyrimidin-7-one, and 2-methylthiopyrazolo(1,5-*a*)*s*-triazin-4-one (6). It was shown that large upfield α shifts and small downfield β shifts were preserved in these complex fused-ring systems when the nucleosides were compared to the anion of the heterocycle. Only one exception was observed in the β shift of 7-amino-2-β-D-ribofuranosyl-*v*-triazolo(4,5-*d*)pyrimidine, where a bridgehead carbon was in-

TABLE III. Carbon-13 Chemical Shifts of *as*-Triazine 1-Oxides

	Compound	*Chemical shift (δ, ppm)*		
		C-3	*C-5*	*C-6*
9		164.9	152.2	118.9
8		151.8	142.4	120.9
7		153.4	138.6	120.9
	R = β-D-ribofuranosyl			
	Δδ 8 - 7	-1.6	+3.8	0.0
	Δδ 9 - 7	+11.5	+13.6	-2.0

volved, but analyses of the carbon-13 chemical shifts alone
were sufficient to establish the glycosylation site.

With the concomitant absence of tautomeric structures it
has been shown (10) that the structures for the isomeric N-1
and N-2 methyl derivatives of 5-amino-3,4-dicyanopyrazole can
be assigned by comparing the carbon-13 chemical shift of the
free base and the N-alkylated species, not requiring a comparison with the corresponding anionic species. However, it is difficult to predict the magnitude of the chemical shift changes
(the N-substituted compound with the free base).

In heterocyclic compounds that possess C_2 symmetry, the glycosylation site may be established by measuring the carbon-13 chemical shifts of the nucleoside alone. Symmetry considerations were applied in the assignment of glycosylation site in naphtho (2,3-*d*)triazole-4,9-dione (19). Of the two possible structures (N-1 and N-2 isomers), the quinone portion of the nucleoside would exhibit C_2 symmetry only for the 2-β-D-ribofuranosylnaphtho (2,3-*d*)triazole-4,9-dione structure (10). Table IV shows the chemical shifts of the quinone carbons of 10. The downfield resonance at 176.6 ppm is assigned to the keto carbons C-4 and C-9. The bridgehead carbons C-a, C-a' and C-b, C-b' can be differentiated from the remaining carbons in the proton-coupled spectra since no protons are bonded to the bridgehead carbons. The C-a and C-a' carbons are assigned to the downfield resonance

TABLE IV. Carbon-13 Chemical Shift of 2-β-D-Ribofuranosylnaph-tho(2,3-*d*)Triazole-4,9-Dione (10)

Compound	Chemical shift (δ, ppm)				
	C-4 C-9	C-a C-a'	C-b C-b'	C-5 C-8	C-6 C-7
10	176.6	145.2	133.6	134.6	126.9

R = β-D-Ribofuranosyl

at 145.2 ppm as they are bonded to nitrogen atoms (20), while the C-b and C-b' resonances can be assigned by elimination. The C-a and C-a' carbons, as well as all the carbon pairs along the C_2 axis in 10, i.e., C-a, C-a'; C-4, C-9; C-5, C-8; and C-6, C-7, have identical chemical shifts. If ribosylation had

occurred at N-1, the C-a carbon would be expected to occur at a
considerably higher field than the C-a' carbon, since it is α
to the ribosylated nitrogen. It is therefore apparent that the
quinone carbons of the triazole nucleoside 10, which have identical chemical shifts, are chemically equivalent and that their
shieldings are not accidentally degenerate. These results establish that ribosylation must have occurred at the N-2 position,
resulting in C_2 symmetry for the quinone portion of 10.

Another example of the convenience of symmetry considerations
is provided in the structure assignment of 1-β-D-ribofuranosyl-
imidazo(4,5-*d*)pyridazine-4,7(5H,6H)-dione (11). In the ionized
heterocyclic base imidazo(4,5-*d*)pyridazine-4,7(5H,6H)-dione (12),
the C-4 and C-7 carbonyl carbons, as well as the bridgehead carbons C-3a and C-7a are expected to be identical from symmetry
considerations. The chemical shifts of the nucleoside (11) are
shown in Table V (21). The C-2 carbon can be readily identified
from the proton-coupled spectrum and the carbonyl carbons can be
assigned downfield from the bridgehead carbons by comparison
with related compounds (22). The bridgehead carbons C-3a and

TABLE V.   Carbon-13 Chemical Shift of 1-β-D-Ribofuranosylimidazo
(4,5-*d*)Pyridazine-4,7(5H,6H)-Dione (11)

		Chemical shift			
		(δ, ppm)			
Compound	C-2	C-3a	C-4[a]	C-7[a]	C-7a
11	142.8	135.9	151.3	152.5	126.5

R = β-D-ribofuranosyl

[a]*Assignment tentative.*

C-7a are found to have a chemical shift difference of 9.4 ppm,
while the C-4 and C-7 carbons, though not unambiguously assigned,
exhibit very similar chemical shifts. Considerations of these
chemical shift differences led to the conclusion that one of

the bridgehead carbons must be α to the glycosylation site. The structure of 11 can therefore be established.

In summary, carbon-13 NMR spectroscopy has been found to be a very useful technique in the structural assignments of the nitrogen at which glycosylation has occurred. Examples of nucleosides that have been successfully studied include those containing imidazoles, 1,2,3- and 1,2,4-triazoles, *as*-triazines, and tetrazoles. Applications in fused-ring heteroaromatic systems have also been demonstrated. In most of these heterocyclic systems, measurement of the α and β substitution shifts alone are sufficient to establish the glycosylation site.

ACKNOWLEDGMENTS

We wish to thank Drs. J. T. Witkowski, P. D. Cook, and G. R. Revankar for providing compounds used inthis study, and for their interests and helpful discussions of this work.

REFERENCES

1. Townsend, L. B., Robins, R. K., Loeppky, R. N., and Leonard, N. J., *J. Am. Chem. Soc. 86,* 5320 (1964).

2. May, J. A. Jr., and Townsend, L. B., *J. Org. Chem. 41,* 1449 (1976).

3. Stothers, J. B., *Carbon-13 NMR Spectroscopy,* Academic Press, New York, 1972, p. 55.

4. Kreishman, G. P., Witkowski, J. T., Robins, R. K., and Schweizer, M. P., *J. Am. Chem. Soc. 94,* 5894 (1972).

5. Szekeres, G. L., Robins, R. K., Dea, P., Schweizer, M. P., and Long, R. A., *J. Org. Chem. 38,* 3277 (1973).

6. Dea, P., Revankar, G. R., Tolman, R. L., Robins, R. K., and Schweizer, M. P., *J. Org. Chem. 39,* 3226 (1974).

7. Bartholomew, D. G., Dea, P., Robins, R. M., and Revankar, G. R., *J. Org. Chem. 40,* 3708 (1975).

8. Cook, P. D., Rousseau, R. J., Mian, A. M., Dea, P., Meyer, R. B. Jr., and Robins, R. K., *J. Am. Chem. Soc. 98,* 1492 (1976).

9. Thurber, T. C., Pugmire, R. J., and Townsend, L. B., *J. Heterocycl. Chem.* 645 (1974).

10. Earl, R. A. Pugmire, R. J., Revankar, G. R., and Townsend, L. B., *J. Org. Chem. 40,* 1822 (1975).

11. Poonian, M. S., Nowoswiat, E. F., Blount, J. F., Williams, T. H., Pitcher, R. G., and Kramer, M. J., *J. Med. Chem. 19,* 286 (1976).

12. Poonian, M. S., Nowoswiat, E. F., Blount, J. F., and Kramer, M. J., *J. Med. Chem. 19,* 1017 (1976).
13. Pugmire, R. J., and Grant, D. M., *J. Am. Chem. Soc. 90,* 697 (1968).
14. Pugmire, R. J. and Grant, D. M., *J. Am. Chem. Soc. 90,* 4232 (1968).
15. Pugmire, R. J., Grant, D. M., Robins, R. K., and Townsend, L. B., *J. Am. Soc. 95,* 2791 (1973).
16. Witkowski, J. T., Robins, R. K., Sidwell, R. W., and Simon, L. N., *J. Med. Chem. 15,* 1150 (1972).
17. Sidwell, R. W., Huffman, J. H., Khare, G. P., Allen, L. B., Witkowski, J. T., and Robins, R. K., *Science 177,* 705 (1972).
18. Prusiner, P., and Sundaralingam, M., *Nature (London), New Biol. 244,* 116 (1973).
19. Pickering, M. V., Dea, P., Streeter, D. G. and Witkowski, J. T., *J. Med. Chem. 20,* 818 (1977).
20. Stothers, J. B., *Carbon-13 NMR Spectroscopy,* Academic Press, New York, 1972, p. 239.
21. Cook, P. D., Robins, R. K., and Dea, P., *J. Org. Chem.,* submitted for publication.
22. Johnson, L. F., and Jankowski, W. C., *Carbon-13 NMR Spectra,* Wiley, New York, 1972.

A "GEOMETRY-ONLY" ^{1}H NMR METHOD FOR
DETERMINATION OF THE ANOMERIC CONFIGURATION OF
RIBOFURANOSYL COMPOUNDS*

MORRIS J. ROBINS AND MALCOLM MACCOSS[†]

Department of Chemistry, The University of Alberta
Edmonton, Alberta, Canada

ABSTRACT

A review of previously described methods for determining the anomeric configuration of ribonucleosides, with apparent limitations and examples of exceptions is outlined. A new method based on the spin-spin coupling of the anomeric proton with H-2' of ribofuranosyl nucleoside-3',5'-cyclic monophosphates is presented. A sharp singlet ($J_{1'-2'} < 0.7$ Hz) for β-anomers and a doublet ($J_{1'-2'} \geq 3.5$ Hz) for α-anomers of the conformationally rigid trans-fused 6- to 5-membered ring system of these biologically significant 3',5'-cyclic nucleotides is observed.

Determination of the anomeric configuration of ribofuranosyl derivatives (2 and 3) has occupied a considerable amount of thought and research effort, since the most widely employed procedures for the synthesis of ribonucleosides (including naturally occurring components of ribonucleic acids (RNAs), nucleoside antibiotics, and a massive array of structural analogs) involve coupling of a derivative (1) of ribose with a heterocycle or ultimate heterocyclic precursor (1). The ribofuranosyl derivatives 2

*This work was generously supported by the National Research Council of Canada (A5890), the National Cancer Institute of Canada, and The University of Alberta.

†Post-Doctoral Fellow, The University of Alberta, 1972-1976. Present Address: Argonne National Laboratory, Argonne, Illinois.

and $\underline{3}$ are not technically anomers when Y is an acyclic carbon
fragment, but the problem of determination of configuration at
the asymmetric center, which was formerly C-1, may be viewed
identically.

$$\mathbf{1} \quad + \quad DS \quad \longrightarrow \quad \mathbf{2}\ (\beta) \quad \text{and}\ /\ \text{or} \quad \mathbf{3}\ (\alpha)$$

y = Displaced group
D = Heterocycle or precursor derivative
S = Proton or other displaced substituent

Early assignments of the anomeric configuration of nucleosides were made on the basis of Baker's 2'-ester participation
"trans-rule" (2) and Hudson's rules of isorotation (3). However, exceptions to both the trans-rule (4) and to Hudson's rules
(5) have been noted some time ago. In fact, exception to Hudson's rules appears quite general for most pyrimidine nucleosides
(vide infra).

Todd and co-workers (6) discovered a *truly* unequivocal
method involving chemical cyclonucleoside ($\underline{5},\underline{7}$) formation between
the β-anomeric heterocyclic moiety and the 5'-carbon of 2',3'-*O*-
isopropylidene nucleosides ($\underline{4a}$, 6). Such intramolecular cyclization to C-5' is sterically impossible with α-anomers. Unfor-
tunately, such "anhydronucleoside" ($\underline{5},\underline{7}$) formation is precluded
in cases such as 3-deazaadenosine ($\underline{4b}$) (7) or pyrimidine nucleosides such as $\underline{8}$ with no 2-substituent.

Ulbricht and co-workers determined ultraviolet optical rotatory dispersion spectra of a large number of nucleosides. They
noted that the "B_{2u}" $\pi{\rightarrow}\pi{*}$ transition (usually contained in the
lowest energy envelope) of purine nucleoside β-anomers generally
exhibits a negative Cotton effect and that of the α-anomers a
positive Cotton effect (8). Pyrimidine nucleosides generally
exhibit a reversed trend (9) with β-anomers positive and α-anomers negative in the long-wavelength Cotton effects of ORD or
CD spectra. However, intrinsic transition dipole vector orientation in the heterocycle, glycosyl-heterocycle rotameric populations, transition coupling with other groups in the molecule,
and solvation effects, can change the sign as well as magnitude
of this B_{2u} Cotton effect (10). The reversed sign observed with
8-bromoguanosine (11) (presumed syn rotamer predominance) and
with 6-azauridine (12) and pseudouridine (5-β-D-ribofuranosylur-
acil) (presumed alteration of intrinsic transition dipole vector
directions) are "exceptions" that illustrate the care with which

this technique should be used in assignment of configuration to conformationally mobile molecules. The normally reversed long-wavelength Cotton effects for purine and pyrimidine nucleosides of like anomeric configuration further illustrate the departures from Hudson's rules to be expected at the sodium D line (*vide supra*).

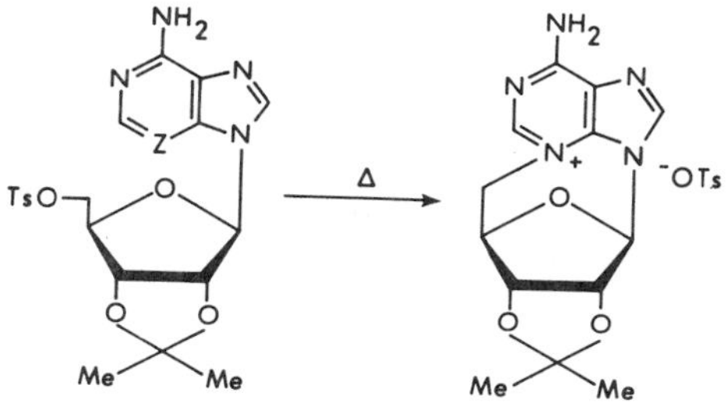

4a Z = N
b Z = CH

5

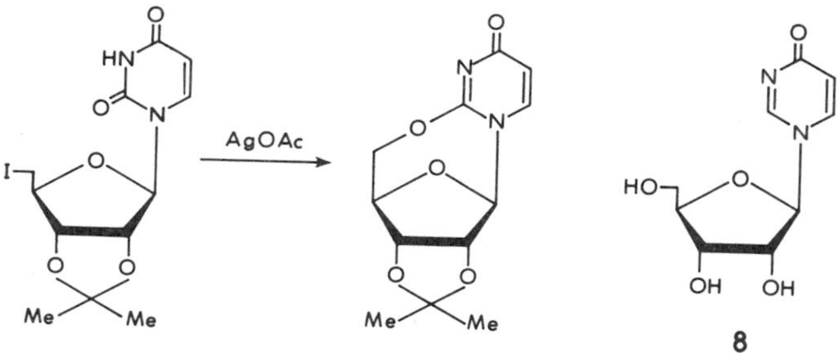

6

7

8

The *cis*-vicinal glycol system of anomeric ribofuranosyl nucleosides (9, 10) has been oxidized with periodate, and the resulting dialdehydes reduced with borohydride (13). The anomeric triols (11, 12) obtained are now enantiomeric, owing to the loss of chirality originally present at C-2', C-3', and C-4' of the precursor nucleoside. The signs and magnitudes of optical rotations of a number of reported triols (11, 12) follow the same trend, and this appears to have been an underutilized procedure for defining anomeric structures. However, in order to be rigorously diagnostic at the present time, both anomers are required and correlation with one known enantiomer should be made.

A number of nuclear magnetic resonance methods have been
investigated, which fall into four general categories. Spin-
spin coupling parameters were first utilized for direct analysis
of H-1' to H-2' dihedral angle limits. In the case of cyclopentane, limiting dihedral angles for the maximally puckered
C_S (envelope) conformation were calculated to be from 0 to 46°
for cis hydrogens and from 74 to 166° for trans hydrogens (14).
Using the Karplus relation (15) the corresponding limiting ran-
ges of $J_{1'-2'} \sim 3.5-8$ Hz for α-ribofuranosyl derivatives (cis hydrogens) and $J_{1'-2'} \sim 0-8$ Hz for the β-anomers (trans hydrogens)
were estimated (16). It was recognized (17) that there were inherent uncertainties in the Karplus relation and a limit of $\lesssim 1$
Hz was suggested for assignment of the β configuration (16,18).
Of course, no assignment is possible for α-anomers since the coupling range is overlapped. It has now come to be generally
accepted that a coupling constant of $J_{1'-2'} \leq 1$ Hz is "unequivocal" proof of β-anomeric configuration (19).

Although this is normally the case, we have observed two
examples that represent formal exceptions to this limit. The
not unexpected flattening and conformational restriction of the
furanose ring by formation of the 2',3'-ribo-epoxide gave 2',3'-
anhydroadenosine (13a) and 2',3'-anhydrotubercidin (13b), with
$J_{1'-2'} < 0.7$ Hz (20). However, spectra of the corresponding
lyxo epoxides, 9-(2,3-anhydro-β-D-lyxofuranosyl)adenine (14a)
and 4-amino-7-(2,3-anhydro-β-D-lyxofuranosyl)pyrrolo(2,3-d)-
pyrimidine (14b), also contain sharp singlets ($J_{1'-2'} < 0.7$ Hz)
for the anomeric proton resonance (20b, 21). Thus although
these lyxo epoxides are unusual structures, owing to the three-
membered ring strain rehybridization at C-2', they do represent
technical exceptions to the $J_{1'-2'} \leq 1$ Hz coupling limit for de-

termining trans hydrogens and emphasize the caution with which "unequivocal" configuration assignments should be made in furanose derivatives.

13 a,b   **14 a,b**

a B = Adenin – 9 – yl
b B = 4 – Aminopyrrolo [2,3 – d] pyrimidin – 7 – yl

Leonard and Laursen employed formation of 2',3'-O-isopropylidene derivatives of ribonucleosides to restrict the conformational mobility of the unblocked furanose ring (22). Although the β-anomers usually show a reduction in the anomeric proton coupling constant, values in the range of $J_{1'-2'}$ = 2-3.5 Hz are not uncommon and result in uncertain assignments (especially if only one anomer is present).

A second general category involves chemical shift or chemical shift difference measurements. Nishimura and Shimizu noted that the anomeric protons of trans (2'-OH to 1'-heterocycle) pentofuranosyl nucleosides resonate at higher magnetic fields than the corresponding cis compounds (23). For the anomeric pair of ribofuranosyl derivatives (15) and (16), this effect can be seen to result from the diamagnetic shielding of the cis 2'-hydroxyl group on H-1' of the trans substituted β-anomer (15) (24). This effect has proven to be very useful and consistent. However, both anomers are required in order to evaluate the relative shifts of the anomeric proton resonances unless very carefully defined shift data are known for one anomer (e.g., the anomeric protons of 15 and 16, where B = thiazolo(5,4-d)pyrimidine-5,7-dion-4-yl were reported to resonate at δ 6.11 and 6.17 respectively (25)). It has recently been stated that "the chemical shift of a β anomer has without exception been found to appear upfield from the corresponding α anomer" (26). However, one technical exception has been noted by Acton and co-workers. An anomeric pair of 2', 3',-5'-tri-O-benzylarabinofuranosyl pyrazole C-nucleoside intermediates were found to have "H-1' doublets identical in both chemical shift and coupling constant" (27), whereas a subsequently cyclized tri-O-benzyl pyrazolo(4,3-d)pyrimidine intermediate and the deblocked product both had the H-1' resonance of the β (cis) arabino anomer at lower field, as expected. Configurational assignments based on this method could be tenuous if anisotropic groups were present or, as noted previously, if only one anomer were available.

Fox and co-workers reported a method based on the difference in chemical shift of the 2'-*O*-acetyl methyl singlet upon
hydrogenation of a double bond in the heterocyclic base (28).
A small upfield shift of trans and a significant downfield shift
of cis 2'-*O*-acetyl protons was observed upon hydrogenation (this
has recently been confirmed for α-uridine (29)). The major technical limitation of this approach is that an easily and specifically reducible double bond in the base is required. Thus although it appears to be consistent for the pyrimidine (28a) and
model indole (28b) systems investigated, it is inapplicable to
purines and related heterocyclic nucleosides. It is also based
on an empirically observed anisotropy, which presumably requires
the "correct" preferred conformational populations. It would
thus be uncertain with a new system, especially if only one anomer were available.

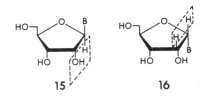

15   16

Montgomery proposed a direct measurement of the chemical
shift of 2'-*O*-acetyl methyl singlets. A series of acetylated
β-ribofuranosyl nucleosides had a peak at δ>2.05, whereas the
corresponding α-anomers had δ<1.95 for this resonance signal (30)
Imbach and Kam have noted an exception with the anomers of 1-
β-D-ribofuranosylbenzotriazole (31). It might be noted that even
in that case the signal (δ∿1-85) for the β-anomer derivative was
downfield from that of the α (δ∿1.60), even though the absolute
limits were displaced to higher fields (31). However, these signals were somewhat solvent dependent and the same cautions concerning this empirical observation apply as to the Fox *et al.*
method (28) (*vide supra*).

Imbach and co-workers have proposed a convenient and useful
approach based on the difference in chemical shift of the two
methyl groups of anomeric 2',3'-*O*-isopropylideneribofuranosyl
nucleosides (29, 31, 32). This proposed criterion (32a), as
refined by experience with further examples (31,32c), states
that α-anomers (<u>17</u>) will have a small chemical shift difference
between isopropylidene methyls (Δδ<0.15), whereas the β-anomers
(<u>18</u>) will exhibit a larger shift difference (Δδ>0.15). Two
technical exceptions to Imbach's rule involve the β uracil O^2→
$C^{5'}$ (<u>7</u>) (33) and C^6→$O^{5'}$ (<u>19</u>) (34) cyclonucleosides, both of which
had reported values of Δδ = 0.10. This chemical shift method is

again based on anisotropy effects resulting from proximity of the base (B) to the endo methyl group of the α-anomers (17). The cyclonucleosides (7 and 19) are trivial examples in terms of anomeric configuration since their formation demands the β orientation at C-1' (6) *(vide supra)*. However, they do emphasize the uncertainties in predicting anisotropy effects, and Imbach has suggested that 5'-substituents should not be present (32c). A more serious limitation involves the apparent requirement for a C-1' substituent with "correct" anisotropy. Moffatt and co-workers have reported an interesting study on the kinetic vs thermodynamic stereochemical mode of condensation of activated ribofuranosyl derivatives with aliphatic compounds to produce *C*-nucleoside precursors (35). "Anomeric" compounds 17 and 18 (B = CH$_2$CN) had Δδ = 0.16 and 0.18 (CDCl$_3$), respectively. Recourse to a single-crystal x-ray determination on the 5-*O*-*p*-bromobenzoyl derivative of 17 (B = CH$_2$CN) was necessitated in order to prove the unexpected "α-anomeric" sterochemistry of the product of thermodynamic control.

17   18

19

We have observed a similar failure of Imbach's rule with the α- and β-anomers of the 2',3'-*O*-isopropylidene-5,6-dihydrouridines. The α- and β-anomeric uridine derivatives (17 and 18, B = uracil-1-yl) have Δδ = 0.094 and 0.199 (DMSO-d_6), respectively, as predicted (32). However, upon saturation of the 5,6-double bond, the resulting 17 and 18 (B = 5,6-dihydrouracil-1-yl) have Δδ = 0.160 and 0.184 (DMSO-d_6), respectively. Since completion of this work, Imbach has reported essentially identical hydrogenation and chemical shift results with these α- and β-uridine derivatives (29). His studies also provided further evidence that the basis for the Δδ criterion is primarily the influence of the anisotropy of the anomeric substituent on the endomethyl of the α-anomers. He therefore noted that this criterion should be limited to only ribofuranosyl compounds having an unsaturated base as the aglycone and containing no 5'-substituents (29).

A third category involves miscellaneous applications of ^{1}H NMR spectroscopy. Cushley and co-workers (36) have directly applied the internal nuclear Overhauser effect (NOE), examined in nucleosides by others (37), to the determination of anomeric configuration. However, ambiguities arise in cases of overlapping sugar proton frequencies. The r^6 distance factor between irradiated and observed protons is dependent upon rotameric populations as well as other variable conformational effects in these flexible furanose molecules. Hall and Preston have observed a difference in the relaxation times (T_1) for α- and β-anomeric sugar protons (38). Although potentially applicable to nucleosides, this measurement requires an advanced level of sophistication in techniques and instrumentation.

A final fourth category involves ^{13}C NMR spectroscopy. Although a number of spectral applications of ^{13}C NMR have appeared, this method has been investigated in only a preliminary fashion with respect to the determination of anomeric structures (35, 39) and it would appear advisable at present to have both anomers available for comparison.

We have been interested in developing an ^{1}H NMR method for determination of the anomeric configuration of ribofuranosyl compounds that would be based solely on the geometry of substitution relative to the C-1'——C-2' bond. Such a method should be independent of anisotropy requirements, rotational conformation populations, solvent effects, and other difficultly predictable empirical considerations. The desirable criteria we considered include (a) formation of a geometrically constrained ring system of uniformly consistent structure, (b) sufficient restriction of the remaining conformational freedom to provide diagnostically consistent ^{1}H NMR spin-coupling parameters, (c) reasonably facile synthetic preparation, and/or (d) significant biological or other "spin-off" interest in the derivatives prepared.

Examination of molecular models indicated that trans fusion of a six-membered ring to the five-membered furanose would provide the structural rigidity and distinctly unique H-1' to H-2' geometries desired. The biologically ubiquitous and extensively studied (40) "second messenger" hormonal nucleoside-3',5'-cyclic monophosphates (20, c-AMP, c-GMP, c-CMP(?) (40c)) have potent biochemical activity, and their rigid trans-fused six- to five-membered ring systems have been established by single-crystal x-ray determinations (41) and NMR spectroscopy (42). Jardetzky observed the singlet resonance of the anomeric proton of adenosine-3',5'-cyclic monophosphate (20, c-AMP, B = adenin-9-yl) at 60 MHz and assigned a limiting coupling of $J_{1'-2'}$ < 1 Hz (42a). Molecular models of the β (20) and α (21) cyclic nucleotide structures indicate H-1' to H-2' dihedral angles of the order of ~105 and <46°, respectively. This corresponds to estimated coupling constants of $J_{1'-2'}$ < 0.7 Hz (β) and $J_{1'-2'}$ > 4 Hz (α), respectively (15, 43). Jardetzky noted that "the conformation

of the five-membered sugar ring in solution is determined solely
by steric requirements" (42) in the case of *c*-AMP. The ICN group
(44) and Moffatt and co-workers (45) have used this observation
for verifying the intact cyclic phosphodiester ring in products
of cyclization reactions of β 5'-mononucleotides and in numerous
heterocyclic transformation products of naturally occurring β
cyclic nucleotides (44, 46). However, the valuable potential of
this rigid fused ring system in determining the anomeric configuration of precursor nucleosides has remained unrecognized.

20

$J_{1'-2'} < 0.7$ Hz

(Sharp singlet)

21

$J_{1'-2'} \geq 3.5$ Hz

(Doublet)

We have examined a number of examples in the literature and
have synthesized several 3',5'-cyclic nucleotide α-anomers. We
find that examination of the well-separated anomeric proton NMR
signal of these straightforwardly prepared biomolecule analogs
indeed represents a rationally defined "geometry-only" method
for determining their anomeric sterochemistry. Over 150 examples
of the β structure (20) have been synthesized and studied by the
ICN group. These contain variously substituted purine (44, 46a-
e,g-n) pyrrolo(2,3-*d*)pyrimidine (46f), imidazole (46e,h),
imidazo(2,1-*i*)purine (46e), and imidazo(4,5-*d*)-*vic*-triazine
e,h), imidazo(2,1-*i*)purine (46e), and imidazo(4,5-*d*)-*vic*-triazine
(46h) heterocyclic base systems (B on 20). Especially significant for this discussion are a number of examples with large substituents at the 2 and/or 8 positions of the purine ring (44a, 46c
g-k,m,n) since these would be expected to exhibit the greatest
glycosyl torsion effects. Adenine N^1-alkylated and alkoxylated
derivatives (46e,n) test the sensitivity of the system to potential conformational distortion by affinity of the negatively
ionized phosphate for the cationic charge on the base. However,
in all cases examined, including the above extensive series (44,

46)* and the pyrimidine cyclic nucleotides evaluated by ourselves
and others (42b,e), a sharp singlet resonance was observed for
the anomeric proton *without exception*. Further support for the
rigidity of the basic trans-fused ring system of 20 is provided
by the cyclic phosphorothioates (22) (48) and the "5'-substitu-
ted" cyclic phosphate derivatives (23) (45). Although sulfur is
larger and has different bonding requirements than oxygen, resonance peaks for H-1' of 22a and 22b were reported to have "*J=ca.*
1 Hz" (48). This represents an upper limit for $J_{1'-2'}$ since significant coupling of phosphorus to H-1' ($^5J_{H}1'-^{31}P \sim$ 0.7 Hz) has
been recently demonstrated in examples of 20 (42e). An estimate
of the value of $J_{1'-2'}$ could be investigated by double resonance
irradiation at H-2'. Compounds 23 illustrate an even more severe
test of conformational rigidity. An axial methyl group is present on the six-membered cyclic phosphate "chair" of the L-talo
(23b) and dimethyl (23c) derivatives. The allo products (23a,d)
contain an equatorial methyl group and 23d an 8-bromo base substituent (favors syn glycosyl rotamer). Again, however, singlet

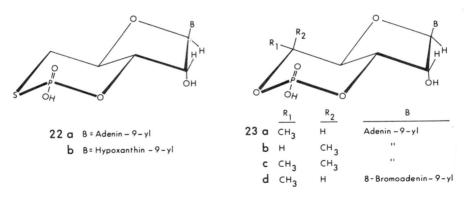

22 a B = Adenin – 9 – yl
   b B = Hypoxanthin – 9 – yl

23	R_1	R_2	B
a	CH_3	H	Adenin – 9 – yl
b	H	CH_3	"
c	CH_3	CH_3	"
d	CH_3	H	8 - Bromoadenin – 9 – yl

resonance peaks were reported for the anomeric protons of 23a-d
(45).
     A final series of β-ribofuranosyl compounds (24) contain O-
2' substituents that could have potential electronic and/or conformational effects. Some 29 examples of 2'-O- acetyl, butyryl,
N-substituted carbamoyl, methyl, 2,4-dinitrophenyl, and p-toluenesulfonyl esters, carbamates, and ethers with different bases
(B) have been evaluated (46b,d,i,l,m, 49, 50). The 2'-O-tosyl
derivatives (50) deserve special mention since they have a

*Spectral data are not reported in some of these quoted
papers. However, we have been informed that virtually every
compound was examined by ^{1}H NMR spectroscopy and that *no exceptio*
to the sharp singlet for H-1' has been observed (47).

strongly electronegative function at C-2' and also possess an
aromatic ring for potential stacking overlap with the base at
C-1' (51). However, a singlet resonance was observed for H-1'
of each of the O-2' derivatives (24) investigated.

24

$$R = -CH_3,$$

$$-NO_2, \quad -OSO_2 - \langle\rangle - CH_3,$$

$$-COR, \quad -CONHR$$

We have now synthesized several α-nucleoside-3',5'-cyclic
phosphates (25) from the corresponding unblocked α-nucleosides
(10). Direct phosphorylations using phosphoryl chloride in tri-
ethyl or trimethyl phosphate (52) proceed in high yields. In
most cases hydrolysis of the initially formed phosphorochlori-
date products and isolation of the α-nucleoside-5'-monophosphates
followed by cyclization to the α-cyclic nucleotides (25) using
N,N'-dicyclohexylcarbodiimide (DCC) in refluxing pyridine (53)
was employed. However, treatment of unblocked nucleosides with
trichloromethylphosphonic acid dichloride followed by cyclization
of the resulting 5'-O-trichloromethylphosphonates in strong base
(54a) or direct cyclization of certain intermediate 5'-O-phos-
phorochloridates in aqueous base (54b) represent new and experi-
mentally more convenient routes to the 3',5'-cyclic nucleotides.

As seen under structures 25a-e the anomeric proton resonances appear as doublets with $J_{1'-2'} \geq 3.5$ Hz. The coupling constants for H-1' of the α-cyclic nucleotides containing an unmodified base are slightly smaller, $J_{1'-2'} \sim 3.5$ Hz, than predicted,
J ($\phi=46°$) ~ 4 Hz (15, 43), for the "maximum distortion angle" in
a cyclopentane ring (14). Two plausible explanations are that
either (a) the steric and parallel dipolar repulsions of the cis
1'-base and 2'-hydroxyl substituents of these α-cyclic nucleotides force an alteration of C-1' to C-2' bond lengths and angles
resulting in H-1' to H-2' dihedral angles of $\phi \sim 50°$, and/or (b)
these effects cause this strained, electronegatively substituted

C-1'──── C-2' "ethane system" to depart (55) from the modified
(larger constants) Karplus relationship (43). Hydrogenation of
α-c-UMP (25c, $J_{1'-2'}$ = 3.5 Hz) gives the 5,6-dihydrouracil derivative (25d, $J_{1'-2'}$ = 5.0 Hz). The 5-Hz anomeric proton coupling constant observed for 25d corresponds to a calculated (43)
H-1'────H-2' dihedral angle of 41°. Hydrogenation of the 5,6-
double bond of uracil could reduce the dipolar character of the
C-1'────N-1 bond cis to O-2'. In addition, the rehybridization

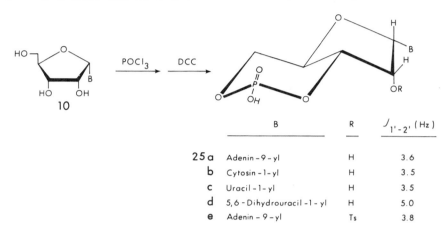

	B	R	$J_{1'-2'}$ (Hz)
25 a	Adenin-9-yl	H	3.6
b	Cytosin-1-yl	H	3.5
c	Uracil-1-yl	H	3.5
d	5,6-Dihydrouracil-1-yl	H	5.0
e	Adenin-9-yl	Ts	3.8

of the now aliphatic base provides a flexible cyclic substituent
at C-1', which can easily be rotated away from O-2' as indicated
by molecular models. These considerations are consistent with
a slight relaxation of the "maximally puckered" ^3E────$^3T_2$ conformation (41) range of the ribofuranose ring of the dihydro compound
(25d) relative to that of the other α-cyclic nucleotides. It is
significant to note that introduction of the electronegative and
aromatic ring-containing 2'-O-p-toluenesulfonyl function (25e)
actually increased ($J_{1'-2'}$ = 3.8 Hz) the anomeric proton coupling
constant slightly, relative to 25a ($J_{1'-2'}$ = 3.6 Hz).
      Several β-D-arabinofuranosyl nucleoside-3',5'-cyclic phos-
phates reported in the literature provide further examples of
cis H-1' to H-2' coupling. Twelve cyclic β-ara-nucleotides (26)
have been found to have anomeric coupling constants of $J_{1'-2'}$ =
6-7 Hz (50,56)*. Five 8-aza-, oxa-, and thia-8→2'-anhydro-β-D-
arabinofuranosylpurine-3',5'-cyclic nucleotides (27) have $J_{1'-2'}$
= 6.2-6.5 Hz (50). It is again on interest to note that these
coupling constants are smaller than predicted (43) for eclipsed
vicinal hydrogens ($J(\phi=0°)$ = 9 Hz). Apparently steric and/or

      *Anomeric couplings not reported were of similar magnitudes
(57).

dipolar repulsions of the cis oriented 1' and 2' substituents result in H-1' to H-2' dihedral angles of $\phi \sim$ 25–35° and/or correspondingly significant electronegativity, bond length, bond angle, etc., effects on the Karplus dependency (55).

26

B =

Z,

Z = NR, O, S

27

It is now apparent that with over 190 variously substituted trans (H-1' to H-2') (β-ribofuranosyl) and some 22 cis (H-1' to H-2') (α-ribofuranosyl and β-arabinofuranosyl) examples of 3',5'-cyclic monophosphates evaluated with *no observed exceptions* to the $J_{1'-2'}$ (trans)$\lesssim$ 1 Hz (singlet) and $J_{1'-2'}$ (cis) $\geq$ 3.5 Hz (doublet) anomeric proton couplings as predicted on the basis of geometrical constraints of the trans-fused six- to five-membered ring system), a generally applicable and rational "geometry-only" method for determination of the anomeric configuration of ribofuranosyl derivatives (nucleosides) is available for the first time.

Of the methods previously described, only the chemical formation of 2',3'-O-isopropylidene-$C^{5'}$→heterocycle cyclonucleosides (6) is *truly* unequivocal for defining anomeric stereochemistry in cases where it is applicable. The anomeric proton shift criterion (23) appears thus far to be reliable if both unblocked anomers of a nucleoside are available; and $J_{1'-2'}$ < 1 Hz is powerful support for the β configuration of a ribofuranosyl compound. Imbach's $\Delta\delta$ criterion is easily applicable and provides supportive evidence for the configuration of "aromatic" nucleosides with no 5'-substituents (29, 31, 32). It may be noted that a limitation of the $\Delta\delta$ criterion observed in going from α($\Delta\delta$ = 0.094) and β($\Delta\delta$ = 0.199) 2',3'-O-isopropylidene-uridines to the corresponding α($\Delta\delta$ = 0.160) and β($\Delta\delta$ = 0.184) 5,6-dihydrouridine derivatives is not a problem in the present case of α($J_{1'-2'}$ = 3.5 Hz) and β($J_{1'-2'}$ < 0.5 Hz (singlet)) uridine-3',5'-cyclic phosphates going to α($J_{1'-2'}$ = 5.0 Hz) and β($J_{1'-2'}$ < 0.7 Hz (singlet)) 5,6-dihydrouridine cyclic nucleotides. It would be of interest to examine the cyclic phosphate of a "ribofuranosyl-acetonitrile" derivative, which previously required a single-crystal x-ray determination to evaluate the "anomeric" configuration (35).

It may be possible to construct 3',5'-cyclic nucleotides in which steric and/or substituent effects could result in anomeric proton couplings outside normal limits. However, the wide variety of diverse examples presently available suggests that rather narrow limiting values exist. We propose that a rationally de-fined "geometry-only" ^{1}H NMR criterion for determination of the anomeric configuration of ribofuranosyl derivatives is β-anomers $J_{1'-2'} \lesssim 1$ Hz (apparent singlet) and α-anomers $J_{1'-2'} \sim 3$ Hz (doublet) for the anomeric proton resonance of their 3',5'-cyclic monophosphates.

Some of the advantages of this method to be noted are as follows:

(a) It is based upon the geometrical orientation of protons on a consistent rigid structure and does not rely on empirical anisotropy effects or conformational fortuities.

(b) It defines either anomer of a pair and is applicable when *only one anomer is available*.

(c) It requires no sophisticated techniques or instrumenta-tion beyond an ^{1}H NMR spectrometer capable of ∿1-Hz resolution.

(d) It entails observation of a single resonance peak in the ordinarily uncomplicated δ∿6 range of the spectrum.

(e) It involves the straightforward preparation of a cyclic nucleotide that is of potentially greater biological interest than its precursor nucleoside.

(f) It is nondestructive and the precursor nucleoside can be recovered by chemical and/or enzymatic hydrolysis of the cyclic phosphodiester group.

REFERENCES

1. For pertinent reviews see: (a) Fox, J. J., and Wempen, I., *Adv. Carbohydr. Chem. 14*, 283 (1959); (b) Montgomery, J. A., and Thomas, H. J., *Adv. Carbohydr. Chem. 17*, 301 (1962); (c) Goodman, L., in *Basic Principles in Nucleic Acid Chemistry*, Vol. 1 (Ts'o, P.O.P., ed.), Academic Press, New York, 1974, pp. 93-208.
2. Baker, B. R., in *Chemistry and Biology of Purines: A Ciba Foundation Symposium* (Wolstenholme, G.E.W., and O'Connor, D. M., eds.), J. & A. Churchill, Ltd., London, 1957, pp. 120-133.
3. Hudson, C. S., *J. Am. Chem. Soc. 31*, 66 (1909); *Adv. Carbo-hydr. Chem. 3*, 1 (1948).
4. Montgomery, J. A., and Hewson, K., *Chem. Commun.* 15 (1969).
5. Lemieux, R. U., and Hoffer, M., *Can. J. Chem. 39*, 110 (1961)
6. (a) Clark, V. M., Todd, A. R., and Zussman, J., *J. Chem. Soc* 2952 (1951); (b) Brown, D. M., Todd, A. R., and Varadar-ajan, S., *J. Chem. Soc.* 868 (1957).

7. Rousseau, R. J., Townsend, L. B., and Robins, R. K., *Biochemistry 5,* 756 (1966).

8. Emerson, T. R., Swam, R. J., and Ulbricht, T.L.V., *Biochem. Biophys. Res. Commun. 22,* 505 (1966).

9. Ulbricht, T.L.V., Emerson, T. R., and Swan, R. J., *Tetrahedron Lett* 1561 (1966).

10. (a) Miles, D. W., Hahn, S. J., Robins, R. K., Robins, M. J., and Eyring, H., *J. Phys. Chem. 72,* 1483 (1968); (b) Miles, D. W., Robins, M. J., Robins, R. K., Winkley, M. W., and Eyring, H., *J. Am. Chem. Soc. 91,* 824, 831 (1969); (c) Miles, D. W., Robins, M. J., Robins, R. K., and Eyring, H., *Proc. Natl. Acad. Sci. US 62,* 22 (1969); (d) Miles, D. W., Inskeep, W. H., Robins, M. J., Winkley, M. W., Robins, R. K., and Eyring, H., *Int. J. Quantum Chem. IIIS,* 129 (1969); (e) Miles, D. W., Inskeep, W. H., Robins, M. J., Winkley, M. W., Robins, R. K., and Eyring, H., *J. Am. Chem. Soc. 92,* 3872 (1970); Miles, D. W., Townsend, L. B., Robins, M. J., Robins, R. K., Inskeep, W. H., and Eyring, H., *J. Am. Chem. Soc. 93,* 1600 (1971).

11. Chantot, J. F., and Guschlbauer, W., *FEBS Lett. 4,* 173 (1969).

12. Ulbricht, T. L. V., Emerson. T. R., and Swan, R. J., *Biochem. Biophys. Res. Commun. 19,* 643 (1965).

13. Wright, R. S., Tener, G. M., and Khorana, H. G., *J. Am. Chem. Soc. 80,* 2004 (1958).

14. Pitzer, K. S., and Donath, W. E., *J. Am. Chem. Soc. 81,* 3213 (1959).

15. Karplus, M., *J. Chem. Phys. 30,* 11 (1959).

16. Lemieux, R. U., and Lineback, D. R., *Ann. Rev. Biochem. 32,* 155 (1963).

17. Lemieux, R. U., Stevens, J. D., and Fraser, R. R., *Can. J. Chem. 40,* 1955 (1962).

18. Rinehardt, K. L., Jr., Chilton, W. S., Hickens, M., and von Phillipsborn, W., *J. Am. Chem. Soc. 84,* 3216 (1962).

19. May, J. A., Jr., and Townsend, L. B., *J. Org. Chem. 41,* 1449 (1976); Townsend, L. B., in *Synthetic Procedures in Nucleic Acid Chemistry,* Vol 2 (Zorbach, W. W., and Tipson, R. S., eds.), Wiley-Interscience, New York, 1973, pp. 330-331.

20. (a) Robins, M. J., Mengel, R., and Jones, R. A., *J. Am. Chem. Soc. 95,* 4074 (1973); (b) Robins, M. J., Fouron, Y., and Mengel, R., *J. Org. Chem. 39,* 1564 (1974); (c) Russell, A. F., Greenberg, S., and Moffatt, J. G., *J. Am. Chem. Soc. 95,* 4025 (1973); (d) Jain, T. C., Russell, A. F., and Moffatt, J. G., *J. Org. Chem. 38,* 3179 (1973).

21. Robins, M. J., Fouron, Y., and Muhs, W. H., *Can. J. Chem. 55,* 1260 (1977).

22. Leonard, N. J., and Laursen, R. A., *J. Am. Chem. Soc. 85,* 2026 (1963).

23. Nishimura, T., and Shimizu, B., *Chem. Pharm. Bull. 13,* 803 (1965).

24. Fay, C. K., Grutzner, J. B., Johnson, L. F., Sternhell, S., and Westerman, P. W., *J. Org. Chem. 38,* 3122 (1973).

25. Schmidt, C. L., Rusho, W. J., and Townsend, L. B., *Chem. Commun.* 1515 (1971).

26. Cook, P. D., Rousseau, R. J., Mian, A. M., Dea, P., Meyer, R. B., Jr., and Robins, R. K., *J. Am. Chem. Soc. 98,* 1492 (1976).

27. Acton, E. M., Fujiwara, A. N., Goodman, L., and Henry, D. W., *Carbohydr. Res. 33,* 135 (1974).

28. (a) Cushley, R. J., Watanabe, K. A., and Fox, J. J., *J. Am. Chem. Soc. 89,* 394 (1967); (b) Cushley, R. J., McMurray, W. J., Lipsky, S. R., and Fox, J. J., *Chem. Commun.* 1611 (1969).

29. Rayner, B., Tapiero, C., and Imbach, J.-L., *Carbohydr. Res. 47,* 195 (1976).

30. Montgomery, J. A., *Carbohydr. Res. 33,* 184 (1974).

31. Imbach, J.-L., and Kam, B. L., *J. Carbohydr. Nucleosides Nucleotides 1,* 271 (1974).

32. (a) Imbach, J.-L. Barascut, J.-L., Kam, B. L., Rayner, B., Tamby, C., and Tapiero, C., *J. Heterocycl. Chem. 10,* 1069 (1973); (b) Imbach, J.-L., Barascut, J.-L., Kam, B. L., and Tapiero, C., *Tetrahedron Lett.* 129 (1974); (c) Imbach, J.-L. *Ann. NY Acad. Sci. 255,* 177 (1975).

33. Otter, B. A., Falco, E. A., and Fox, J. J., *J. Org. Chem. 34,* 1390 (1969).

34. Verheyden, J. P. H., and Moffatt, J. G., *J. Org. Chem. 35,* 2319 (1970).

35. Ohrui, H., Jones, G. H., Moffatt, J. G., Maddox, M. L., Christensen, A. T., and Byram, S. K., *J. Am. Chem. Soc. 97,* 4602 (1975).

36. Cushley, R. J., Blitzer, B. L., and Lipsky, S. R., *Biochem. Biophys. Res. Commun. 48,* 1482 (1972).

37. (a) Hart, P. A., and Davis, J. P., *J. Am. Chem. Soc. 93,* 753 (1971); Schirmer, R. E., Davis, J. P., Noggle, J. H., and Hart, P. A., *J. Am. Chem. Soc. 94,* 2561 (1972); (c) Son, T.-D., Guschlbauer, W., and Guéron, M., *J. Am. Chem. Soc. 94,* 7903 (1972); and references therein.

38. Hall, L. D., and Preston, C., *J. Chem. Soc., Chem. Commun.* 1319 (1972).

39. Sugiyama, H., Yamaoka, N., Shimizu, B., Ishido, Y., and Seto, S., *Bull. Chem. Soc. Japan 47,* 1815 (1974).

40. See for example: (a) Robison, G. A., Butcher, R. W., and Sutherland, E. W., *Cyclic AMP,* Academic Press, New York, 1971; (b) *Advances in Cyclic Nucleotide Research,* Contin-

uing Volumes (Greengard, P., Robison, G. A. (and other) eds.), Raven Press, New York, 1972-; (c) Bloch, A., *Biochem. Biophys. Res. Commun.* 58, 652 (1974); Bloch, A., Dutschman, G., and Maue, R., *Biochem. Biophys. Res. Commun.* 59, 955 (1974).

41. See: (a) Sundaralingam, M., and Abola, J., *J. Am. Chem. Soc.* 94, 5070 (1972); (b) Yathindra, N., and Sundaralingam, M., *Biochem. Biophys. Res. Commun.* 56, 119 (1974); and references therein.

42. (a) Jardetzky, C. D., *J. Am. Chem. Soc.* 84, 62 (1962); (b) Smith, M., and Jardetzky, C. D., *J. Mol. Spectrosc.* 28, 70 (1968); (c) Blackburn, B. J., Lapper, R. D., and Smith, I.C.P., *J. Am. Chem. Soc.* 95, 2873 (1973); (d) Lavallee, D. K., and Zeltmann, A. H., *J. Am. Chem. Soc.* 96, 5552 (1974); (e) Kainosho, M., and Ajisaka, K., *J. Amer. Chem. Soc.* 97, 6839 (1975); (f) Lapper, R. D., Mantsch, H. H., and Smith I.C.P., *J. Am. Chem. Soc.* 95, 2878 (1973).
97, 6839 (1975); (f) Lapper, R. D., Mantsch, H. H., and Smith, I.C.P., *J. Am. Chem. Soc.* 95, 2878 (1973).

43. Abraham, R. J., Hall, L. D., Hough, L., and McLauchlan, K. A., *J. Chem. Soc.* 3699 (1962).

44. See for example: (a) Muneyama, K., Bauer, R. J., Shuman, D. A., Robins, R. K., and Simon, L. N., *Biochemistry* 10, 2390 (1971); (b) Simon, L. N., Shuman, D. A., and Robins, R. K., in *Advances in Cyclic Nucleotide Research*, Vol. 3 (Greengard, P., and Robison, G. A., eds.), Raven Press, New York, 1973, pp. 225-353; (c) Schweizer, M. P., and Robins, R. K., in *Conformation of Biological Molecules and Polymers, Proceedings of the Fifth Jerusalem Symposium On Quantum Chemistry and Biochemistry* (Pullman, B., and Bergmann, E., eds., Academic Press, New York, 1973, pp. 329-343.

45. Ranganathan, R. S., Jones, G. H., and Moffatt, J. G., *J. Org. Chem.* 39, 290 (1974).

46. (a) Meyer, R. B., Shuman, D. A., Robins, R. K., Bauer, R. J., Dimmit, M. K., and Simon, L. N., *Biochemistry* 11, 2704 (1972); (b) Miller, J. P., Shuman, D. A., Scholten, M. B., Dimmit, M. K., Stewart, C. M., Khwaja, T. A., Robins, R. K., and Simon, L. N., *Biochemistry* 12, 1010 (1973); (c) Miller, J. P., Boswell, K. H., Muneyama, K., Simon, L. N., Robins, R. K., and Shuman, D. A., *Biochemistry* 12, 5310 (1973); (d) Boswell, K. H., Miller, J. P., Shuman, D. A., Sidwell, R. W., Simon, L. N., and Robins, R. K., *J. Med. Chem.* 16, 1075 (1973); (e) Meyer, R. B., Jr., Shuman, D. A., Robins, R. K., Miller, J. P., and Simon, L. N., *J. Med. Chem.* 16, 1319 (1973); (f) Miller, J. P., Boswell, K. H., Muneyama, K., Tolman, R. L.,

Scholten, M. B., Robins, R. K., Simon, L. N., and Shuman, D. A., *Biochem. Biophys. Res. Commun.* *55*, 843 (1973); (g) Muneymma, K., Shuman, D. A., Boswell, K. H., Robins, R. K., Simon, L. N., and Miller, J. P., *J. Carbohydr. Nucleosides Nucleotides,* *1*, 55 (1974); (h) Meyer, R. B., Jr., Shuman, D. A., and Robins, R. K., *J. Am. Chem. Soc.* *96* 4962 (1974); (i) Boswell, K. H., Christensen, L. F., Shuman, D. A., and Robins, R. K., *J. Heterocycl. Chem.* *12*, 1 (1975); (j) Christensen, L. F., Meyer, R. B., Jr., Miller, J. P., Simon, L. N., and Robins, R. K., *Biochemistry 14*, 1490 (1975); (k) Meyer, R. B., Jr., Uno, H., Robins, R. K., Simon, L. N., and Miller, J. P., *Biochemistry 14*, 3315 (1975); (l) Meyer, R. B., Jr., Uno, H., Shuman, D. A., Robins, R. K., Simon, L. N., and Miller, J. P., *J. Cycl. Nucleotide Res.* *1*, 159 (1975); (m) Miller J. P., Boswell, K. H., Mian, A. M., Meyer, R. B., Jr., Robins, R. K., and Khwaja, T. A., *Biochemistru 15.* 217 (1976); (n) Uno, H., Meyer, R. B., Jr., Shuman, D. A., Robins, R. K., Simon, L. N., and Miller, J. P., *J. Med. Chem.* *19, 419* (1976).

47. Meyer, R. B., Jr., personal communication.
48. Shuman, D. A., Miller, J. P., Scholten, M. B., Simon, L. N., and Robins, R. K., *Biochemistry,* *12* 2781 (1973).
49. Jastorff, B., and Freist, W., *Bioorg. Chem.* *3*, 103 (1974).
50. (a) Mian, A. M., Harris, R., Sidwell, R. W., Robins, R. K., and Khwaja, T. A., *J. Med. Chem.* *17*, 259 (1974); (b) Khwaja, T. A., Boswell, K. H., Robins, R. K., and Miller, J. P., *Biochemistry 14*, 4238 (1975).
51. Christensen, L. F., and Broom, A. D., *J. Org. Chem. 37*, 3398 (1972).
52. Yoshikawa, M., Kato, T., and Takenishi, T., *Tetrahedron Lett.* 5065 (1967).
53. Smith, M., Drummond, G. I., and Khorana, H. G., *J. Am. Chem. Soc. 83*, 698 (1961).
54. (a) Marumoto, R., Nishimura, T., and Honjo, M., *Chem. Pharm Bull. 23*, 2295 (1975); (b) Tazawa, I., Tazawa, S., Alderfer, J. L., and Ts'o, P.O.P., *Biochemistry 11*, 4931 (197
55. (a) Karplus, M., *J. Am. Chem. Soc. 85*, 2870 (1963); (b) Jackman, L. M., and Sternhell, S., *Applications of Nucle Magnetic Resonance Spectroscopy in Organic Chemistry,* 2nd ed., Pergamon Press, London, 1969, pp. 281-289, 292-294.
56. (a) Long, R. A., Szekeres, G. L., Khwaja, T. A., Sidwell, R. W., Simon, L. N., and Robins, R. K., *J. Med. Chem. 15* 1215 (1972); (b) Revankar, G. R., Huffman, J. H., Allen, L. B., Sidwell, R. W., Robins, R. K., and Tolman, R. L., *Med. Chem. 18*, 721 (1975).
57. Long, R. A., personal communication.

INVESTIGATION OF NUCLEOSIDE ALKYLPHOSPHONATES

M. N. PREOBRAZHENSKAYA, S. YA. MELNIK, T. P. NEDOREZOVA,
I. D. SHINGAROVA, AND D. M. OLEINIK

Cancer Research Center of the USSR Academy of Medical
Sciences, 115478, Moscow

ABSTRACT

By the interaction of 1-adamantylphosphonodichloride (APD) with uridine, 5-bromouridine, 5-fluorouridine, 6-azauridine, 5'-fluoro-5'-deoxyuridine, adenosine, or 9-β-D-ribofuranosyl-6-methylmercaptopurine, corresponding diastereoisomeric at phosphorus nucleoside 2',3'-O-(1-adamantyl)phosphonates were obtained. The five-membered phosphonates were formed starting both from 5'-O-blocked nucleosides and from 5'-O-unblocked nucleosides. The configuration of cyclophosphonates was established by the use of PMR data. The interaction of APD with thymidine gave 5'- and 3'-O-(1-adamantyl)chlorophosphonates, 5'- and 3'-O-di-P_1,P_2-(1-adamantyl)-P_2-chloropyrophosphonates, eight-membered 3',5'-O-di-P_1,P_2-(1-adamantyl)pyrophosphonate, and 5'-chloro-5'-deoxythymidine. Model five- or six-membered cyclo-1-adamantylphosphonates were prepared from ADP and ethyleneglycol or 1,3-propandiol. Triphenylmethylphosphonodichloride and 5'-O-acetyluridine yielded diasteroisomeric uridine 2',3'-O-tritylphosphonates. The compounds obtained were studied by PMR, ^{31}P NMR, CD spectroscopy, electrophoresis, and mass-spectrometry methods.

So far only limited studies have been made of nucleoside alkylphosphonates. These compounds are of interest as analogs of nucleotides and also as depot forms of nucleosides, potentially capable of passing through the blood-brain barrier. Nucleoside adamantylphosphonates are of particular interest since certain adamantane derivatives are known to display high biological activity. Up to the present nucleoside cycloalkylphosphonates have remained practically unstudied.

SYNTHESIS OF PYRIMIDINE NUCLEOSIDE 2',3'-*O*-1-ADAMANTYLPHOSPHONATES

At the first stage of our investigation we studied (1,2) the interaction of pyrimidine nucleosides with 1-adamantylphosphono-dichloride (APD) (3). The interaction of uridine (1), 5-bromo-uridine (2), 5-fluorouridine (3) or 6-azauridine (4) with APD in pyridine at 37°C during several days or at 80 to 100°C during a period of 12 to 20 hr, after TLC separation, gave mixtures of nucleoside 2',3'-*O*-1-adamantylphosphonates, differing in the configuration at the asymmetric phosphorus atom, in the following total yields: 5a and 5b, 52%; 6a and 6b, 16%; 7a and 7b, 30%; 8a and 8b, 37%.

The mixture of 5-bromouridine 2',3'-*O*-1-adamantylphosphonate (6a and 6b) was chromatographically separated on Silufol plates in a system of CHCl$_3$:MeOH (10:1) (system A). The ratio of the faster and slower moving isomers (a:b) was approximately 2:1. Starting from 4, besides isomers 8a and 8b, which could be re-solved using TLC techniques on silica gel in a system of CHCl$_3$: iso-PrOH (15:1), a third product, compound 9, was also obtained; the resolved isomers 8a and 8b also contained this substance (9) as an admixture. In all cases when resolving the isomers, multi-ple development utilizing the solvents was employed.

Structures 1-16

The adamantylphosphonate mixture (a and b) was acetylated without separation by an acetic anhydride-pyridine mixture
to give in each case two acetyl derivatives, which were resolved
by TLC techniques on silica gel in system A. Acetates with a
higher rf value (10a,11a,12a or 13a) are formed in a greater
quantity than the corresponding acetates with a lower rf value
(10b,11b,12b or 13b). The yields at the acetylation and resolution stages are 10a and 10b, 44 and 17%; 11a and 11b, 50 and 20%;
12a and 12b, 43 and 22%; 13a and 13b, 22 and 15%.

Acetylation of separated cyclophosphonates 8a or 8b (each
containing 9 as an admixture) leads to acetates 13a or 13b, respectively, with an admixture of substance 14 (acetate of substance 9). Thus, acetylation is not accompanied by isomerization.

APD was also reacted with 5'-O-acetyluridine (15) (4) and
5'-O-acetyl-5-bromouridine (16). 16 was in turn produced from
5'-O-acetyl-2',3'-O-isopropylidene-5-bromouridine (5). Starting
from 15 or 16, isomeric adamantylphosphonates 10a and 10b or 11a
and 11b were also produced in approximately the same ratio (10a
and 10b, 21.5 and 9%; 11a and 11b, 7 and 3.5%).

The interaction of 5'-fluoro-5'-deoxyuridine (17) (6) with
APD in pyridine also gives two isomeric adamantylphosphonates:
on TLC in system A a faster moving isomer (18a) was separated
with a yield of 50.5% and a slower moving isomer (18b) with a
yield of 19.6%.

Structure 17-18b

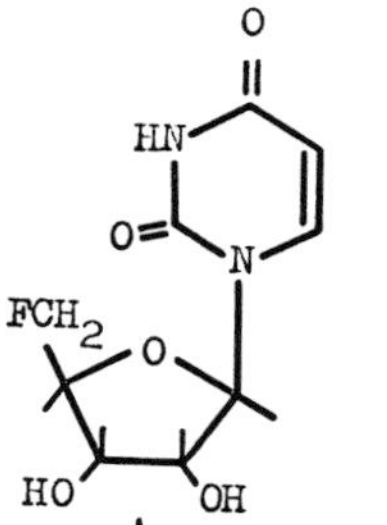

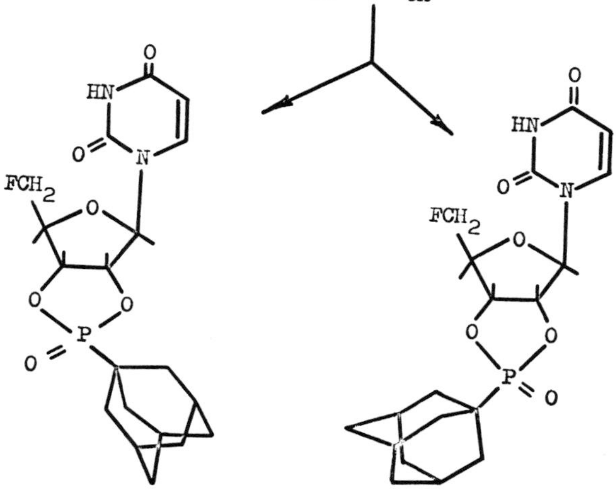

18a

18b

at C_2, and C_3, are split because of interaction with the ^{31}P atom with values of $^3J_{HH}$ and $^3J_{HP}$ constants for both isomers being close to the values of these constants for the corresponding isomeric 5'-O-acetyl-2;3'-O-1-adamantylphosphonates (see below).

Structure 17-18b of the individual adamantylphosphonates 6a and 6b, 18a, and 18b, and of individual acetates (10,11,12,13a and b), was determined by electrophoresis, ir, uv, CD spectroscopy, NMR spectroscopy, and mass spectrometry.

None of these compounds nor mixtures of isomers 5a, b, 7a, b,8a,b, or substances 9 and 14 migrate on paper electrophoregrams in a phosphate-alkaline buffer (pH 7.8) or in borate buffer (pH 9.2). This fact confirms the lack of POO⁻ or cis-diol groups in the structure. Absorption maxima in the uv spectr coincide with those of the initial nucleosides. The CD spectra of 6a and 6b are generally similar to 5-bromouridine CD spectrum, and are positive curves with maxima at 297 nm, and with slight shoulders at 285 nm (6a) or 290 nm (6b) (Fig. 1).

Mass spectra of both 6a and 6b display molecular ion peaks at m/e 502. The peaks of $(M-H_2O)^{+}$, $(M-B)^{+}$, and $(M-B-H_2O)^{+}$ ions are present at m/e 484, 313, and 295 respectively. All peaks of the fragments containing Br have the characteristic isotopic distribution. The presence of peaks corresponding to 1-adamantylphosphonic acid (APA) and adamantane were observed for mass-spectra of 18a and 18b, (see Fig. 2).

In the PMR spectra of 6a or 6b the position of the proton signals of the CH_2OH group at 4.12 ppm (having the form of a doublet) indicates that the CH_2OH group is not substituted. The PMR spectra of 18a and 18b are given in Table I. Proton signals

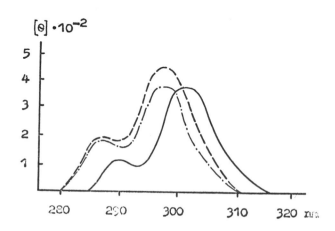

Figure 1. CD spectra of 5-bromouridine (6) (——), 6a (----), 6b(-.-.-.) in ethanol.

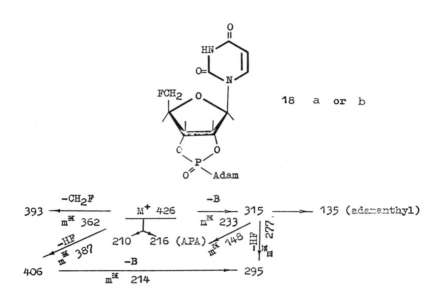

Figure 2

The mass spectra of the acetates 10-13a,b (fragmentation of a and b(isomers)practically coincide (for example, see Fig. 3). The parameters of the PMR spectra of the acetates 10-13a,b were determined by means of the double resonance method (Table I). In the case of 10a and 10b the H-C-O-P interaction was confirmed by heteronuclear ^{1}H-(^{31}P) decoupling. As in the case of the non-acetylated phosphonates the position of the signals of the pyrimidine protons and $C_{1'}$H for both isomers are close and the $C_{2'}$H and $C_{3'}$H signals in the b isomers are deshielded as compared with a isomers. All a and b isomers of 10-13 display close values of $J_{H_1'H_2'}$ and $J_{H_2'H_3'}$ constants. The most characteristic features of the PMR spectra are differences in the corresponding $^3J_{HP}$ constants for both isomers; for stereochemically similar phosphonates the values of $J_{H_2'P}$ and $J_{H_3'P}$, as well as the values of $J_{H_3'H_4'}$ constants, practically coincide. For predominant a isomers $^3J_{HP}$ constants are close to the values of the constants for 2',3'-UMP (Ba) (7).

Spin coupling constants J_{CP} in ^{13}C-NMR spectra of isomers 13a and 12a are different from the constants obtained for 2', 3'-UMP (Ba) (Table II). Assignment of the ^{13}C resonances was made with reference to previous work on cyclic nucleotides (7).

^{31}P-NMR spectra of 11a and 11b indicate the presence of one P atom in the molecule; signals are observed at 57.4 ppm for 11a and at 56.1 ppm for 11b (relative to H_3PO_4), the difference in

TABLE I. PMR Spectra of Pyrimidine Nucleoside 2',3'-O-1-Adamantylphosphonates

Compound	Chemical shifts (ppm)							Coupling constants (Hz)					Solvent temperature (°C)
	AcO	C_5,HH / C_4,H	C_3,H	C_2,H	C_1,H	C_5H	C_6H	$J_{1',2'}$	$J_{2',3'}$	$J_{3',4'}$	$J_{2',P}$	$J_{3',P}$	
10a	2.03	4.10–4.52	5.18	5.47	5.94	5.65	7.72	2.0	7.0	5.0	7.0	15.0	d_6-DMSO 30
10b	2.00	4.14–4.50	5.25	5.55	5.87	5.65	7.76	2.0	6.5	3.5	2.0	6.5	d_6-DMSO 30
11a	1.98	4.20–4.70	5.28	5.56	6.00	--	7.92	2.0	6.5	5.0	6.5	14.0	Py-d_5 70
11b	2.08	4.30–4.70	5.42	5.70	6.06	--	7.96	2.5	6.5	3.5	2.5	6.5	Py-d_5 70
12a	2.04	4.20–4.50	5.20	5.48	5.76	--	7.80	2.0	7.0	5.0	7.0	14.0	CD_3OD 20
12b	2.04	4.20–4.48	5.26	5.54	5.75	--	7.80	2.0	6.5	3.6	3.0	6.5	CD_3OD 20
13a	2.04	4.16–4.60	5.20	5.46	6.25	7.38	--	1.7	6.5	4.5	6.5	14.5	CD_3OD 20
13b	2.15	4.20–4.40	5.42	5.70	6.34	7.54	--	1.0	6.5	3.0	2.5	6.5	CD_3OD 20
18a	--	4.55[a]–4.40	5.08	5.34	5.57	5.57	7.42	2.0	7.0	5.0	7.0	14.0	CD_3OD 20
18b	--	4.60[a]–4.40	5.22	5.44	5.78	5.58	7.44	2.0	6.5	3.5	3.0	6.5	CD_3OD 20
2',3'-UMP (Ba)[b]								3.0	6.9	5.5	6.9	11.5	D_2O (pH 7.2)
14	2.05	4.14–	5.43	5.56	6.16	7.33	--	3.5	5.5	3.0	1.5	5.5	CD_3OD 20

[a] $J_{H,F}$ 48 Hz.
[b] See (7).

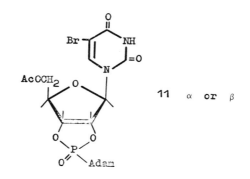

11   α   or   β

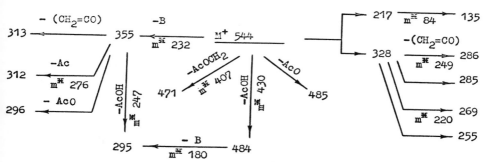

APA-adamanthylphosphonic acid

Figure 3

the width of the ^{31}P multiplet signal for 11a and 11b is approximately 11 Hz, which corresponds to the difference between the
sums of the constants $(J_{H_2,P} + J_{H_3,P})$11a - $(J_{H_2,P} + J_{H_3,P})$ 11b
∿11 Hz.

From the PMR spectrum of the acetate (14), it follows that
this compound is a monoacetate. From the $^3J_{HP}$ values it follows
that the Adam-POO- fragment is connected to the nucleoside in
positions 2' and 3'. Integration data show that the molecule
contains two adamantyl residues therefore 9 and 14 can be
assigned as 2',3'-O-di-P$_1$P$_2$-(1-adamantyl)pyrophosphonates. The
mass spectrum of 14 shows a peak corresponding to the fragment
which forms at the breakage of the pyrophosphonic acid from the
molecular ion (m/e 251), and peaks associated with decomposition
of the acetyl group of this fragment or with splitting of aglycon
from it (m/o 139). The CD-spectrum of 14 is similar to that of
6-azauridine (see Fig. 4).

TABLE II. ^{13}C-NMR Spectra of Predominant Isomers <u>12a</u> and <u>13a</u> in CD$_3$OD

	Chemical shifts (ppm; J_{CP}, Hz)				
	$C_{1'}$	$C_{2'}$	$C_{3'}$	$C_{4'}$	$C_{5'}$
12a	*94.1*	*84.6*	*81.9*	*85.7*	*63.6*
	(3.8)	*(<1.5)*	*(3.8)*	*(1.9)*	*(<1.5)*
13a	*93.0*	*85.2*	*82.9*	*86.0*	*64.0*
	(3.8)	*(<1.5)*	*(3.8)*	*(∿2)*	*(<1.5)*
2',3'--UMP (7)	*93.74*	*81.74*	*78.26*	*86.25*	*62.12*
(Ba) (7) in H$_2$O	*(6.8)*	*(2.5)*	*(0.7)*	*(2.5)*	*(0.2)*

CONFIGURATIONS AND CONFORMATIONS OF 2',3'-*O*-1-ADAMANTYLPHOS-PHONATES OF PYRIMIDINE NUCLEOSIDES

      Comparison of the PMR data for the known isomeric 6-membered cyclophosphonates reveals that in the case when cis-di-axial orientation of the C-H and P=O groups in the $\overset{\text{H}\quad\text{O}}{\underset{\text{C-O-P}}{}}$ moiety is possible this proton is more deshielded than the proton of the isomer where such orientation is impossible (8). Recently, Bentrude and Tan (9) used the relative chemical shifts of the ring protons cis and trans to the phosphorus lone electron pair in the isomeric 4-methyl-1,3,2-dioxaphospholanes for the determination of the configuration of these isomers. The C-H and C-CH$_3$ protons cis to the phosphorus lone electron pair are downfield to those with the transorientation. In our case in b isomers the protons at C$_{2'}$ and C$_{3'}$ are shifted downfield, which allows us to conclude that in b isomers the P=O group is cis oriented to these protons, i.e., is in an exo position (S-configuration at phosphorus). a isomers show the P=O group in an endo position (R-configuration at phosphorus).

      The $^3J_{HH}$ and $^3J_{HP}$ coupling constants are approximately the same for all compounds through the series (a or b). The above coupling constants of a isomers are similar to those found in the 2',3'-cyclonucleotides, which have been previously studied 3'-CMP show a preference in the direction of the C$_{3'}$-endo (C$_{2'}$-exo)-puckered form. We have also used the relationship $^3J_{HP}=$ 16.3 cos$^2\theta$-4.6 cosθ to estimate the H-C-O-P dihedral angles in the a and b isomers (Table III). As in the case of 2',3'-cyclonucleotides (7), the H-C$_{3'}$-O-P dihedral angle values for both isomers (a and b) are larger than for H-C$_{2'}$-O-P, where the C$_{3'}$-endo-

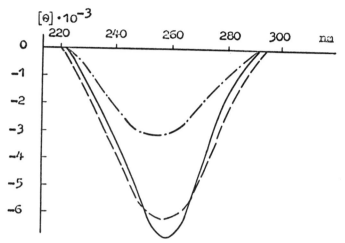

Figure 4. CD spectra of <u>13a</u> (——), <u>13b</u> (————), <u>14</u> (-.-.)
in ethanol.

9 R = H

14 R = Ac

19   a isomer

20   b isomer

Structures 19-20

($C_{2'}$-exo) conformation apparently predominate. It can be noticed that the $H-C_{3'}-O-P$ angle value increases for a isomers and the $H-C-O-P$ angle values for b isomers decrease in comparison with the 2',3'-cyclonucleotides. An examination of molecular models indicates that the increase of $H-C_{3'}-O-P$ dihedral angle values in α isomers is due to the P atom endo puckering in the dioxaphos pholane ring; with the bulky adamantyl group occupying the pseudo equatorial position (19). In b isomers the decrease of $H-C_{3'}-O-P$ and especially of the $H-C_{2'}-O-P$, dihedral angle values may be explained by possible P atom exo puckering, where the bulky ada-mantyl group also occupies the pseudo equatorial position (20). The decrease of $H-C-O-P$ dihedral angle values may also be due to the change of the conformational equilibria states of the ribose ring and may be connected with the decrease of $C_{3'}$-endo puckering

TABLE III. Dihedral Angles Calculated from $^3J_{HP}$ Coupling Constants and Predicted for Rigid Conformation of 2',3'-Cyclonucleotides

Compound	$P-C-C_{2'}H$	$P-O-C_{3'}-H$
a *isomer*	122	∿150
b *isomer*	∿106	122
2',3'-UMP	122	136
(Ba) (7)		
$C_{3'}$-*endo (7)*	105	135
($C_{2'}$-exo)		
$C_{3'}$-*exo (7)*	135	105
($C_{2'}$-endo)		

CD spectra of the isomeric acetates (10-13) practically coincide and are close to those for the corresponding nucleosides Thus one may conclude that the pyrimidine nucleoside 2',3'-O-adamantylphosphonates retains the anti conformation around the glycosidic bond (for example, see Fig. 4).

PURINE NUCLEOSIDE 2',3'-O-1-adamantylphosphonates

The next stage of our investigation was to study the inter-action of APD with purine nucleosides. APD and adenosine (21) in pyridine at 80-100°C during 15 hr gave a mixture of products that were difficult to separate and from which we succeeded in isolating only one adenosine, 2',3'-O-1-adamanthylphosphonate (22) (15%). The structure of 22 was confirmed by mass-spectro-metric and PMR-spectroscopic data (Table IV). The CD curve is similar to that of adenosine. Starting from 5'-O-acetyladenosine (23) (11), a mixture of isomeric phosphonates (24a and 24b) was obtained with a total yield of 30% in approximately an equal ratio. The mixture was separated on silica gel in system A. Acetylation of 22 gave 24a.

Structure 21-24b

In Table IV PMR data for 22 and also for 24a and 24b are presented, as well as those for 2',3'-AMP(Na) (7). The values of corresponding $^3J_{HH}$ AND $^3J_{HP}$ coupling constants for sterochemically similar isomers in purine and pyrimidine series are close. Similar transformations were carried out for 9-β-D-ribofuranosyl-6-methylmercaptopurine and its 5'-O-trityl derivative.

The formation of five-membered adamantylphosphonates in the interaction of ADP with nucleosides proved to be unexpected. No reaction with APD occurred under similar conditions when using uridine or 5-bromouridine 2',3'-O-isopropylidene derivatives. Cyclo-1-adamantylphosphonates of nucleosides split to corresponding 2'(3')-1-adamantylphosphonates of nucleosides in tris-HCl buffer (pH 9.0) at room temperature during several hours. In acidic solutions they are more stable.

Verheyden and Moffatt (12) suggested that 5'-O-acetyluridine
2',3'-O-methylphosphonates were unstable and easily split, producing a mixture of 5'-O-acyluridine 2'- and 3'-O-methylphospho-
nates. Apparently, the stability of five-membered adamantylphosphonates is due to steric hindrance offered by the bulky adamantyl
residue. Easy ring closure of the cis-diol group into the five-
membered cyclic ring under the effect of APD may also be determine
by the size of the X-P-X valence angle in Adam-POX$_2$ derivatives,
which is evidently somewhat smaller as compared to the derivatives
of lower alkylphosphonates or phosphates.

It may be assumed that other alkylphosphonodichlorides that
have a sufficiently bulky residue can also form five-membered
cyclophosphonates. We have shown that uridine (1) or 5'-O-acetyl-
uridine (15), when treated with triphenylmethylphosphonodichloride
(13), also forms the corresponding 2',3'-tritylphosphonates. It
is of interest to study the dependence between the stability of
five-membered rings and the size of the alkyl group in alkylphosphonates.

INTERACTION OF APD WITH THYMIDINE .

Six-membered nucleoside 3',5'-O-cyclophosphonates hold a
unique position among cyclic nucleotides. It was of interest to
investigate the possibility of synthesizing 3',5'-O-alkylphos-
phonates of nucleosides or 2'-deoxynucleosides. We studied the
interaction of thymidine (25) and APD in pyridine at 55°C during
10 days (14). The complex mixture of reaction products was separat
using TLC techniques on silica gel in a system of EtOAc:iso-PrOH:
H$_2$O 20:1:6, into fractions A,B,C,D (rf=0.40; 0.33; 0.28; 0.25,
respectively). Each fraction was then chromatographed on silica
gel plates in system A. Fraction A was resolved into four substances (26,27,28,29). This yielded 6% 5'-chloro-5'-deoxythymi-
dine 3'-O-1-adamantylchlorophosphonate (26) (rf=0.22 in system
A), about 3.7% thymidine 3'-O-1-adamantylchlorophosphonate (27)
(rf=0.24), about 2.2% thymdine 3'-O-di-P$_1$P$_2$-(1-adamantyl)-P$_2$-
chloropyrophosphonate (28) (rf=0.27) and 14% of its isomer, differing by the configuration at P (29) (rf= 0.20). After purification of fraction B in system A, thymidine 5'-O-1-adamantylchlor-
ophosphonate (30) (rf=0.20) was obtained in a yield of 8.4%, and
from fraction C 4.5% of thymidine 5'-O-di-P$_1$P$_2$-(1-adamantyl)-P$_2$-
chloropyrophosphonate (31) was obtained (rf=0.17). For the substance separated from fraction D the structure was presumed to
be that of 8-membered thymidine 3',5'-di-O-P$_1$P$_2$-(1-adamanthyl)
pyrophosphonate (32) (rf=0.12,∿2.4%). The structures of the substances obtained were determined by PMR, ^{31}P-NMR spectroscopy,

and mass-spectrometric methods. The total yield of the reaction products for the reacted thymidine was ∿40%. None of the substances isolated (26-32) migrated under the condition of paper electrophoresis using phosphate-alkaline (pH 7.8) and borate (pH 9.5) buffers.

Chlorodeoxythymidine (26) was identified by spectral methods and its properties proved coincident with those of the substance described by Verheyden and Moffatt (15).

Structures 26-35

The parameters of the PMR spectra of the thymidine derivatives obtained are presented in Table V. The positions of the signals were determined with the aid of the double resonance method. In 3'-O-derivatives (27,28) the signals of the C_5,HH protons exist as a doublet that turns into a singlet upon decoupling of the C_4,H signal. Upfield positions of the doublet and absence of niticeable spin coupling with ^{31}P indicate that the CH_2OH group is not substituted. Signals of the C_3'H group are shifted downfield as compared with those of thymidine. The values of the J_{PH} constants confirm the position of the P-containing substituent at C_3,-O. The PMR spectrum of (29) is very close to that of 28, but apparently (29) is a mixture of several isomers of 28 (at P) with very close parameters, as noted in the PMR spectrum.

TABLE IV. PMR Spectra of Adenosine and 5'-O-Acetyladenosine 2',3'-O-1-Adamantylphosphonates (22,24a, and 24b).

Compound	Chemical shifts (ppm)							Coupling constants (Hz)					Solvent temperature (°C)
	C_2^H C_8^H	$C_1,^H$	$C_2,^H$	$C_3,^H$	$C_4,^H$	$C_5,,^{HH}$	Adam Aco	$J_{1'2'}$	$J_{2'3'}$	$J_{3'4'}$	$J_{2'P}$	$J_{3'P}$	
22	8.24 8.42	6.38	5.82	5.36	4.36	3.66	1.64- -2.04	3.0	7.0	4.0	10.5	11.5	d_6-DMSO,100
24a	8.26 8.34	6.38	5.91	5.44	4.19--4.64		1.64- -2.04	3.0	7.0	4.5	9.0	14.0	d_6-DMSO,100
24b	8.24 8.32	6.34	6.04	5.48	4.24	4.48	1.64- -2.04	2.5	7.0	3.5	3.5	6.0	d_6-DMSO,100
2',3'-AMP(Na)[7]								4.4	6.8	3.9	10.6	7.6	D_2O,30

In the PMR spectra of 5'-*O*-substituted thymidines (30) and
(31) C_5,HH signals are complex multiplets shifted downfield as
compared with 27 or 28. Determination of the $^3J_{HP}$ constants
presents difficulties since in 30 the C_5,HH signal is overlapped
by the C_4,H signal, and in 31 it is overlapped by the C_3,H signal.
C_3,H signals in 30 and 31 are shifted relatively upfield as compared to 27 and 28, just as would be expected for the proton of
the CHOH group. The C_3,H signal in 30 given by decoupling of
the C_2,HH signal, has the form of a doublet with the spin coupling
3 Hz, which evidently, corresponds to interaction with C_4,H. Thus,
the P-containing substituent in 30 and 31 is found at C_5'HH. In
32 the character of the multiplicity and the position of proton
signals at C_3, and C_5,, show that the P-containing substituent
is connected with the nucleoside at C_3, and C_5,.

In the ^{31}P-^1H spectrum of 5'-*O*-1-adamantylchlorophosphonate
30 (in d_6-DMSO) two ^{31}P signals are observed that do not interact
and are closely spaced at 52.7 and 52.0 ppm. Their intensity
is approximately equal, which indicates that 30 is a mixture of
two isomers differing in configuration at P. Evidently, 3'-*O*-
1-adamanthylchlorophosphonate (27) is also a mixture of two diastereoisomers. The ^{31}P-NMR spectrum of 31 (in d_6-DMSO) contains
two groups of interacting signals that correspond to three pyrophosphonate isomers δ40.45 and 27.50 ppm, J_{POP}53 Hz; δ40.61 and
27.34 ppm: J_{POP} 54 Hz; δ40.71 and 27.34 ppm, J_{POP} 50 Hz. The
^{31}P-NMR spectrum of 29 indicates this compound to be a complex
mixture of several (evidently 4) isomers, ^{31}P signals make up
two groups of interacting multiplets at 42.6-39.1 and 27.5-25.1 ppm.

Acetylation of 29 gave a mixture of monoacetates (33) (according to the ir-and PMR-spectroscopy data), which was separated
into two fractions (by TLC). The first fraction, according to
the ^{31}P-NMR data (in *d*-pyridine) is a mixture of two thymidine
5'-*O*-acetyl-3'-*O*-di-P_1P_2-(adamanthyl)-P_2-chlorophosphonates with
^{31}P signals at 41.49 and 26.85 ppm, J_{POP} 48 Hz, and at 41.79 and
27.14 ppm, J_{POP} 52 Hz. The second fraction contains an individual *O*-acetate (33), with ^{31}P signals at 41.45 and 26.92 ppm,
J_{POP} 48 Hz. Thus, substances 27-32, containing asymmetric P
atoms, are mixtures of diastereoisomers whose PMR spectra (except
28 and 29) coincide and ^{31}P signals observed at different but
rather close ppm values.

The mass spectrum of 5'-phosphonate (30) displays peaks of
melecular ion M^+(m/e 458) and peaks of $(M-B)^+$, $(M-B-H_2O)^+$, and
$(M-OH)^+$ at m/e 333,315,441, respectively, with the corresponding
isotopic distribution, and also peaks of $(M-HCl)^+$ and $(M-HCl-B)^+$
at m/e 422 or 297. In the mass spectrum of acetate 35, and also
in the mass spectrum of acetate 34, there are molecular ion peaks
at m/e 500, peaks of $(M-B)^+$ at m/e 375, and peaks of $(M-B-CH_3COOH)^+$

at m/e 315. All peaks containing Cl have characteristic isotopic distributions. In the mass spectrum of 32 a peak of M^+ and $(M-B)^+$ are seen; peaks which form upon splitting of diadamantyl-pyrophosphonic acid from the molecule (m/e 620,495, and 206). Keeping 32 in an aqueous DMSO gives a substance that migrates on electrophoregrams in the TEAB buffer (pH 7.6). Evidently, opening of the cycle takes place.

To elucidate to what extent the formation of 6-membered adamanthylphosphonate is more hindered than that of the 5-membered one, we studied the interaction of APD with ethylene glycol (36) or propanepiol-1,3(37) in pyridine at 55°C over a period of 3 days. 36 gave the 5-membered adamantylphosphonate (38) in a yield of 24%, and 2-chloroethanol O-di-P_1P_2-(1-adamantyl)-P_2-chloropyrophosphonate (39) in a yield of 21%. 37 gave the 6-membered adamantylphosphonate (40) in a yield of 12% and 1-propanediol-1,3 O-di-P_1P_2-(1-adamantyl)-P_2-chloropyrophosphonate (41) in a yield of 2%. No other reaction products were detected by the TLC method. Acetylation of 41 gave a corresponding monoacetate (42).

Structures 36-42

Thus, the simplest and conformationally mobile models indicate that the formation of 6-membered adamantylphosphonates proceed with substantially lower yields than the formation of similar 5-membered rings. Nucleoside 3',5'-O-cyclophosphates are known to be very strained, and therefore the formation of their adamantylphosphonate analogs encountered considerable steric hindrance.

TABLE V. PMR Spectral Data of Thymidine Adamantylphosphonates and Pyrophosphonates (27-32)

Com-pound	$C-CH_3$	C_6H	C_1,H	C_2,H^a	C_3,H	C_4,H	C_5,HH	Solvent temperature (°C)
					Chemical shift (ppm)			
27	1.78	7.76	6.28	2.4	5.02 J_{H_3,H_4}, 2Hz $J_{H_3,P}$ 8Hz	4.12	3.68	d_6-DMSO,100
28	1.82	7.78	6.30	2.4	5.04 J_{H_3,H_4}, 3Hz $J_{H_3,P}$ 6 Hz	4.14	3.73	d_6-DMSO,100
30	1.81	7.55	6.24	2.2	4.35	4.19	-3.91	d_6-DMSO,100
31	1.75	7.48	6.21	2.2	b	4.03	b	d_6-DMSO,100
32	1.82	7.64	6.22	2.2	5.60	4.03	4.30-4.90	d_6-DMSO,100

^{a}The signals are overlapped by signals of adamantane protons.

^{b}Signals C_3,H and C_5,HH are overlapped and have chemical shift 4.21-4.45 ppm.

REFERENCES

1. Preobrazhenskaya, M. N., Melnik, S. Ya., Oleinik, D. M.,
    Shepeleva, E. S., Turchin, K. F., and Sanin, P. I.,
    Bioorganicheskaya Khim. 2, 627 (1976).

2. Melnik, S. Ya., Nedozezova, T. P., and Preobrazhenskaya, M.
    N., *J. Carb. Nucleosides Nucleotides* 3, 129 (1976).

3. Stetter, H., and Last, W. D., *Chem. Ber.* 102, 3364 (1969).

4. Brown, D., and Todd, A., *J. Chem. Soc.* 2388 (1956).

5. Inoue, H., and Ueda, T., *Chem. Pharm. Bull.* 19, 1743 (1971).

6. A gift of Professor P. Langen and Dr. G. Kowollik (Central
    Institute of Molecular Biology of Academy of Science, GDR)

7. Lapper, R., and Smith, I., *J. Am. Chem. Soc.* 95, 2880 (1973).

8. Bentrude, W., Tan, H.-W., and Yee, K., *J. Am. Chem. Soc.*
    94, 3264 (1972).

9. Bentrude, W., and Tan, H.-W., *J. Am. Chem. Sox.* 98, 1850
    (1976).

10. Lavallee, D., and Coulter, C, H., *J. Am. Chem. Soc.* 95,
    576 (1973).

11. Michelson, A., Szabo, L., and Todd, A., *J. Chem. Soc.* 1546
    (1956).

12. Verheyden, J., and Moffatt, J., *J. Org. Chem.* 35, 2868 (1970)

13. Hardy, D. and Hatt, H., *J. Chem. Soc.* 3778 (1952).

14. Melnik, S. Ya., Shingarova, I. D., Yartzera, I. V. and Preo-
    brazhenskaya, M. N.

15. Verheyden, J. and Moffatt, J., *J. Org. Chem.* 37, 2289 (1972)

THE *TERT*-BUTYLDIMETHYLSILYL GROUP AS A
PROTECTING GROUP IN OLIGONUCLEOTIDE SYNTHESIS

WILFRIED KÖHLER, WILHELM SCHLOSSER,
GEETA CHARUBALA, AND WOLFGANG PFLEIDERER

Fachbereich Chemie, Universität Konstanz,
Konstanz, West Germany

ABSTRACT

N-Benzoylcytidine (1) and *N*-benzoylguanosine (14) have been silylated by *tert*-butyldimethylsilylchloride (TSDMSiCl) to the various corresponding mono-, di-, and tri-TBDMSi-ethers, whose structures are characterized by chemical and spectrophotometric methods. ^{13}C-NMR spectroscopy proved to be particularly useful for
rapid determination of the positions of the TBDMSi groups in the
mono- and disilylated nucleotides. Isomerizations of 2'- and
3'-silyl derivatives take place, especially in alcohols. These
interconversions are base catalyzed. Deblocking experiments under various reaction conditions are described. The usefulness
of the TBDMSi group as an OH protecting group in oligonucleotide
synthesis is demonstrated by the synthesis of CpC via the triester approach.

The strategy of chemical syntheses with polyfunctional molecules
is mainly dependent on the availability of apporpriate protecting
groups (1,2). In nucleoside, nucleotide, and oligonucleotide
chemistry the right protection of the sugar hydroxyl groups is
still an interesting challenge since no perfect solution for all
the various problems has yet been found. Besides the different
types of acid- and base-labile protecting groups, deblocking under other conditions such as photochemical cleavages (3) or de-
protection in aprotic solvents provide new solutions to the
problems and enhance the chemical possibilities to a large
extent.

The *tert*-butyldimethylsilyl (TBDMSi) group, introduced to
organic chemistry by Stork (4) and being made more popular by
Corey (5), adds great versatility to the commonly used protecting groups due to its removal with fluoride ions without affecting other acid- or base-labil blocking groups. The use and utility of the TBDMSi group in nucleoside chemistry has been described and investigated by Ogilvie *et al.* (6-9) applying known
conditions preferentially to deoxynucleosides for the first time.
Furthermore, a series of trisubstituted silyl groups (9,10) provides an additional range in selectivity on blocking of hydroxyl
groups and deblocking of the resulting silyl ethers.

   Our interest in this field has mainly been directed to the
functionalization of ribonucleosides by the TBDMSi group, since
very little is known in this respect aside from some preliminary
investigations with uridine (8). Treatment of *N*-benzoyl-cytidi-
dine (1) with *tert*-butyldimethylsilylchlorid (TBDMSiCl) and imidazole in DMF in a 1: 2,5':5 ratio led at room temperature to
a mixture of 43% 2',5'-di'(5), 31% 3',5'-di-(8), and 4% 2',3',5'
-tri-*tert*-butyldimethylsilyl-*N*-benzoylcytidine (11), respectively. Their separation could be achieved by preparative thick-
layer or column chromatography on silica gel.

   A selective monosubstitution at the 5'-position (2) took
place with a 1:1,1: 2,2 ratio of the reactants at -30° in 65%
yield. Starting from 2 a similar product distribution was obtained as on direct silylation of 1 when using 1.4 eq. of
TBDMSiCl at room temperature. The structural assignments of 5
and 8 are based on their acylations to the corresponding 3'-
(6) and 2'-acetyl derivatives (9), whereby 6 has also been obtained by silylation of authentic *N*-benzoyl-3'-*O*-acetyl-cytidine
(3) (11). Analogous silylation reactions with *N*-benzoyl-guano-
sine (14) afforded a much higher nucleoside: TBDMSiCl ratio
(1:4.4), which led, at 3°C, to a 73% yield, of 5'-*O*-*tert*-butyl-
dimethylsilyl-*N*-benzoylguanosine (15) and at 30°C lead to a
mixture of 4% of 15 and 34% of 2',5'- (16) and 3',5'-disubstitu-
ted derivatives (18). An 8.8 molar excess of silylating agent
finally resulted in an almost quantitative yield of 2',3',5'-
tri-*O*-*tert*-butyldimethylsilyl-*N*-benzoylguanosine (20).

   NMR studies of the silylated ribonucleotides indicated that
structural assignments by PMR spectra on the basis of Rees's
rules (12) may lack unambiguity as seen from the chemical shifts
of the pairs 5, 8 and 7, 10. Unlike Reese's findings for other
pairs of 2'- and 3'-isomers, the 1'-H resonance is at lower field
in the silylated *N*-benzoylcytidine series, for the 3'- than the
2'-isomers, whereas the coupling constants $J_{1',2'}$ show the expected features. A good correlation is found for the OH signals
based on 1 as the reference substance (Table I).

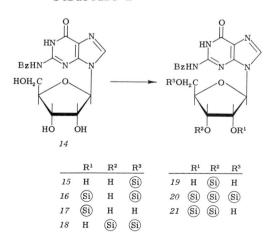

Structure 1

Structure 2

TABLE I H-NMR Spectra of Silylated N-Benzoylcytidine Derivatives (δ_{TMS} values in ppm)

N-Benzoyl-cytidine

5'	3'	2'		Solvent	1'-H	2'-OH	3'-OH	5'-OH	$J_{1',2'}$ (Hz)
OH	OH	OH	(1)	DMSO-D$_6$	5.88	5.5	5.1	5.2	3
O Si	OH	OH	(2)	DMSO-D$_6$	5.86	5.6	5.1	–	0
OH	O Si	OH	(10)	DMSO-D$_6$	5.86	5.3	–	5.2	3.6
OH	OH	O Si	(7)	DMSO-D$_6$	5.79	–	4.9	5.2	2.4
O Si	O Si	OH	(8)	DMSO-D$_6$	5.87	5.5	–	–	2.4
O Si	OH	O Si	(5)	DMSO-D$_6$	5.83	–	5.0	–	2.0
OH	O Si	O Si	(12)	DMSO-D$_6$	5.84	–	–	5.2	4
O Si	O Si	O Si	(11)	CCl$_4$	5.60	–	–	–	0
OMMTr	O Si	OH		DMSO-D$_6$	5.88	5.4	–	–	1.5
OMMTR	O Si	OH		DMSO-D$_6$	5.81	–	5.1	–	0
O Si	OAc	O Ai	(6)	CDCl$_3$	5.98	–	–	–	2.8
O Si	O Si	OAc	(9)	CDCl$_3$	6.22	–	–	–	3.6

On the other hand the silylated *N*-benzoylguanosine derivatives behave normally according to Reese's rules, whereby the chemical shifts of the OH protons serve as additional indicators in the structural determinations (Table II).

13-CONMR spectra so far offer the best physical method in the structure elucidations of the *tert*-butyldimethylsilyl ethers since substitution at the sugar OH group is associated with a substantial downfield shift of the basic ring carbon in comparison with the parent substance (1) (Table III).

The chemical properties of the various *tert*-butyldimethylsilylated nucleosides are dominated by the cleavage reactions of the O-Si bonds. The large difference in rates of acid hydrolysis between a 2' or 3' and a 5' substituent offers a selective method for removal of the latter. 5 and 8 could be converted to 2'- (7) and 3'-*tert*-butyldimethylsilyl-*N*-benzoylcytidine (10), respectively, in 72% and 44% yield after heating at 60°C for 4 hr in 80% acetic acid. Under similar conditions 11 was converted to the 2'.3'-disilylether 12. The relatively low yield of 38% could be increased starting from 5'-*O*-monomethoxytrityl-*N*-benzoylcytidine (4), which after silylation to 13 and detritylation lead to 12 in an overall yield of 60%.

The deblocking of the 5'-*O*-silyl groups in the *N*-benzoylguanosines series afforded a somewhat longer reaction time but was also selective in 80 - 90% acetic acid, as seen from the interconversion of 16, 18, and 20 into 17, 19, and 21, respectively. Furthermore, a selective desilylation in 5'-position could be achieved with HF in ethanol, whereas treatment of 2',5'-di-*tert*-butyldimethylsilyl-*N*-benzoylcytidine (5) with NH_4F in acetone/water resulted in a selective deblocking of the 2'-silyl group. This selectivity is mainly due to a neighboring group participation effect of the 3'-hydroxy function and the reduced nucleophilicity of the F^- ion due to its solvation in water. As can be seen from the deblocking experiments with tetra-*N*-butylammonium fluoride in aprotic solvents, an unsubstituted vicinal OH group again facilitates the cleavage of an adjacent silyl group to a large extent. A more quantitative picture of this effect is shown in Table IV.

Another interesting behavior of 2'- and 3'-*O*-*tert*-butyldimethylsilyl ribonucleosides has been recognized during purification of 5 and 8. After separation of the two isomers by preparative thick-layer chromatography on silica gel the chromatographically pure compounds showed mutual interconversion on recrystallization from methanol, giving raise to an isomeric mixture again with 5 as the predominant component. The equilibrium constants of the system 5 $\rightleftarrows$ 8 have been determined at three different temperatures in methanol and in the presence of a trace of triethylamine,which catalyzes the isomerization. The isomeric mixture was analyzed quantitatively by high-pressure liquid chromatography.

TABLE II. ^{1}H-NMR Spectra of Silylated N-Benzoylguanosine Derivatives in DMSO-D_6 (δ_{TMS} Values in ppm)

N-Benzoyl-guanosine

5'	3'	2'		1'-H	2'-OH	3'-OH	5'-OH
OH	OH	OH	(14)	5.88	5.42	5.12	5.00
O Si	OH	OH	(15)	5.88	5.48	5.12	-
OH	O Si	OH	(19)	5.86	5.34	-	5.04
OH	OH	O Si	(17)	5.88	-	5.08	5.08
O Si	O Si	OH	(18)	5.86	5.44	-	-
O Si	OH	O Si	(16)	5.94	-	5.10	-
OH	O Si	O Si	(21)	5.92	-	-	-
O Si	O Si	O Si	(20)	5.90	-	-	-

TABLE III. ^{13}C-NMR Spectra of Silylated N-Benzoylcytidine Derivatives in DMSO-D_6 (δ_{TMS} values in ppm)

N-Benzoyl-cytidine

5'	3'	2'		C-1'	C-2'	C-3'	C-4'	C-5'
OH	OH	OH	(1)	90.2	74.6	68.7	84.3	59.9
O Si	OH	OH	(2)	90.5	74.8	68.2	83.4	61.6
OH	O Si	OH	(10)	90.0	74.2	70.9	84.9	60.0
OH	OH	O Si	(7)	90.3	76.5	68.2	84.0	59.6
O Si	O Si	OH	(8)	90.2	74.1	69.9	83.6	61.4
O Si	OH	O Si	(5)	90.1	76.8	68.1	83.4	61.6
OH	O Si	O Si	Si(12)	89.5	75.6	70.3	84.3	59.3
O Si	O Si	O Si	Si(11)[a]	91.2	77.1	70.5	84.5	62.0

[a] in acetone-D_6

TABLE IV

N-Benzoyl-cytidine	Methanol	Ethanol	Acetone	Tetrahydrofurane	Dimethylformide
5'-O-TBDMSi-(2)	2 days[a]	1 day[a]	4 hr	10 min	10 min
3'-O-TBDMSi-(10)	2 hr	1.5 hr	2 min	<15 sec	<15 sec
2'-O-TBDMSi-(7)	2 hr	1.5 hr	2 min	<15 sec	<15 sec

[a] ∿50% cleavage after given time.

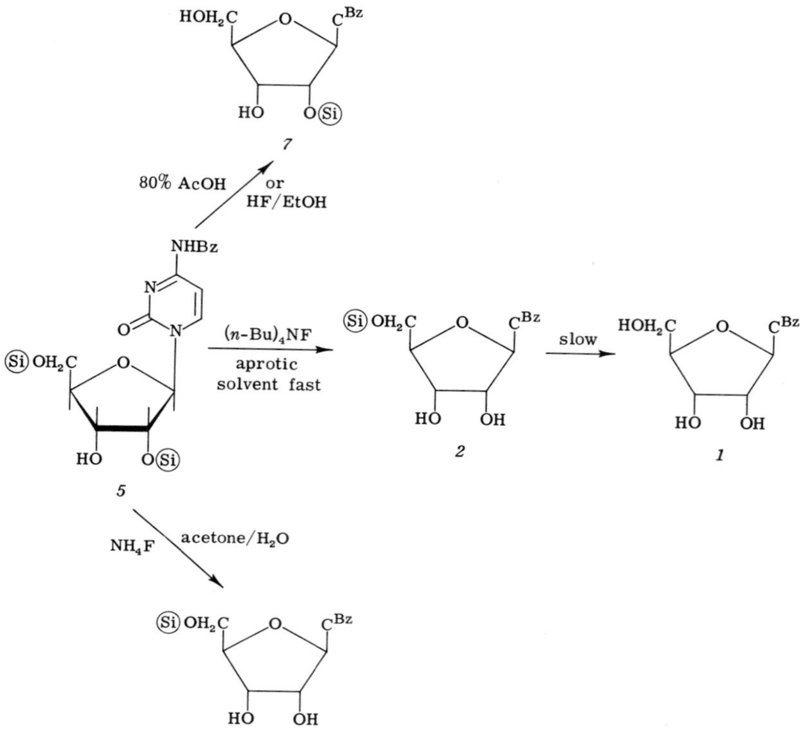

Scheme 1

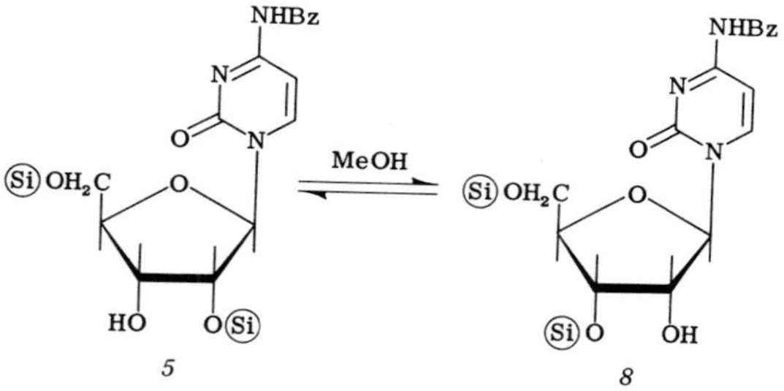

K = C_5/C_8	Temp.
0.44	0°
0.51	25°
0.58	50°

Scheme 2

Since the isomerization is base catalyzed and should involve d-orbital participation of the Si-atom, some rate measurements independent of different bases have been undertaken. As expected an increase in base strength enhances the isomerization rate (Table V). It was furthermore of interest for the preparative

TABLE V   *Catalysis of Isomerization*
(0.01 M Solutions of 5, Room Temperature)

Solvent	Base	Concentration of base (M)	Time until $(K=C_5 / C_8=0.3)$
CH_3OH	–	–	30 hr
CH_3OH	$(C_2H_5)_3N$	0.01	5 min
CH_3OH	$CH_3 O Na$	0.001	15 sec

work with *tert*-butyldimethylsilyl blocked ribonucleosides to investigate the influence of the chemical nature of the solvent on the rate of isomerization. It was found that the isomerization takes place, particularly in alcohols such as methanol, ethanol, and *n*-propanol, whereas the rate of isomerization in iso-propanol is very slow and addition of small amounts of acetic acid exerts a strong stabilization effect. In pure anhydrous aprotic solvents no isomerization could be detected within 44 hr at room temperature (Table VI).

The usefulness of the TBDMSi group in oligonucleotide synthesis had first been checked by some comparative phosphorylation studies with 5'-*O*-monomethoxytrityl-(22) and 5'-*O*-*tert*-butyldimethylsilyl-*N*-benzoyldeoxyadenosine (23) as model substances using a variety of phosphorylating agents. Phenyl- and cyanoethylphosphoric acid in pyridine/TPS gave a somewhat better yield with 22 than with 23, but in principle no striking difference in their chemical behavior could be observed. Diphenylphosphochloridate reacted very smoothly with 23 to give the corresponding diphenyl triester, which could be isolated without any difficulties in high yield by single silica gel chromatography.

The synthesis of a dinucleoside phosphate with TBDMSi blocking groups has been achieved according to the triester approach (13,14) starting from 2',5'-di-*O*-*tert*-butyldimethylsilyl-*N*-benzoylcytidine (5). Thus 5 was phosphorylated with β-cyanoethyl phosphate to 24 and condensed with 12 to give the fully blocked triester 25 in 22% yield. Deblocking worked very well in a two-step procedure by treatment (1) with methanolic ammonia and (2) with (*n*-butyl)4 NF in absolute THF. The resulting cytidylyl-(3'-5')cytidine (CpC), which separated out on acidification with

TABLE VI   *Rate of Isomerization in Different Solvents*
            (0.005 M solutions, Room Temperature)

Solvent	Reaction time (hr)	$K = C_{\underline{5}} / C_{\underline{8}}$
Methanol	43	0.4
Ethanol	44	0.3
n-Propanol	44	0.25
iso-Propanol	45	0.03
Methanol + 0.1% AcOH	96	0.01
Ethanol + 0.1% AcOH	96	0.01
Dioxane	44	0
Acetonitrile	44	0
Acetone	44	0
Chloroform	44	0
Pyridine	44	0
Benzene	44	0
Cyclohexane	44	0

acetic acid in 90% yield, was chromatographically pure and was completely degraded by both snake venon and spleen phosphodiesterases.

Structure 24-25

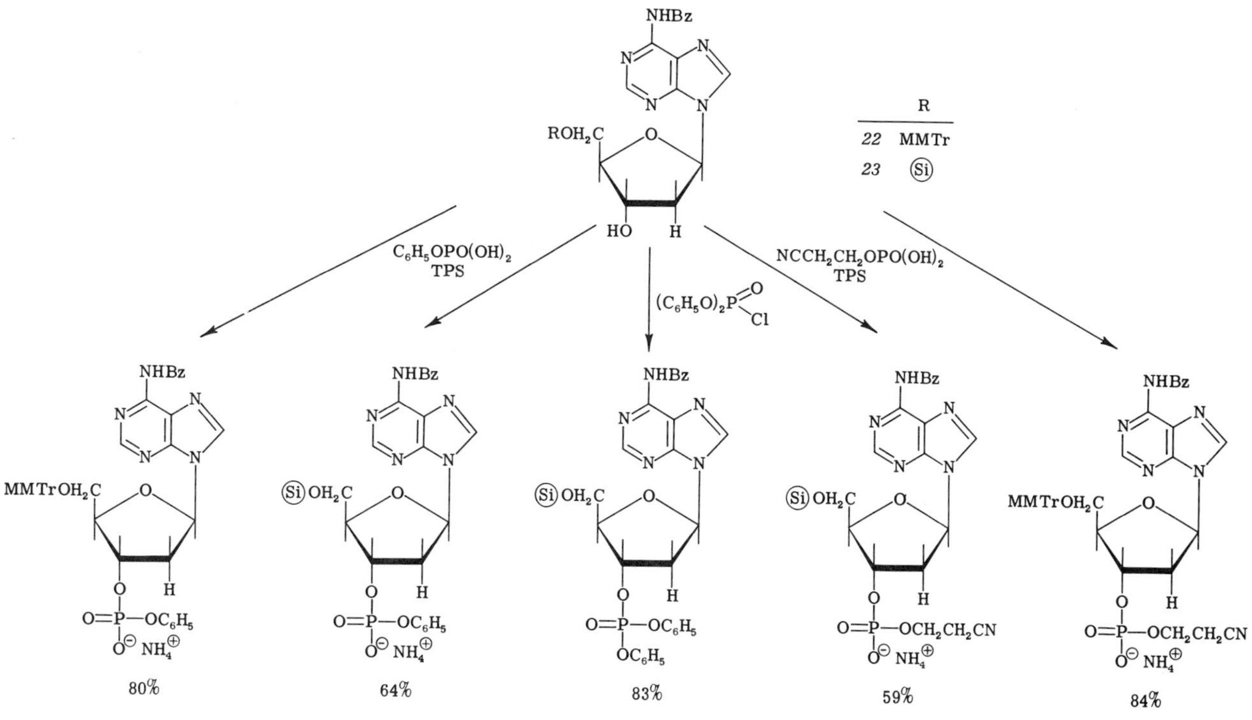

Scheme 3

REFERENCES

1. J. F. W. McOmie, *Protective Groups in Organic Chemistry,* Plenum Press, 1973.
2. S. Coffey, *Rodd's Chemistry of Carbon Compounds,* Vol. 1F, 2nd ed., Elsevier Publishing Co., New York, 1967.
3. D. G. Bartholomew and A. D. Broom, *Chem. Commun. 38* (1975).
4. G. Stork and P. F. Hudrlik, *J. Am. Chem. Soc. 90,* 4462 (1968).
5. E. J. Corey and A. Venkateswarlu, *J. Am. Chem. Soc. 94,* 6190 (1972).
6. K. K. Ogilvie and D. J. Iwacha, *Tetrahedron Lett. 317* (1973).
7. K. K. Ogilvie, *Can. J. Chem. 51,* 3799 (1973).
8. K. K. Ogilvie, K. L. Sadana, E. A. Thompson, M. A. Quilliam, and J. B. Westmore, *Tetrahedron Lett. 2861* (1974).
9. K. K. Ogilvie, E. A. Thompson, M. A. Quilliam, and J. B. Westmore, *Tetrahedron Lett.* 2865 (1974).
10. S. Hanessian and P. Lavallee, *Can. J. Chem. 53,* 2975 (1975).
11. D.P.L. Green, T. Ravindranathan, C. B. Reese, and R. Saffhill, *Tetrahedron 26,* 1031 (1970).
12. H. P. M. Fromageot, B. E. Griffin, C. B. Reese, J. E. Sulston, and D. R. Trentham, *Tetrahedron 22,* 705 (1966).
13. R. L. Letsinger and K. K. Ogilvie, *J. Am Chem. Soc. 89,* 4801 (1967).
14. K. K. Ogilvie, S. L. Beaucage, and D. W. Entwistle, *Tetrahedron Lett.* 1255 (1976).

ADVANCES IN THE SYNTHESIS OF *C*-GLYCOSYL NUCLEOSIDES

JOHN G. MOFFATT, HANS P. ALBRECHT, GORDON H. JONES,
DAVID B. REPKE, GÜNTER TRUMMLITZ, HIROSHI OHRUI,
AND CHHITAR M. GUPTA

Institute of Molecular Biology, Syntex Research,
Palo Alto, California

ABSTRACT

A number of basic approaches for the synthesis of a variety of
C-glycosyl nucleosides are described. One approach involves the
synthesis of certain protected derivatives of 2,5-anhydro-D-
allose. The aldehyde function of the latter compounds can be
elaborated into a wide range of heterocyclic systems, and examples of pyrazoles, isoxazoles, maleimides, and oxadiazoles are
provided. The condensation of 2,3-*O*-isopropylidene sugars with
stabilized phosphoranes leads to the formation of *C*-glycosides
via reversible cyclization of an intermediate olefin. In each
case the kinetic and thermodynamic products were those in which
the "aglycone" and isopropylidene groups were, respectively,
trans and *cis* oriented. The synthesis of a triazine ring onto
glycoside is described via annulation of a pyrimidine ring onto
a 3-amino-4-ribofuranosyl-pyrazole. The latter compound was
prepared via Curtius rearrangement of the corresponding acyl
azide. A new approach to the synthesis of appropriate pyrazole
C-glycosides involves condensation of 2,3,5-tri-*O*-benzyl-D-
ribose with a propiolic acid Grignard reagent followed by cycloaddition of diazomethane and acid-catalyzed cyclization.

In recent years a number of *C*-glycosyl nucleosides such as
pseudouridine, pyrazomycin, formycin, oxazinomycin, and showdomycin have been isolated from natural sources (1). Since many
of these compounds have been found to possess significant antibacterial, antiviral, or antineoplastic activities, the development of suitable synthetic methods for preparing such *C*-glyco-
sides has become a popular pursuit. Such methods have proved to
be considerably more complex than those that have been devised
for the synthesis of *N*-glycosyl nucleosides (2) and can be con-
ceptually divided into two major pathways. The first of these
(route a) comprises the direct condensation of a suitably activated carbohydrate moiety (e.g., 1 or an acyclic equivalent)
with a carbanion (2) derived from an appropriate heterocycle in
order to directly form the requisite *C*-glycosyl bond present in
3. This method has been used for the synthesis of pseudouridine
(3) and certain of its sugar analogs (4), as well as 1-deazauri-
dine (5) and 5-ribosylcytosine (6). Considerable progress has
also been made in the direct, Lewis-acid catalyzed glycosylation
of aromatic and certain heteroaromatic systems (7). The second,
and as yet considerably more flexible, route (b) involves the
elaboration of the requisite heterocyclic systems (3) from
suitable, functionally substituted anhydrosugars (4). The suc-
cess of the latter method rests upon the availability of appropriate compounds of type 4. This approach has been pioneered
by the work of Acton *et al.* (8) and of Farkas and Sorm (9), who
have described syntheses of appropriately blocked β-D-ribofuran-
osyldiazomethanes, which could be converted into formycin B (6)
and oxoformycin (7). The Czech group has also prepared a β-D-
ribofuranosylpyruvate via ozonolysis of a 1-β-D-ribofuranosyl-
2,4,6-trimethoxybenzene, and have used this substance in synthesis of showdomycin (10) and pyrazomycin (11).

Based upon the above successes, syntheses of many different
compounds of type 3 have been developed in which the functional
group CHX is formyl (12), carboxyl, or carboxyl derivatives (13),
either mono- or difunctionally substituted methyl (14) or an
acetylenic grouping (15). It is clear that all of these functional substituents are well suited for further elaboration into
a variety of heterocyclic systems. It may also be noted that
many of these same types of reactions have been explored by Tronchet and his colleagues and have been reviewed (16). These
studies have, however, largely been carried out by starting with
derivatives of 5-aldehydofuranoses and hence generally lead to
structures that are not conventional *C*-glycosyl nucleosides in
that they still possess a derivatized reducing sugar.

Our participation in this general area has been largely
based upon the premise that a formyl group provides a particic-
ularly versatile synthon from which a broad spectrum of heterocyclic systems can be derived. Hence, our initial objective

was the development of a versatile synthesis of a variety of
differently substituted 2,5-anhydro-D-alloses (*8-10*). This has
been achieved using the readily available 2,3,5-tri-*O*-benzoyl-β-
D-ribofuranosylcyanide (*5*) (*17*) as a starting material already
containing the elusive carbon-carbon linkage with the desired
β configuration. Preliminary efforts to subject the nitrile
group to reductive hydrolysis using Raney nickel and sodium hy-
pophosphite in aqueous pyridine and acetic acid (*18*) were accompanied by elimination of benzoates and formation of a furfural
derivative (*12*). If, however, the reaction was conducted in the
presence of *N,N'*-diphenylethylenediamine, a reagent that has
proved to be very valuable for the derivatization of nucleoside
5'-aldehydes (*19*), the aldehyde was immediately trapped as the
imidazolidine derivative (*6*) that was isolated in crystalline
form in 74% yield. Typically, such imidazolidine derivatives are
very stable to bases but sensitive to acids. Hence, the benzoyl
groups can be readily hydrolyzed from *6* and the resulting triol
can be treated with benzyl bromide and sodium hydride to form
the tribenzyl ether (*7*) in 84% yield (*12,15d*). Cleavage of the
imidazolidine derivatives from *6* or *7* can be readily accomplished
by treatment with *p*-toluenesulfonic acid monohydrate in methylene
chloride-acetone at 0°C giving 2,5-anhydro-3,4,6-tri-*O*-benzoyl-

D-allose (*8*) and its tri-*O*-benzyl analog (*9*) in pure form (*12*).
Alternatively, selective hydrolysis of the secondary benzoates
from *5* followed by acetonide formation and reductive hydrolysis
(as above) generated the 3,4-*O*-isopropylidene derivative (*10*).
It may be noted that Montgomery *et al.* (*20*) have independently
investigated the reductive hydrolysis of 5-*O*-benzoyl-2,3-*O*-iso-
propylidene-β-D-ribofuranosylcyanide and have observed epimerization of the aldehyde function. The absence of any such epimerization when the product is trapped *in situ,* as the imidazolidine derivative, further emphasizes the utility of the "Wanzlick
reagent" (*21*) in aldehyde chemistry.
    With the availability of the variously protected 2,5-anhydro-
D-allose derivatives (*8-10*), we initiated the synthesis of a variety of heterocyclic *C*-glycosides. One objective was the synthesis of the nucleoside antibiotic showdomycin (*17a*) (*1*) and
some of its derivatives. The extreme sensitivity of showdomycin
towards alkaline conditions precluded the use of benzoate pro-

tecting groups and dictated the use of the tribenzyl ether (9) as a starting material. The conversion of the aldehyde group of (9) into a bifunctional unit suitable for cyclization was brought about by reaction with sodium cyanide and hydrogen peroxide in aqueous dioxane containing sodium carbonate in order to give a pair of epimeric hydroxy amides (11) (22). This particular reaction is patterned after one that we have previously employed for synthesis of the nucleoside skeleton of the polyoxin antifungal agents (19a) and avoids the isolation of the labile cyanohydrins, which show a marked tendency to revert to the aldehyde upon attempted purification. Since various attempts at the oxidation of 11 to the corresponding keto amide were unsuccessful, the amide was converted to the hydroxy esters (12) by treatment with methanol in the presence of anhydrous Dowex 50 (H$^+$) resin. Once again, oxidation of 12 to the keto ester 13 was rather difficult, particularly in view of the instability of the latter product. Oxidation could, however, be rapidly accomplished using the DMSO-DCC method (23) in the presence of dichloroacetic acid. Without attempted purification, crude 13 was immediately reacted with carbamoylmethylenetriphenylphosphorane (14). The resulting meleamic acid ester underwent spontaneous cyclization, giving

the crystalline maleimide derivative 16a in an overall yield of 43% from 12. Completion of the synthesis then only required debenzylation of 16a, which was readily achieved by treatment with boron trichloride in methylene chloride at -78°C, giving crystalline showdomycin (17a) that was in every way identical to the natural product (22).

The above synthesis could also be adapted for the synthesis of ring-substituted showdomycins (e.g., 17b). Thus, reaction of

13 with the dimethylphenylphosphorane (*15*) followed by debenzylation of the resulting *16b* with boron trichloride gave 3-methyl-
showdomycin (*17b*) (24). It is interesting to note that the introduction of the methyl group in *17b* leads to a marked reduction
in antibacterial activity. Recently, the above sequence has been
adapted by Just *et al.* (25) for the synthesis of the biologically
inert carbocyclic analogs of D,L-showdomycin and D,L-pyrazomycin.

The interesting spectrum of biological activities exhibited
by the pyrazole *C*-glycoside pyrazomycin (Pyrazofurin) (26) prompted us to undertake the elaboration of the aldehyde group in compounds such as *8-10* into functionalized pyrazoles. To this end
8 was converted in high yield into the acrylate derivative *18*
through reaction with carbomethoxymethylenetriphenylphosphorane
(27). Subsequent 1,3-dipolar cycloaddition of diazomethane to
18 gave the pyrazoline *19* as a mixture of tautomers in essentially quantitative yield, and dehydrogenation to the crystalline
pyrazole (*20a*) was achieved in an overall yield of 86% from *18*
by reaction with chlorine in carbon tetrachloride. Treatment
of *20a* with methanolic ammonia, methanolic sodium methoxide, and
aqueous sodium hydroxide converted *20a* into 3(5)-carboxamido-4-
(β-D-ribofuranosyl)-pyrazole (*21a*) and its related methyl ester
(*21b*) and free acid (*21c*), respectively. The acrylate *18* was
also subjected to 1,3-dipolar cycloaddition of ethyl diazoacetate giving, after dehydrogenation with chlorine, the 4-β-D-ri-
bofuranosylpyrazole-3,5-diester (*20b*), which was converted into
the 3,5-dicarboxamide analogous to *21a* upon treatment with ammonia. All of the above cycloadditions appeared to proceed by the
conventional mode of attack, inserting nitrogen β to the carbomethoxy group (28). Even using more hindered dipoles such as *t*-

butyl diazoacetate, we obtained no evidence for the formation of
3-glycosylpyrazoles arising from inverse addition. Such products
would have been rather useful since they can be used as intermediates for the synthesis of formycin derivatives (8,9).

We have also developed a specific pathway leading to the 3-
β-D-ribofuranosylpyrazole system present in pyrazomycin and formycin (29). Thus, condensation of *8* with 1-chloroacetonylidene-
triphenylphosphorane (*22*) led smoothly to the α-chlorovinyl ketone (*23a*) in 78% yield. Treatment of this substance with hydrazine, according to the procedure of von Auwers (30), led directly to the desired pyrazole *24a*. Debenzoylation of *24a* using
methanolic ammonia or sodium methoxide was accompanied by the
formation of several byproducts and isolation of the desired 5-
methyl-3-(β-D-ribofuranosyl)pyrazole (*24c*)could only be accomplished by ion-exchange chromatography. This difficulty was
circumvented by a repetition of the above synthesis starting with
the tribenzyl ether (*9*). The resulting chloroketone (*23b*), which
was obtained as a 4:1 mixture of the Z and E isomers in 87% yield,
was readily cyclized with hydrazine to the pyrazole *24b*. While
catalytic hydrogenolysis of *24b* was somewhat capricious, debenzylation could be readily achieved by treatment with sodium in
liquid ammonia, which afforded readily purified *24c* in high yield.
This sequence can doubtless be extended to the synthesis of more
highly functionalized pyrazole systems.
     The use of 1,3-dipolar cycloaddition reactions can also be
applied to the synthesis of some functionalized isoxazole *C*-gly-
cosides (31). Thus, the readily prepared oxime (*25a*) derived
from *8* was treated with chlorine in ether at -60°C to form the

chlorooxime (26) or its nitroso tautomer 27. The chlorination of the oximes of several aldehydosugars has previously been examined in some detail by Tronchet et al. (32). Without purification, the chlorinated product was treated with triethylamine to generate the nitrile oxide 28, which underwent cycloaddition in the presence of ethyl propiolate to give 3-(2,3,5-tri-O-benzoyl-β-D-ribofuranosyl)-5-ethoxycarbonylisoxazole (29a) in 67% overall yield from 25a. While 29a could be slowly, but cleanly,

converted to the triol amide 30a upon treatment with methanolic ammonia the isoxazole ring appeared to be unstable during attempted debenzoylation with sodium methoxide. Hence, in order to prepare the ethoxycarbonylisoxazole (30b), it was necessary to repeat the above sequence starting with the tri-O-acetyl oxime 25b. Hydrolysis of the acetyl groups from the resulting 29b could then be achieved using ethanolic hydrogen chloride to give the desired 5-ethoxycarbonyl-3-β-D-ribofuranosylisoxazole (30b)

The nitrile oxides (28) could also be trapped by reaction with dimethylacetylenedicarboxylate giving the 4,5-dimethoxycarbonylisoxazoles 31a,b. Acidic deacetylation of 31b then readily gave the free nucleoside 31c. The oxime derived from 10 was also chlorinated, converted to the nitrile oxide, and reacted with ethyl propiolate. It is interesting to note that in this case the product, isolated in 62% yield, appeared on NMR analysis to be a roughly 5.5:1 mixture of the expected isoxazole 32 and the isomer 33, arising from inverse cycloaddition. Upon treatment with 90% trifluoroacetic acid and then methanolic ammonia, the mixture of 32 and 33 readily gave 30a in 71% yield. Since no evidence of inverse cycloaddition was observed during reactions with the triesters, 28 (R=Bz or Ac), it would be of interest to examine the addition of diazomethane to the acrylate derived from *10* in greater detail (27) in order to ascertain any unique role of the isopropylidene group. Recently, De Las Heras *et al.* (15c) have shown that addition of diazomethane to the propiolate *34* proceeds only in the normal fashion, while a similar reaction with ethyl diazoacetate leads to partial inverse addition.

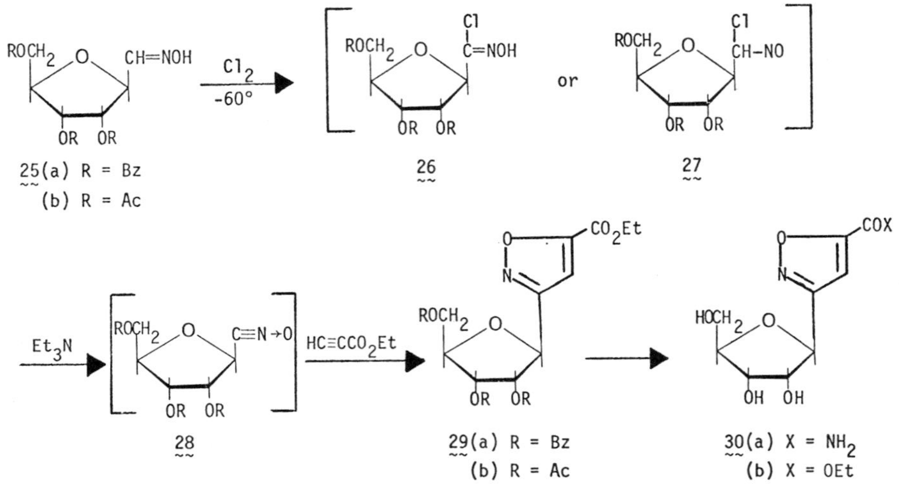

25(a) R = Bz
  (b) R = Ac
26
27

28
29(a) R = Bz
   (b) R = Ac
30(a) X = NH₂
   (b) X = OEt

We have also prepared one example of a 5-ribosylisoxazole via oxidative cyclization of an α,β-unsaturated ketoxime (33). To this end *8* was condensed with acetonylidenetriphenylphosphorane to give the pure *trans* enone *35a*. The latter was converted to the oxime *35b* and then treated with iodine and potassium iodide to form *36a*. Debenzoylation then readily afforded the desired 3-methyl-5-β-D-ribofuranosylisoxazole (*36b*) (31).

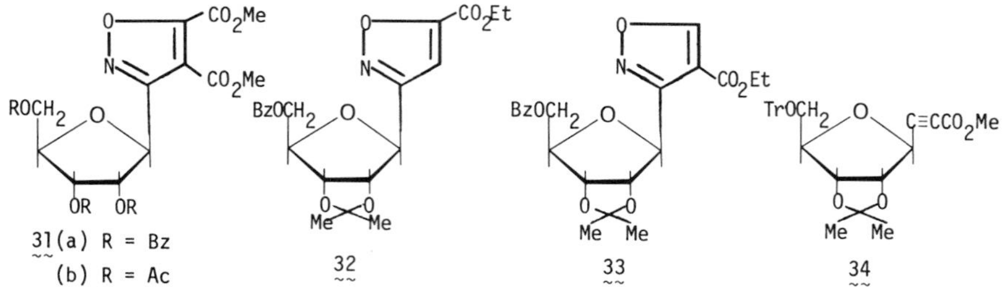

31(a) R = Bz
  (b) R = Ac
  (c) R = H
32
33
34

2,3,5-Tri-*O*-benzoyl-β-D-ribofuranosylcyanide (5) itself serves as a precursor for the synthesis of a 1,2,4-oxadiazole *C*-glycoside (29) via the well-known acylation of amidoximes (34). Treatment of 5 with hydroxylamine in methanol at 50°C gave the amidoxime 37, but only in a yield of 34%. This compound can be prepared much more efficiently by reaction of the chlorooxime 26 with ammonia, the overall yield from the oxime 25a being 92%. The reaction of 37 with acetic anhydride under reflux for 12 hr led to the direct formation of the 1,2,4-oxadiazole 38a, which could be converted to the free nucleoside 38b with methanolic ammonia. Alternatively, the amidoxime 37 could be reacted with acetaldehyde at room temperature to form the Δ^2-1,2,4-oxadiazo-line 39 as a mixture of diasterisomers. Dehydrogenation of 39 with chlorine in carbon tetrachloride rapidly gave 38a identical to the product from the acetic anhydride reaction. Presumably, these reactions can be modified so as to introduce a variety of functionalized substituents rather than the methyl group.

Structures 35-36

The reactions described above point out the versatility of derivatives of 2,5-anhydro-D-allose as intermediates in the synthesis of a wide range of heterocyclic *C*-ribofuranosides. We were, however, also interested in the development of routes to anhydro-sugars in which the functional group was separated from the tetrahydrofuran ring by a methylene group. Such compounds could not only themselves be elaborated into "homo *C*-glycosyl nucleo-sides," but could also serve as precursors for difuntional de-rivatives. The search for routes to the latter type of com-pound has independently been carried on by others, and it has been shown that ribofuranosyl halides bearing nonparticipating groups at C_2 can indeed be condensed with malonates, in the presence of sodium hydride, to form 2-ribofuranosylmalonates (14c,35). This has also been extended to condensations with a few β-ketoesters (14d,35).

Our approach to the synthesis of the desired monofunctional compounds was based upon the reactions of otherwise protected reducing sugars with stabilized phosphoranes (14a). Previous work by Zhdanov *et al.* (36) has shown that unprotected, and

partially protected, sugars react with stabilized phosphoranes
to form anomeric mixtures of *C*-glycosides. This reaction was
considered to proceed via concerted attack by a distal hydroxyl
group accompanying loss of triphenylphosphine oxide from the
intermediate phosphonium betaine (36).

In our hands, the reactions of 2,3-*O*-isopropylidene-5-*O*-trityl-
D-ribose *(40a)* with carbomethoxymethylenetriphenylphosphorane
(41a) or cyanomethylenetriphenylphosphorane *(41b)* proceeded readily in acetonitrile under reflux to give separable mixtures of
the α- and β-D-ribofuranosyl *C*-glycosides *(45* and *43)* in almost
quantitative yield. In each case the more polar isomer predomi-
nated in a ratio of about 3:1. Upon prolonging the reaction itself, or preferably by treatment of the crude product with sodium
methoxide, the distribution of isomers was found to drastically
change, the less polar isomer then being favored in a ratio of
roughly 3:1. Similar patterns were observed using 2,3-*O*-isopro-
pylidene-D-ribose *(40b)* and in these cases the stereoselectivity
of the initial reaction was even more striking, with the major
isomers predominating in ratios of 22:1 and 50:1. Once again,
base-catalyzed equilibration led to the the emergence of the
other isomer as the more thermodynamically stable and predominant
product.

Neither optical rotations nor ^{1}H-NMR spectra of the pure
products allowed a convincing assignment of "anomeric" configuration to the isomers of type *43* and *45*. An examination of the
^{13}C-NMR spectra, however, allowed us to unequivocally reach the

rather surprising conclusion that the kinetic products of the reaction have, in fact, the β-D-*ribo* configuration (43), while the more thermodynamically stable products are the seemingly more hindered α-D-*ribo* isomers (45) (14a). These assignments were based upon the well-known upfield shifts exhibited by carbon atoms that are subjected to steric crowding, particularly by a neighboring oxygen function, in a 5-membered ring (37). In the present case, the chemical shifts of C_2, C_3, and C_4 in the more thermodynamically stable and more sterically hindered isomers (45) appeared 1.8-3.8 ppm upfield of those in the kinetic products 43. The veracity of these assignments was ultimately confirmed by x-ray crystallography of the *p*-bromobenzoyl derivative of 45 (X=CN, R=H). It may also be noted that several other consistent patterns concerning the ^{13}C-chemical shifts of the isopropylidene groups in compounds of type 43 and 45 allow assignments of anomeric configuration (14a).

The course of the above reactions clearly involves an initial reaction of the stabilized phosphorane with the hemiacetal function in 40 to form the hydroxyolefin 42. In polar solvents such as acetonitrile this species is very short lived and undergoes spontaneous conjugate addition of the free hydroxyl group, giving a preponderance of the kinetically controlled β-*C*-glycoside (43). In the presence of a base, however, abstraction of a proton from C_2 of 43 can lead to β elimination of the ring oxygen generating the olefin anion 44, which can ultimately recyclize and thereby lead to the more thermodynamically stable α-*C*-glycoside (45).

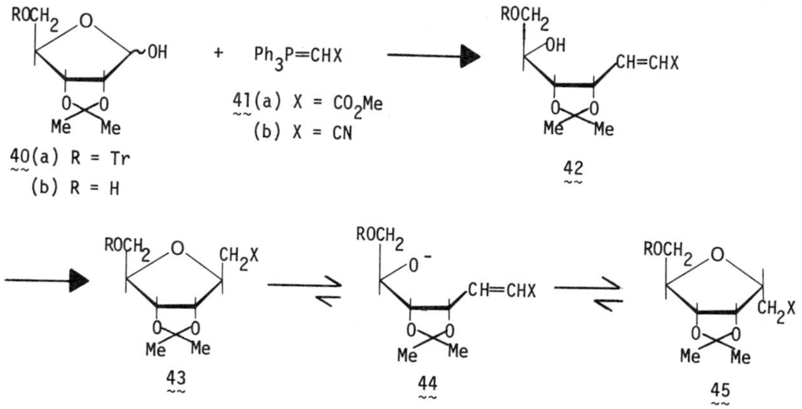

The steric course outlined is not unique to 2,3-*O*-isopropy-lidene-D-ribose derivatives, but appears to be consistent for 2,3-*O*-isopropylidene sugars in general. Thus the analogous re-actions of stabilized phosphoranes with 2,3;5,6-di-*O*-isopropyl-idene-D-allose (*46*) and 2,3;5,6-di-*O*-isopropylidene-D-mannose (*49*) lead, in each case, to a kinetic product in which the CH$_2$X moiety is introduced *trans* to the acetonide (*47* and *50*). Equi-libration of these compounds leads to the more thermodynamically stable products in which there is a *cis* relationship between those functions (*48* and *51*).

As might be expected, the *C*-glycosyl malonates (*52-55*), which can be prepared from the glycosyl chlorides and diethyl sodiomalonate (38,39) are also subject to anomeric equilibration (14a). In each case the major isolated product from these highly basic reactions proved to be the more thermodynamically stable isomer (*53,55*) with the malonate and isopropylidene functions in a *cis* relationship. It had previously been assumed (38), on purely steric grounds, that the thermodynamic product derived from 2,3-*O*-isopropylidene-5-*O*-trityl-β-D-ribofuranosyl chloride was the seemingly less hindered isomer *52*.

The 2,3-O-isopropylidene group plays a unique role in this process (14a). Thus, the corresponding reaction of 2,3,5-tri-O-benzyl-D-ribose (56) with 41a in acetonitrile leads to the isolation of the *cis* (57) and *trans* (58) isomers of an acyclic olefin in yields of 60% and 34% and without any indication of spontaneous cyclization. Brief treatment of either pure isomer with sodium methoxide, however, leads to rapid cyclization; the *cis* isomer (57) gives pure β-C-glycoside (59), while the *trans* olefin (58) gives an inseparable 3:2 mixture of 59 and 60. In the absence of a 2,3-O-isopropylidene group, these compounds show only a modest tendency towards equilibration, the pure β isomer (59) showing only about 25% conversion to 60 after 24 hr in 0.1N sodium methoxide. Similar treatment of the 3:2 mixture of 59 and 60 led only to a small increase in the proportion of 59 (2:1). Thus, in the absence of the 2,3-O-isopropylidene group the tendency towards cyclization is markedly depressed and there appears to be some thermodynamic preference for the β *(trans)* configuration.

Generally similar results have been independently arrived at by Buchanan *et al.* (40) from the reaction of 56 and ethoxy-carbonylmethylenetriphenylphosphorane, except that cyclization of both isomers corresponding to 57 and 58 led to roughly 1:1 mixtures of the α- and β-C-glycosides. It has also been reported that treatment of the 4,5,7-tri-O-benzoyl ethyl ester analog of 60 with sodium ethoxide leads to epimerization to the β-C-glycoside (41).

The literature provides ample examples of the role of the isopropylidene function in inducing cyclization of the intermediate olefins (42, 44). In general, the presence of one oxygen-containing five-membered ring appears to favor the formation of a second fused five-membered ring. Thus, for example, it is know that D-mannose-2,3-cyclic carbonate (42) and 3,6-anhydro-

D-glucose (43) exist to an extent of greater than 99% in the
furanose form and that acidic equilibration of methyl 3,6-anhy-
dro-D-glucopyranoside leads predominantly to the furanose isomer (44). The formation of only furanose C-glycosides from *40b*
and the lack of spontaneous cyclization of *57* and *58* are further
examples of this effect. It should be pointed out that the cyclization reaction is solvent dependent, with the reaction of
40a with *41a* in chloroform, rather than acetonitrile, leading
to acyclic olefins similar to *57* and *58* (45).

Analysis of the ^{1}H-NMR spectra of the various furanose C-
glycosides described above shows that the kinetic and thermodynamic products tend to exist in O-*exo* and O-*endo* envelope conformations, respectively (14). By adopting the O-*endo* conformation the major steric interactions between the *cis* oriented
isopropylidene and "aglycone" moieties are minimized. Recently,
fairly detailed conformational studies based upon analysis of
the ^{1}H-NMR spectra of *43* and *45* (X=CN,R=Tr) have indeed suggested
that the nonbonding interactions in the preferred conformation
of the α-anomer are, in fact, less than those present in the β-
isomer (46). In view of the somewhat different patterns of cyclization shown by *57* and *58*, the sterochemistry of the intermediate olefins (*42* and *44*) might also play a role. This point
remains unexplored at this time.

Finally, it should be pointed out that two striking exceptions to the above generalities have been noted in the literature
Thus, base catalyzed cyclization with urea of the more thermodynamically stable ribofuranosyl malonate (now known to be the α-
anomer *53* (14a)) leads to what is almost certainly the barbituric
acid β-C-glycoside, *61* (38,14a). Similarly, prolonged treatment
of the α-C-glycoside *62* with ammonia leads to complete "anomerization," giving the 2',3'-O-isopropylidene derivative of pyrazomycin (*63*) (14d). A unifying feature of these two exceptions is
that in both cases the final product is enolic in nature. This
suggests that polar interactions can perhaps play an overriding
role in the determination of relative stabilities. This is somewhat akin to the fact that a number of 2,3-O-isopropylidenefurano
syl halides and glycosides, both of which differ from the C-glycosides of types *43* and *45*, etc., by the existence of an additional anomeric dipole, prefer an *exo (trans)* configuration (14a,
46). Clearly, a full understanding of the factors controlling
anomeric configuration in furanose derivatives of various types
deserves further attention.

The work described so far leads the way for the synthesis
of a wide range of C-glycosides derived for 5- and 6-membered
heterocycles. The problem of devising routes to comparable derivatives of bicyclic systems more closely allied to the purines
is considerably more difficult. While this challenge has been
met with the successful syntheses of formycin B (6) and oxofor-

mycin (7), the problem of preparing monocyclic *C*-glycosides bearing suitable functionality for the annulation of a second fused ring remains formidable. Clearly, the preparation of 2- and 8-*C*-glycosylpurines, in which the carbon directly attached to the sugar is flanked by two nitrogen atoms, is less demanding and has been more easily exploited (13a,47).

Considerable recent interest has centered on the synthesis of purine nucleoside analogs in which there is a bridgehead nitrogen atom (48). It seemed to us that *C*-glycosides derived from this type of heterocycle might be readily available via .elaboration of a triazine ring using the ring nitrogen and carbonyl function present in the previously described pyrazole derivatives such as *20a*. The route we envisioned, however, required the conversion of the carbomethoxy function to a hydrazide, and the benzoyl groups were found to be incompatible with this step. Hence we chose to prepare the tribenzyl ether (*64*), which was obtained quite readily by starting with 2,5-anhydro-3, 4,6-tri-*O*-benzyl-D-allose (*9*). This compound was converted to the acrylate (*63*) and subjected to 1,3-dipolar cycloaddition of diazomethane and dehydrogenation with chlorine as previously outlined with the analogous tribenzoate (*8*).

The dehydrogenation step in the above sequence was, however, somewhat capricious and the yield dropped considerably during larger scale reactions, presumably due to side reactions between the benzyl ethers and chlorine. A convenient way to avoid this problem was to prepare methyl 3-(2,3,5-tri-O-benzyl-β-D-ribofuranosyl)propiolate (65), which would undergo direct cycloaddition of diazomethane, giving 64 without the necessity of dehydrogenation. Recent work by Buchanan *et al.* (15b) has shown that synthesis of 65 can be achieved in 21% yield via condensation of 2,3,5-tri-O-benzyl-D-ribofuranosylbromide with the silver derivative of methyl propiolate. Unfortunately, however, the major product from this reaction is the α-isomer of 65, which was isolated in 42% yield. Since completion of this aspect of our work, De Las Heras *et al.* (15c) have reported the related preparation of methyl 3-(2,3-O-isopropylidene-5-O-trityl-β-D-ribofuranosyl)propiolate and have successfully achieved the cycloaddition of diazomethane giving the desired pyrazole ester 67a. We too had previously prepared a related pyrazole protected with an isopropylidene group (67b) (27). We preferred, however, to proceed with the benzyl ether (64), which, because of the considerations presented above, was expected to show less tendency towards epimerization to the α-C-glycoside during subsequent steps.

65

66

67(a) R = Tr
   (b) R = Bz

Because of the difficulties sometimes experienced during the dehydrogenation step leading to 64, we attempted a new synthesis of 65 that would hopefully give a higher yield than that described by Buchanan (15b). To this end, 2,3,5-tri-O-benzyl-D-ribose (68) (49) was reacted with the Grignard derivative of methyl propiolate, but only poor results were obtained, probably due to instability of the Grignard reagent. On the other hand, condensation of 68 with the reagent derived from propiolic acid and two equivalents of ethylmagnesium bromide proceeded smoothly and, following acid-catalyzed conversion of the crude product to the methyl ester, a single acetylenic diol (69) was obtained in 73% yield (15d). This reaction finds close precedent in the condensation of 68 with ethynylmagnesium bromide, which has been enequivocally shown to lead predominantly to the D-*altro*-ethynylpen

titol derivative (50). This result is consistent with the rules
of asymmetric induction (51), and, based upon these considerations, we consider our product to also have the D-*altro* configuration (*69*).

Much precedent exists for the cyclization of 1,4-diols,
particularly when one hydroxyl is allylic in nature, upon treatment with acids (52,3c) or with sulfonyl chlorides in pyridine
(50,53). Unfortunately, our attempts to effect cyclization of
69 by these methods were fruitless and led mainly to nonacetyl-
enic products. In view of this failure we also attempted cyclization of 69 via activation of a hydroxyl group through reaction
with methyltriphenoxyphosphonium iodide (15d). It was anticipated
that selective activation of the propargylic alcohol would occur
leading to an oxyphosphonium salt that would undergo S_N2 attack
by the C_7-hydroxyl, giving 65. In practice, however, 65 and its
α-isomer (*66*) were obtained in yields of only 19 and 2%, respectively. The major acetylenic product proved to be the C_7 epimer
of *66*, based upon a variety of chemical transformations and spectral data. The formation of this compound must be explained by
the preferential activation of C_7OH rather than the expected C_3-
OH followed by S_N2 cyclization.

Treatment of 69 with diazomethane, however, led smoothly to
the formation of the crystalline pyrazole *70a*, which was isolated
in 76% yield. A small amount (7%) of the related *N*-methylpyra-
zole (*70b*) resulting from alkylation of *70a* by excess diazomethane was also isolated. Quite unlike the results with 69, treatment of *70a* with *p*-toluenesulfonic acid in benzene led to facile
cyclization giving *64*, identical to that described via the acrylate *63*, in 85% yield. This method provides a reliable route to
64 in yields at least as good as those available by the alternate
scheme. It is interesting that the cyclization of 69 proceeds
stereospecifically with apparent inversion of configuration at
the allylic center. This same situation was previously observed
during the synthesis of pseudouridine (3c) and suggests a direct
S_N2 displacement of the protonated allylic alcohol, rather than
intervention of an allylically stabilized carbonium ion. In the
present case, the D-*allo* isomer of 69 was not avaible for comparative study. It could be shown, however, that neither 64 nor
its α-anomer, which is readily available from 66, underwent any
anomeric equilibration under the acidic conditions used for cyclization.

In order to place a nitrogen atom at the desired site
in the ultimate purine-related product (76) it was necessary to
convert the carbomethoxy group of 64 into an amine. To this
end 64 was reacted with hydrazine in methanol under reflux
giving the hydrazide *71a* in 86% yield. The latter was, in
turn, converted to the acyl azide *71b*, which, without purification, was subjected to a Curtius rearrangement by heating in *t*-

butanol. The resulting *t*-butyl urethane *72*, which was isolated
in 71% yield, was converted to 3(5)-amino-4-(2,3,5-tri-*O*-benzyl-
β-D-ribofuranosyl)pyrazole (*73a*) by treatment with aqueous trifluoroacetic acid. It is interesting to note that in a recent
brief communication De Las Heras *et al.* (54) have reported a
closely related synthesis of the 2',3'-*O*-isopropylidene-5'-*O*-
trityl analog of *73a*. These authors have reported that this compound undergoes anomeric equilibration in both aqueous and non-
aqueous solvents, even under neutral conditions. In our experience the benzyl ether *73a* is rather unstable and tends to give
several spots upon TLC examination after storage. There was,
however, no clear indication of anomerization and *73a*, as directly obtained, gave satisfactory ^{1}H- and ^{13}C-NMR spectra,
clearly indicating the presence of only the pure β compound.
This observation supports the contention that protection via the
tribenzyl ether offers a distinct advantage over the use of an
isopropylidene function. It may be noted that the crystalline
N-methyl derivative *73b*, prepared as above from *70b*, appears to
be completely stable.

Cyclization of *73a* to 2,4-dioxo-8-(2,3,5-tri-*O*-benzyl-β-D-
ribofuranosyl)-1H,3H-pyrazolo(1,5-*a*)-1,3,5-triazine (*75*) was
readily achieved by treatment with phenoxycarbonylisocyanate *(74)*
(55), a reagent that we have previously found to be valuable
with other amino heterocycles (56). Related cyclizations using
ethoxycarbonylisothiocyanate and related compounds have recently
been reviewed (57). Debenzylation of *75* was efficiently accomplished by treatment with boron trichloride in methylene chloride
at -78°C. The resulting purine-related *C*-glycoside *76*, an analog
of xanthosine in which N^9 has been translocated to an angular
polition, was obtained in crystalline form in 73% yield. Debenzylation of *75* was also attempted using sodium in liquid ammonia,
but this reaction led to the formation of a chromatographically
distinguishable isomer in addition to the desired *76*. Examination of the mixture by borate electrophoresis at pH 9.2 showed
that *76* had a mobility similar to that of β-pseudouridine, while
the second isomer from the reaction with sodium in liquid ammonia behaved like α-pseudouridine. Presumably, under the strongly
basic conditions of the reaction using sodium in ammonia a proton was abstracted from N^1 of the heterocycle leading to concerted opening and reclosing of the furanose ring, resulting in anomeric equilibration. A similar mechanism has been invoked to
explain the base-catalyzed anomerization of pseudouridine (58).
In the present case no apparent furanose-pyranose equilibration
accompanies this process. De Las Heras *et al.* (54) have recently
reported the preparation of a fully protected derivative of the
2-thio analog of *76* but, as yet, removal of the protecting groups
has not been described. Clearly, the aminopyrazole *73a* is a versatile intermediate from which to build a variety of purine-related *C*-glycosides containing bridgehead nitrogens.

In this paper we have surveyed a rather large body of work on the synthesis of C-glycosyl nucleosides that has been done in these laboratories during the past few years. Clearly, the availability of key intermediates such as *8,43* (X=CO₂Me, CN), and *65* has made possible the synthesis of a wide range of C-glycosyl nucleosides. This type of work is now being widely explored in a number of laboratories and it is to be hoped that these efforts will lead to the development of pharmacologically useful agents.

REFERENCES

1. For a general review see Suhadolnik, R. J., *Nucleoside An-
    tibiotics,* Wiley-Interscience, New York, 1970.
2. For reviews see, e.g. (a) Goodman, L., in *Basic Principles
    in Nucleic Acid Chemistry,* Vol. I Ts'0 P.O.P., Academic
    Press, New York, 1974, p. 93; (b) Watanabe, K. A., Hollen
    berg, D. H., and Fox, J. J., *J. Carbohydr. Nucleosides
    and Nucleotides 1,* 1 (1974).
3. (a) Shapiro, R., and Chambers, R. W., *J. Am. Chem. Soc. 83,*
    3920 (1961); (b) Brown, D. M., Burdon, M. G., and
    Slatcher, R. P., *J. Chem. Soc.* 1051 (1968); (c) Lerch,
    U., Burdon, M. G., and Moffatt, J. G., *J. Org. Chem. 36,*
    1507 (1971).
4. Asbun, W. A., and Binkley, S. B., *J. Org. Chem. 33,* 140
    (1968).
5. Mertes, M. P., Zielinski, J., and Pillar, C., *J. Med. Chem.
    10,* 320 (1967).
6. David, S., and Lubineau, A., *Carbohydr. Res. 29,* 15 (1973).
7. (a) Ohrui, H., Kuzuhara, H., and Emoto, S., *Agric. Biol.
    Chem. 36,* 1651 (1972); (b) Kalvoda, L., *Collect. Czech.
    Chem. Commun. 38,* 1679 (1973).
8. Acton, E. M., Ryan, K. J., Henry, D. W., and Goodman, L.,
    Chem. Commun. 986 (1971).
9. Farkaś, J., and Šorm, F., *Collect. Czech. Chem. Commun. 37,*
    2798 (1972).
10. Kalvoda, L., Farkaś, J., and Šorm, F., *Tetrahedron Lett.*
    2297 (1970).
11. Farkaś, J., Flegelová, Z., and Šorm, F., *Tetrahedron Lett.*
    2279 (1972).
12. Albrecht, H. P., Repke, D. B., and Moffatt, J. G., *J. Org.
    Chem. 38,* 1836 (1973).
13. See, e.g. (a) Dinh, T. H., Kolb, A., Gougette, C., Igolen,
    J., and Dihn, T. S., *J. Org. Chem. 40,* 2825 (1975) and
    references therein; (b) Just, G., and Ramjeesing, M.,
    Tetrahedron Lett. 985 (1975); (c) Fuertes, M., Garcia-
    Lopez, M. T., Garcia-Munoz, G., and Madronero, R., *J.
    Carbohydr. Nucleosides and Nucleotides 2,* 277 (1975).
14. See, e.g. (a) Ohrui, H., Jones, G. H., Moffatt, J. G.,
    Maddox, M. L., Christensen, A. T., and Byram, S. K., *J.
    Am. Chem. Soc. 97,* 4602 (1975); (b) Chu, C. K., Watanabe,
    K. A., and Fox, J. J., *J. Heterocycl. Chem 12,* 817
    (1975); (c) Hanessian, S., and Pernet, A. G., *Can. J.
    Chem. 52,* 1280 (1974); (d) De Bernardo, S., and Weigele,
    M., *J. Org. Chem. 41,* 287 (1976).

15. See, e.g., (a) Buchanan, J. G., Edgar, A. R., and Power,
     M. J., *J. Chem. Soc., Perkin Trans 1* 1943 (1974); (b)
     Buchanan, J. G., Edgar, A. R., Power, M. J., and Williams,
     G. C., *Chem. Commun.* 501 (1975). (c) De Las Heras, F. G.,
     Tam, S. Y. K., Klein, R. S., and Fox, J. J., *J. Org.
     Chem. 41*, 84 (1976); (d) Gupta, C. M., Jones, G. H., and
     Moffatt, J. G., *J. Org. Chem. 41,* 3000 (1976).
16. Tronchet, J. M. J., *Biologie Médicale 4*, 83 (1975).
17. Bobek, M., and Farkaś, J., *Collect. Czech. Chem. Commun.
     34,* 1684 (1969).
18. Backeberg, O. G., and Staskun, B., *J. Chem. Soc.* 3961 (1962).
19. See, e.g. (a) Damodaran, N. P., Jones, G. H., and Moffatt,
     J. G., *J. Am. Chem. Soc. 93,* 3812 (1971); (b) Ranganathan,
     R. S., Jones, G. H., and Moffatt, J. G., *J. Org. Chem.
     39* 290 (1974) and unpublished work from this laboratory.
20. Montgomery, J. A., Hewson, K., and Laseter, A. G., *Carbo-
     hydr. Res. 27*, 303 (1973).
21. Wanzlick, H. W., and Löchel, W., *Chem. Ber. 86,* 1463 (1953).
22. Trummlitz, G., and Moffatt, J. G., *J. Org. Chem. 38,* 1841
     (1973).
23. Pfitzner, K. E., and Moffatt, J. G., *J. Am. Chem. Soc. 87,*
     5661, 5670 (1965).
24. Trummlitz, G., Repke, D. B., and Moffatt, J. G., *J. Org.
     Chem. 40,* 3352 (1975).
25. Just, G., and Kim, S., *Tetrahedron Lett.* 1063 (1976).
26. Gutowski, G. E., Sweeney, M. J., De Long, D. C., Hamill,
     R. L., Gerzon, K., and Dyke, R. W., *Ann. NY Acad. Sci.
     255,* 544 (1975).
27. Albrecht, H. P., Repke, D. B., and Moffatt, J. G., *J. Org.
     Chem. 39*, 2176 (1974).
28. See, e.g., Bastide, J., El Ghandous, N., and Henri-Rousseau,
     O., *Bull Soc. Chim. Fr.* 2290 (1973) and references therein.
29. Repke, D. B., Albrecht, H. P., and Moffatt, J. G., *J. Org.
     Chem. 40,* 2481 (1975).
30. von Auwers, K., and Broche, H., *Ber. 55*, 3880 (1922).
31. Albrecht, H. P., Repke, D. B., and Moffatt, J. G., *J. Org.
     Chem. 40,* 2143 (1975).
32. Tronchet, J. M. J., Barbalat-Rey, F., and Le-Hong, N., *Car-
     bohydr. Res. 29,* 297 (1973).
33. Büchi, G., and Vederas, J. C., *Am. Chem. Soc. 94,* 9128 (1972).
34. Eloy, F., and Lenaers, R., *Chem. Rev. 62,* 155 (1962).
35. Ohrui, H., and Fox, J. J., *Tetrahedron Lett.* 1951 (1973).
36. Zhdanov, Y. A., Alexeev, Y. E., and Alexeeva, V. G., *Adv.
     Carbohydr. Chem. Biochem. 27,* 227 (1972).
37. (a) Christl, M., Reich, H. J., and Roberts, J. D., *J. Am.
     Chem. Soc. 93,* 3463 (1971); (b) A. S. Perlin, in *Inter-
     national Review of Science. Organic Chemistry, Series
     Two,* Vol. 7 (G.O. Aspinall, ed.), Butterworths, London,
     1976, p. 1; (c) additional references in Ref. 14a.

38. Ohrui, H., and Fox, J. J. *Tetrahedron Lett.* 1951 (1973).
39. Hanessian, S., and Pernet, . G., *Can J. Chem. 52,* 1266 (1974).
40. Buchanan, J. G. Edgar, A. R., Power, M. J., and Theaker, P. D., *Carbohydr. Res. 38,* C22 (1974).
41. Hanessian, S., Ogawa, T., and Guidon, Y., *Carbohydr. Res. 38,* C12 (1974).
42. Perlin, A. S., *Can. J. Chem. 44,* 539 (1966).
43. Angyal, S. J., *Angew. Chem. Int. Ed. Engl. 8,* 159 (1969).
44. Haworth, W. N., Owen, L. N., and Smith, F., *J. Chem. Soc.* 88 (1941).
45. Unpublished experiments by Dr. G. H. Jones.
46a. Ohrui, H., Emoto, S., Jones, G. H., and Moffat, J. G., Abstracts of the VIII International Symposium on Barbohydrate Chemistry, Kyoto, Japan, Aug. 1976, Abstract $4B_2$-8.
46b. Ohrui, H., and Emoto, S., *J. Org. Chem. 42,* 1951 (1977).
47. El Khadem, H. S., and El Ashry, E.S.H., *Carbohydr. Res. 32,* 339 (1974).
48. Bartholomew, D. G., Huffmann, H. J., Matthews, T. R., Robins, R. K., and Revankar, G. R., *J. Med. Chem. 19,* 814 (1976) and references therein.
49. Barker, R., and Fletcher, H. G., *J. Org. Chem. 26,* 4605 (1961).
50. Buchanan, J. G., Edgar, A. R., and Power, M. J., *J. Chem. Soc., Perkin Trans. 1* 1943 (1974).
51. (a) Cram, D. J., and Wilson, D. R., *J. Am. Chem. Soc. 85,* 1245 (1963). (b) Karabatsos, G. J., *J. Am. Chem. Soc. 89,* 1367 *(1967).*
52. Hudson, B. G., and Barker, R., *J. Org. Chem. 32,* 3650 (1958).
53. See, e.g. (a) Rabinsohn, Y., and Fletcher, H. G.. *J. Org. Chem. 32,* 3452 (1967); (b) Defaye, J., and Horton, D., *Carbohydr. Res. 14,* 128 (1970).
54. De Las Heras, F. G., Chu, C. K., Tam, S.Y.K., Klein, R. S., Watanabe, K. A., and Fox, J. J., *J. Heterocycl. Chem. 13,* 175 (1976).
55. Speziale, A. J., Smith, L. R., and Fedder, J. E., *J. Org. Chem. 30,* 4306 (1965).
56. Prisbe, E. J., Verheyden, J.P.H., and Moffatt, J. G., Abstracts of the Centennial Meeting of the American Chemical Society, San Francisco, August-September 1976, CARB 90.
57. Esmail, R., and Kurzer, F., *Synthesis* 301 (1975).
58. Chambers, R. W., *Progr. Nucl. Acid Res. Mol. Biol. 5,* 349 (1966).

SYNTHESIS OF C-GLYCOSYL THIAZOLES

M. FUERTES, M. T. GARCÍA-LÓPEZ, G. GARCÍA-
MUÑOZ, AND M. STUD

Instituto de Química Médica
Juan de la Cierva
Madrid, Spain

ABSTRACT

Condensation of 2,3,5-tri-O-benzoyl-β-D-ribofuranosyl thiocarboxamide (1) with α-chloroketo compounds yielded the corresponding
2-C-glycosyl thiazole nucleosides (4a and 7) as the major products, along with the 2-(thiazol-2-yl)-5-benzoyloxymethylfuran
derivatives (5a and 8). Reaction of 1 with ethyl bromopyruvate
gave the 2-C-glycosyl thiazole nucleoside 12 as the only resulting compound. A similar series of reactions was carried out with
2-thiocarboxamide-5-benzoylosymethylfuran (2) and α-haloketones.
Finally, treatment of methyl 6-deoxy-6-diazo-2,3-O-isopropylidine-
β-D-ribo-hexofuranosid-5-ulose (24) with thiourea afforded the
4-C-glycosyl thiazole 26.

Of the several synthetic procedures described in the literature
for obtaining thiazole derivatives, the reaction of thioamides
and related compounds with α-halocarbonyl derivatives has been
the most extensively used. In a recent preliminary communication
(1) we have reported on the synthesis of a 2-C-glycosyl thiazole
nucleoside and also the synthesis of several acyclic sugar 4-
thiazolyl nucleosides analogs. Now we wish to give a full account
of this and related work.

The starting material in our synthesis of 2-C-glycosyl thiazole nucleosides (Scheme 1), the 2,3,5-tri-O-benzoyl-β-D-ribofur-
anosyl thiocarboxamide, was obtained in 20% yield as an amorphous
solid by reaction of 2,3,5-tri-O-benzoyl-β-D-ribofuranosyl cyanide (2) with hydrogen sulfide in ethanol containing triethylamine. This furan derivative, resulting from the elimination of

two benzoyloxy groups, was also separated from the reaction in 14% yield. Similar base-catalyzed eliminations of these pro- tecting groups have been previously reported by Moffatt and his research group (3).

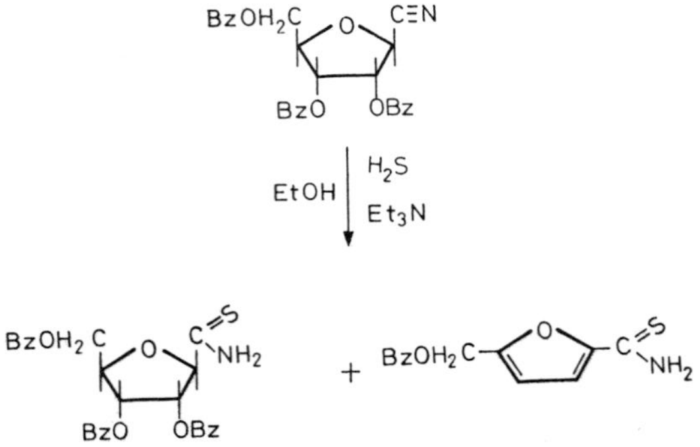

Scheme 1

two benzoyloxy groups, was also separated from the reaction in 14% yield. Similar base-catalyzed eliminations of these protect- ing groups have been previously reported by Moffatt and his re- search group (3).

The assignment of the anomeric configuration of this thio- carboxamide was made on the basis of the known configuration of the nitrile used as starting material, since the doublet corres- ponding to the anomeric proton in this compound showed a coupling constant of 5 Hz. This assignment was further supported by the consistent application of Imbach's criterion (4) on a 2,3-*O*-iso- propylidene thiazole-*C*-nucleoside obtained from this thiocarbox- amide, as described below.

Reaction of the thiocarboxamide (1) with chloroacetone in ethanol afforded a mixture of 2-(2,3,5-tri-*O*-benzoyl-β-D-ribo- furanosyl)-4-methylthiazole (4a) in 32% and the furan derivative (5a) in 15% yield. Debenzoylation of these products with meth- anolic ammonia gave the corresponding deblocked compounds 4b and 5b. Similarly, reaction of the furan thiocarboxamide (2) with chloroacetone gave a 35% yield of 5a identical with the compound obtained in the foregoing reaction (Scheme II).

The thiocarboxamide (1) reacted smoothly with ethyl oxaloch- loroacetate to give a mixture of the blocked *C*-glycosyl nucleo- side (7) and the elimination product (8), which were isolated by preparative layer chromatography in yields of 31 and 27%, respec- tively. Compound 8 was also obtained from the reaction of furan thiocarboxamide (2) with ethyl oxalochloracetate.

Treatment of compounds 7 and 8 with methanol saturated with ammonia gave the corresponding deblocked dicarboxamides in low yields. In the case of compound 7, thin-layer chromatography of the crude reaction showed a complex mixture of products. Separation by preparative thin-layer chromatography yielded the expected deblocked β-anomer 9 in 25% yield (Scheme III).

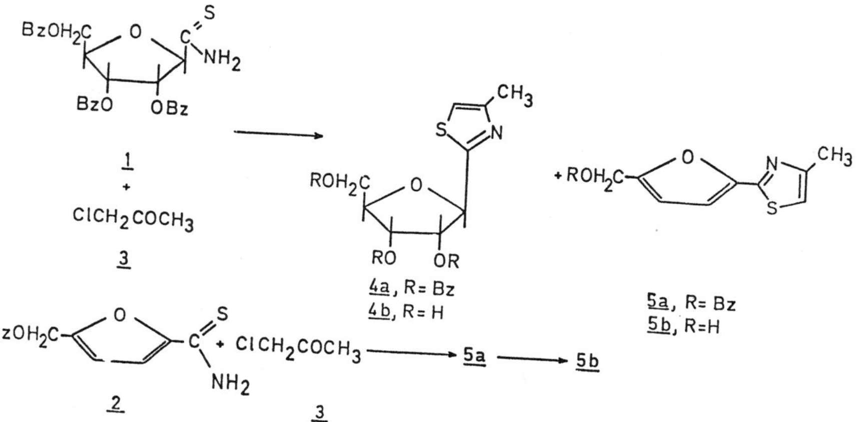

Scheme II

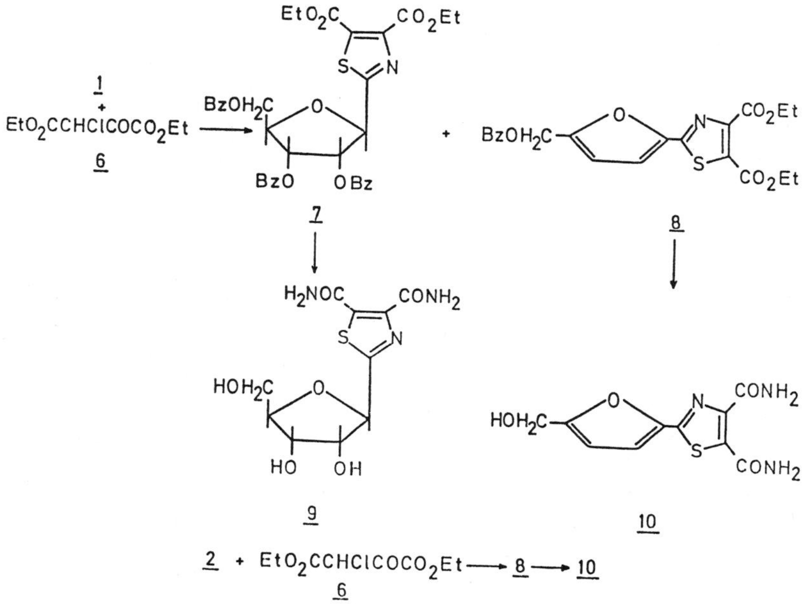

Scheme III

In a similar fashion the thiocarboxamide (1) when treated
with ethyl bromopyruvate at reflux in thanol solution gave the
protected C-glycosyl nucleoside 12 in 55% yield as a syrup
(Scheme IV). Although a furan derivative was expected, no compound of this type was found. Treatment of 12 with methanolic
ammonia afforded 13 as an amorphous solid in 81% yield. This
latter compound was then converted to the 2,3-O-isopropylidene
derivative 14. Its NMR spectrum showed the protons of the isopropylidene methyl groups as two singlets at δ= 8.50 and 8.68.
This difference of 0.18 ppm has been shown to be consistent with
the β configuration.

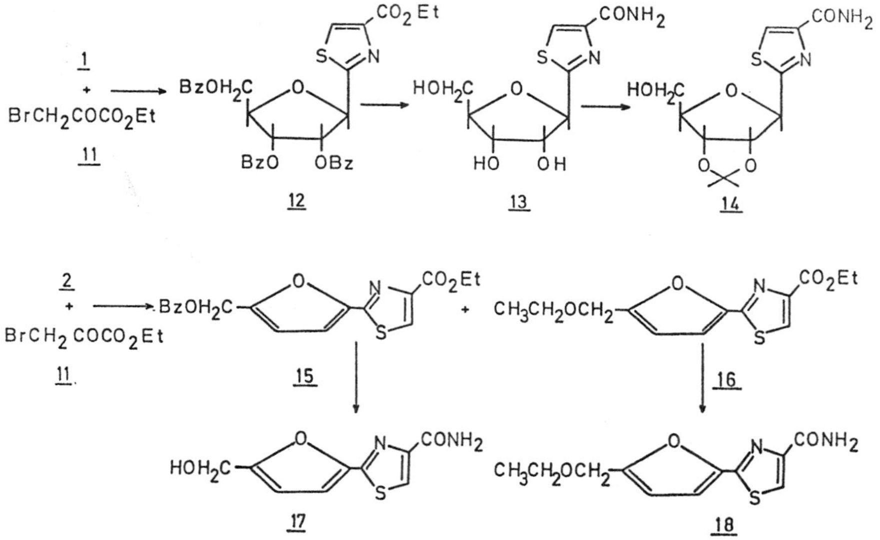

Scheme IV

As before, the use of the thiocarboxamide (2) in the synthesis of nucleoside-related compounds having a furyl moiety was
evaluated. Reaction of 2 with ethyl bromopyruvate in refluxing
ethanol gave, besides the expected compound 15 in 30% yield,
another compound in 27% yield that was identified as 16. The
NMR spectrum of 16 clearly indicated the absence of benzoyl protons and the presence of signals corresponding to an ethoxy
group, in addition to the signals of the carboethoxy thiazole
substituent. Further evidence for the structure assignment of
16 stems from its conversion to the carboxamide 18, the NMR
spectrum of which retains the characteristic pattern for the
ethoxymethyl moiety. The formation of 16 can be explained by
taking account of the stability of the carbonium ion resulting
from the acidic media originated on the reaction. Attack of the
carbonium ion by ethanol affords 16. Chemical evidence for this

assumption was supported by the formation of 16 from 15, when this last compound was treated with ethanol-hydrogen bromide at reflux temperature. Attempts to find compounds similar to 16 in the reaction of 2 with chloroacetone or ethyl oxalochloroacetate were unsucessful.

In order to avoid the formation of furan derivatives by loss of the benzoyl groups in the condensation of 1 with α-haloketones, we synthesized the glycosyl thiocarboxamide (22) having an iso-propylidene protecting group (Scheme V). The glycosyl nitrile

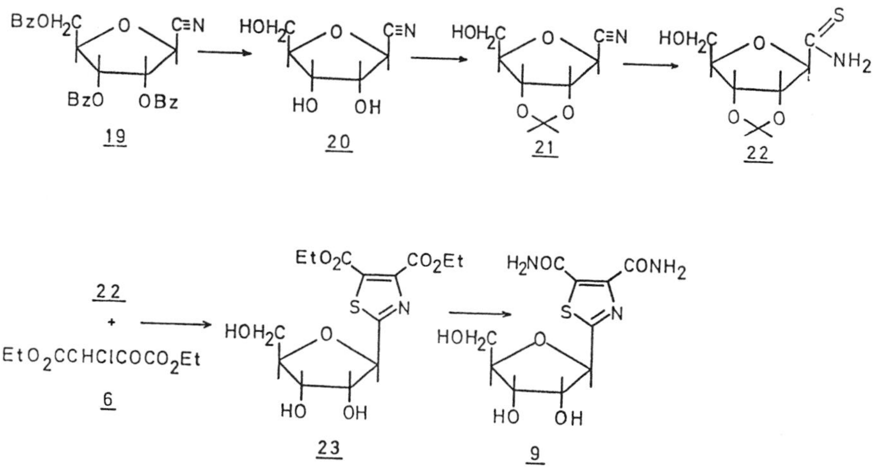

BzOH$_2$C C≡N   HOH$_2$C C≡N   HOH$_2$C C≡N   HOH$_2$C C–S / NH$_2$

BzO OBz   HO OH   O O   O O
19   20   21   22

22
+   EtO$_2$C CO$_2$Et   H$_2$NOC CONH$_2$
EtO$_2$CCHClCOCO$_2$Et   HOH$_2$C S–N   HOH$_2$C S–N
6   HO OH   HO OH
                          23   9

Scheme V

(19) was debenzoylated at room temperature with methanolic amm-onia to give the deblocked derivative 20, which was treated with ethyl orthoformate and acetone in the presence of hydrochloric acid to yield 2,3-O-isopropylidene-β-D-ribofuranosyl cyanide (21). Treatment of this product with hydrogen sulfide in ethanol con-taining triethylamine afforded 22.

Reaction of thiocarboxamide (22) with ethyl oxalochloroace-tate was then examined. As we expected only one thiazole deriv-ative was formed in a process involving the concomitant removal of the protecting isopropylidene group, due to the acidity of the reaction medium. In any case the obtained yield of 23 was not very high, only about 32%. Subsequent ammonlysis of the diester (23) gave the crystalline dicarboxamide derivative (9) which was identical with the compound obtained from 7.

It should be pointed out that compounds 7,9, and 23 provide a series of valuable intermediates for synthesis, via cyclization, of new purinelike nucleosides.

Finally, since it has been reported that the reaction of α-diazoketones with thioamides gives thiazole derivatives (5), we extended our studies to the synthesis of the 4-C glycosyl thia-

zole derivative 26 by reaction in refluxing ethanol of the dia-
zoketose 24 (6) with thiourea (Scheme VI). An analytically pure
sample of 26 was obtained via its picrate derivative since all
the attempts we made to obtain it from the repeatedly chromatographed reaction product were unsuccessful. The synthesis of
compound 26 failed when the reaction was performed starting from
the corresponding α-haloketose, probably due to the known instability of this compound (6).

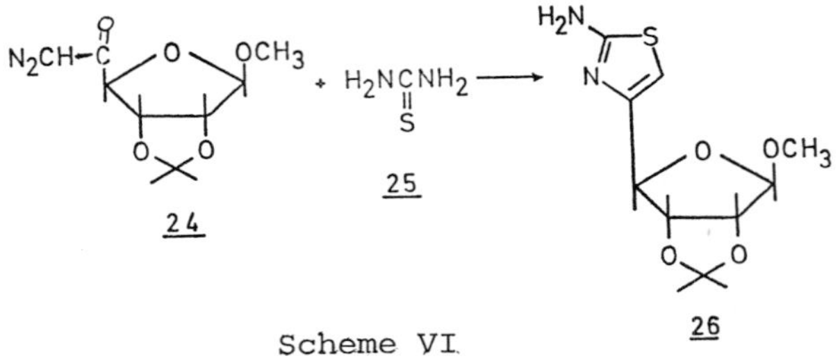

Scheme VI

REFERENCES

1. M. Fuertes, M. T. Garcia-Lopez, G. Garcia-Muñoz, and R. Mad-
      roñero, *J. Carbohydr. Nucleosides Nucleotides 2,* 277 (1975)
2. M. Bobek and J. Farkas, *Coll. Czech. Chem. Commun. 34,* 247
      (1969).
3. H. P. Albrecht, D. B. Repke, and J. G. Moffatt, *J. Org. Chem.*
      39, 2176 (1974).
4. (a) J. L. Imbach, J. L. Barascut, B. L. Kam, B. Rayner, C.
      Tamby, and C. Tapiero, *J. Heterocycl. Chem. 10,* 1069 (1973)
      (b) J. L. Imbach, J. L. Barascut, B. L. Kam, and C. Tapiero
      Tetrahedron Lett. 129 (1974); (c) J. L. Barascut, C. Temby,
      and J. L. Imbach, *J. Carbohydr. Nucleosides Nucleotides 1,*
      177 (1974).
5. L. C. King and F. M. Miller, *J. Am. Chem. Soc. 58,* 367 (1949).
6. A. Hampton, F. Perini, and P. J. Harper, *Carbohydr. Res. 37,*
      359 (1974).

A NEW METHOD FOR THE SYNTHESIS OF 2'-DEOXY-2'-
SUBSTITUTED PURINE NUCLEOSIDES. SYNTHESIS OF THE
ANTIBIOTIC 2'-DEOXY-2'-AMINOGUANOSINE

MORIO IKEHARA, TOKUMI MARUYAMA, AND HIROKO MIKI

*Faculty of Pharmaceutical Sciences, Osaka University
Suita, osaka, Japan*

ABSTRACT

8,2'-Anhydro-8-oxy-9-β-D-arabinofuranosyladenine (IIIa) was
treated with sodium azide in DMF to give an azido compound (IV),
which was deaminated, chlorinated, and derivatized to the 6,8-
di-methylmercapto compound (VIII). Compound VIII gave N^6-dimethyl-
2'-deoxy-2'-aminoadenosine (X) by treatment with dimethylamine
and Raney nickel. The dichloro compound (VI) also gave the 6,8-
dithio purine, which was transformed to 2'-deoxy-2'-azidoinosine
(XIb) by thiolation and oxidation with hydrogen peroxide. Alternatively, IIIa was converted to 3',5'-diacetylarabinofuranosyla-
denine and mesylated at 2'-hydroxyl. Heating of this compound
with sodium azide gave a 3'-azidoxylofuranosyladenine (XVb).
Compound IIIa was then protected with tetrahydropyranyl groups
at the 3' and 5'-hydroxyls and transformed to 2'-mesylarabinosyl-
adenine (XIIIc). Compound XIIIc was then subjected to reaction
with sodium azide and an azido compound (XV) was obtained. Deblocking of XV gave 2'-deoxy-2'-azidoadenosine (XVI) which showed
PMR signals of $H_{1'}$ and $H_{3'}$ shifted towards low field and C^{13} signals shifted 9 ppm upfield relative to adenosine. Hydrogenation
of the azido group of XVI gave 2'-deoxy-2'-aminoadenosine (XVIII),
which was identical with an authentic sample. This method was
then applied to 8,2'-anhydro-8-oxy-9-β-D-arabinofuranosylguanine
(XVIII), which was transformed to N^2,3',5'-tetrahydropyranyl-2'-
mesylarabinofuranosylguanine (XX). The reaction of XX with sodium azide in acetamide gave an azido compound (XXI). After deblocking, this compound was hydrogenated over palladium charcoal to give 2'-deoxy-2'-aminoguanosine (II), which was found to
be identical with the antibiotic 2'-deoxy-2'-aminoguanosine.

It is of considerable interest to synthesize purine nucleosides containing amino groups in the sugar moiety related to the antibiotics puromycin (I) and 3'-deoxy-3'-aminoadenosine (1). To this end several investigators have reported on the synthesis of several of these types of nucleosides (2). Quite recently Nakanishi *et al.* (3) found an antibiotic exhibiting antibacterial and antileukemic activities in the culture broth of *Aerobacter sp KY 3071* and elucidated its structure as 2'-deoxy-2'-aminoguanosine (II), which was the first natural nucleoside having the 2'-amino ribose structure. There are two reports (4,5) dealing with the chemical synthesis of 2'-deoxy-2'-aminoadenosine but none with the synthesis of 2'-deoxy-2'-aminoguanosine.

We attempted, therefore, to synthesize 2'-amino nucleosides in the ribo configuration starting from naturally occurring adenosine and guanosine. Since we have already investigated the cleavage reaction of cyclonucleoside bonds with a variety of nucleophiles (6,7), we employed sodium azide for opening up the cyclo linkage to yield the desired ribo configuration.

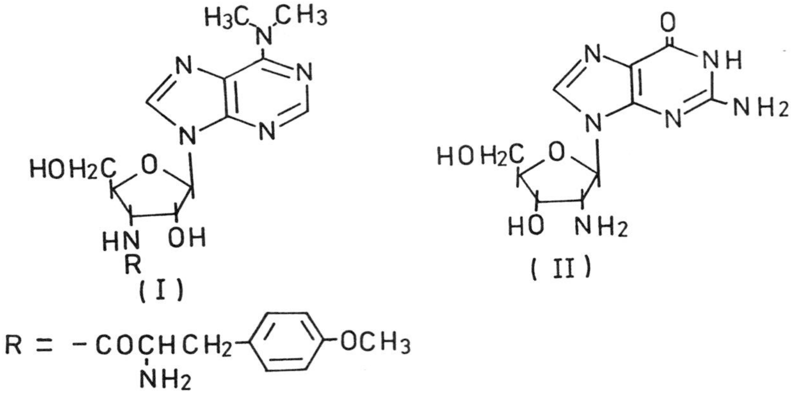

Chart 1

When 8,2'-anhydro-8-oxy-9-β-D-arabinofuranosyladenine (IIIa) (8) was heated with sodium azide in DMF at 150°C for 3 hr, a nucleoside (IV) having the 8-oxyadenine chromophore and an azide band in its ir spectrum was obtained in a yield of 57%. In order to change the 8-oxy function, compound IV was deaminated with nitrous acid to the 6,8-dioxy compound, which was subsequently treated with acetic anhydride to give 6,8-dioxy-9-β-(3',5'-di-O-acetyl-2'-deoxy-2'-azido-D-ribofuranosyl)purine (V). We have previously found a method for chlorinating 6,8-dioxyfunctions in 8-oxyinosine derivatives (9). Accordingly, compound V was heated at reflux temperature with phosphoryl chloride in tri-n-butylamine for 9 hr. The 6,8-dichloro compound (VI) was obtained in a yield of 18%. The 6-monochloro compound (VII) was also detected in the reaction mixture by TLC.

Compound VI was then allowed to react with aqueous sodium methylmercaptide in dioxane. The 6,8-dimethylmercapto compound (VIII) was obtained in a yield of 52%. The intactness of the ribose portion was confirmed by NMR decoupling experiments. As

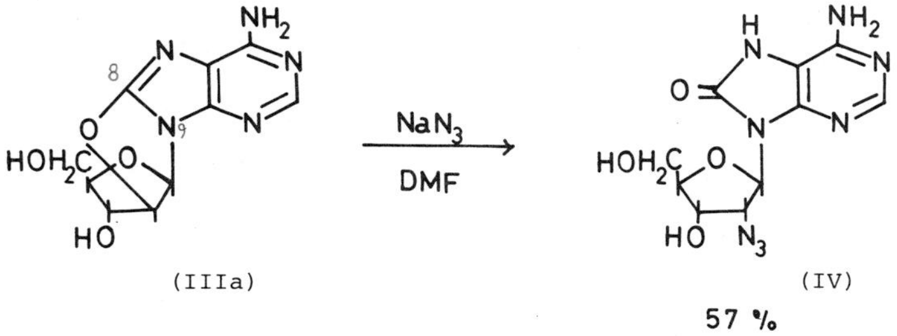

Chart 2

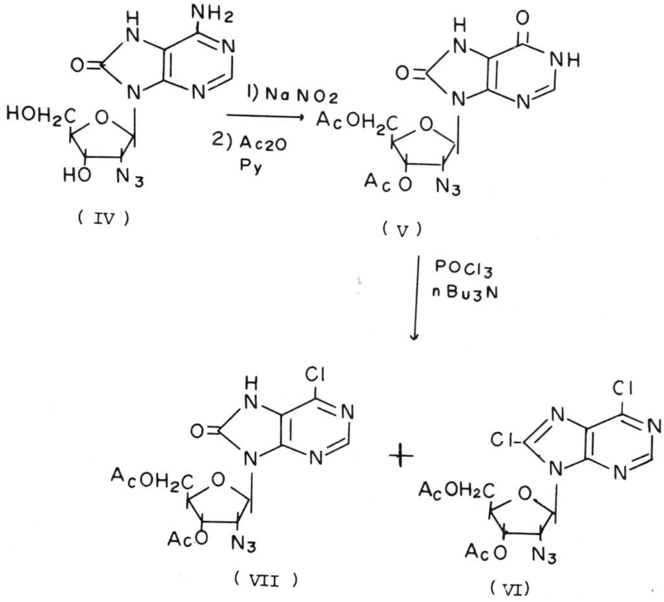

Chart 3

shown in Fig. 1, H_3, and H_1, signals appeared lower field relative to the ribo counterpart. If we irradiated at the position
of H_3,, the 3'-OH signal changed from a doublet to a singlet and
the H_4, signal from a quartet to triplet. When the position of

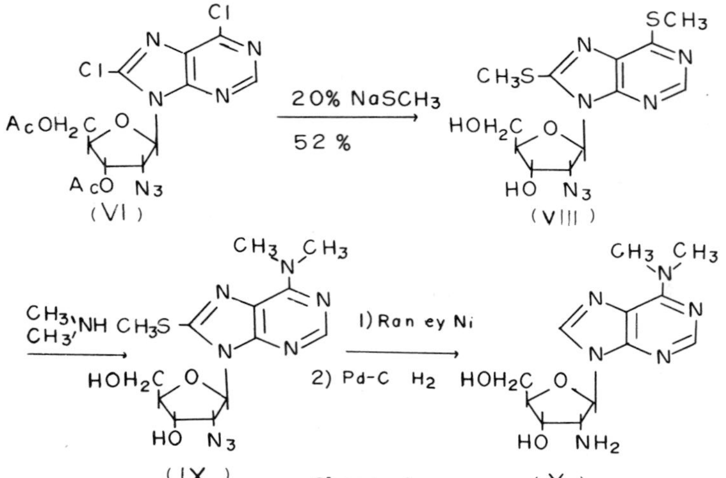

Chart 4

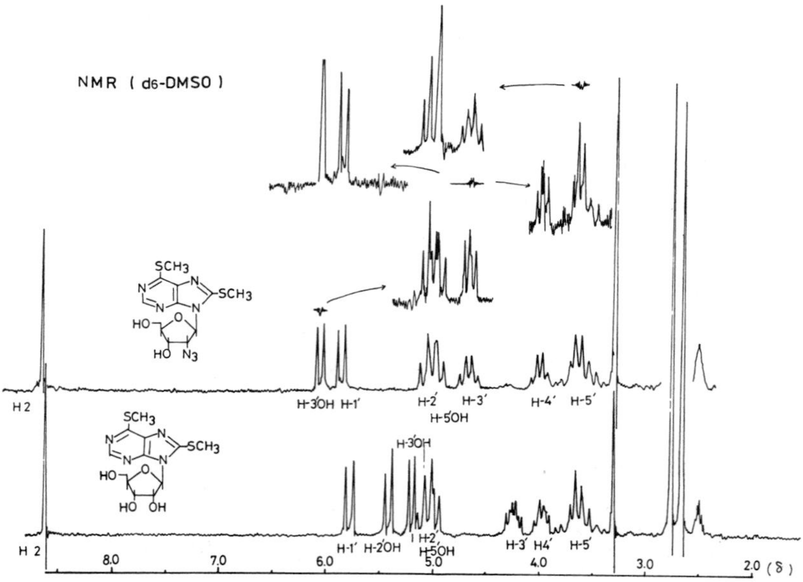

NMR (d6-DMSO)

Figure 1. NMR spectra of 6,8-dimethylmercapto-9-β-(2'-de-
oxy-2'-azido-D-ribofuranosyl) purine and 6,8-dimethylmercapto-
9-β-D-ribofuranosylpurine.

Figure 1

$H_{5'}$ was irradiated the 5'-OH changed from a triplet to a singlet. Furthermore, the 3'-OH and 5'-OH signals were diminished by D_2O. This evidence clearly showed that the azido group was in the 2'-position.

Compound VIII was then reacted with dimethylamine at 100°C for 12 hr to give a N^6-dimethyl-8-methylmercapto compound (IX) in a yield of 44%. Raney nickel dethiolation and reduction gave N^6-dimethyl-2'-deoxy-2'-aminoadenosine (X) as needles with 152°C melting point. This compound is a positional isomer of the puromycin aminonucleoside.

In order to obtain an inosine-type compound, the 6,8-dichloro compound (VI) was thiolated by heating with H_2S in pyridine and then treated with hydrogen peroxide at 4°C overnight (10). The reaction proceeded as that shown in the case of 6,8-dimercaptopurine riboside (9), and di-O-acetyl-2'-deoxy-2'-azidoinosine (XIa) was obtained in a yield of 35%. This compound was easily deacylated to give 2'-deoxy-2'-azidoinosine (XIb), which was identical to a sample obtained from 2'-deoxy-2'-azidoadenosine obtained below. This method for obtaining 2'-azido compounds, however, involves too many steps and an alternative route was next investigated, namely the inversion of arabinosuranosyl purine nucleosides to the ribo-azido compounds.

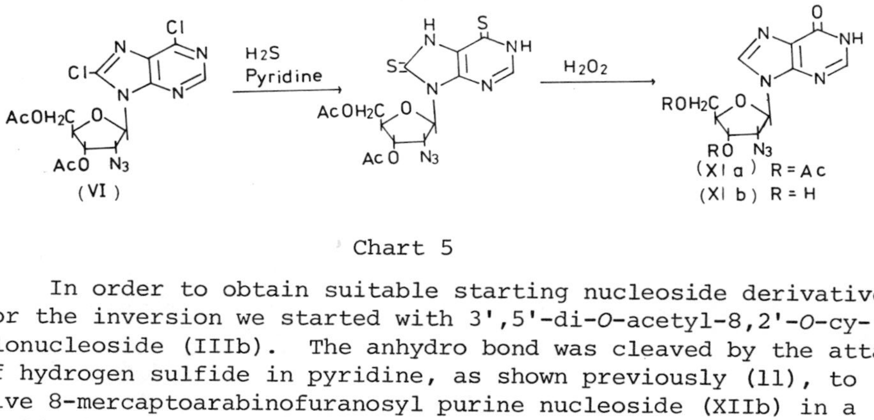

Chart 5

In order to obtain suitable starting nucleoside derivatives for the inversion we started with 3',5'-di-O-acetyl-8,2'-O-cyclonucleoside (IIIb). The anhydro bond was cleaved by the attack of hydrogen sulfide in pyridine, as shown previously (11), to give 8-mercaptoarabinofuranosyl purine nucleoside (XIIb) in a yield of 87%. Compound XIIb was then dethiolated with Raney nickel to afford 3',5'-di-O-acetyl-9-β-D-arabinofuranosyladenine (XIIIb). In order to introduce a strong leaving group, compound XIIIb was mesylated at the 2'-OH group. The mesylated arabinoside (XIVb) showed an ir band at 1180 cm^{-1}, confirming that a mesyl group had been introduced. Elemental analysis also supported this structure. Deacylation of XIVb with methanolic ammonia gave 2'-mesyl-arabinofuranosyladenine (XIVa).

When compound XIVb was heated with sodium azide in DMF at
150°C for 1 hr, the 3'-azidoxylofuranosyl derivative (XVb) was
obtained in a yield of 60%. This may involve an acetoxonium ion
intermediate as shown in XIIb-XVa. The compound (XVa) was found
to be identical to that obtained by Robins *et al.* (12). Similar
treatment of unprotected 2'-mesylarabinosyladenine (XIVa) also
gave the same xylo type compound, presumably through an epoxide
intermediate.

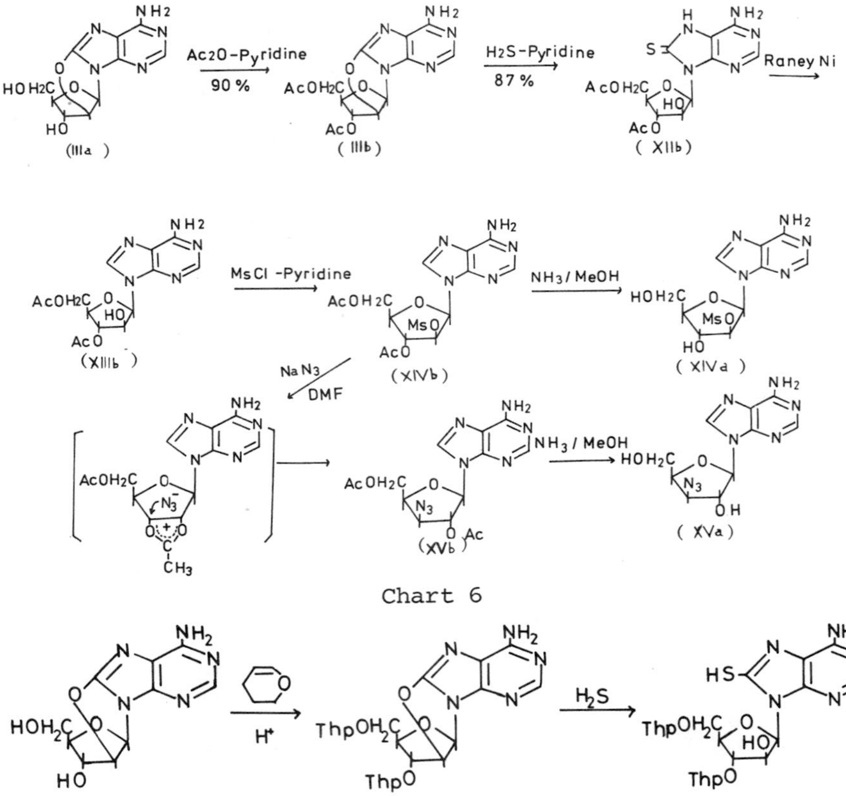

Chart 6

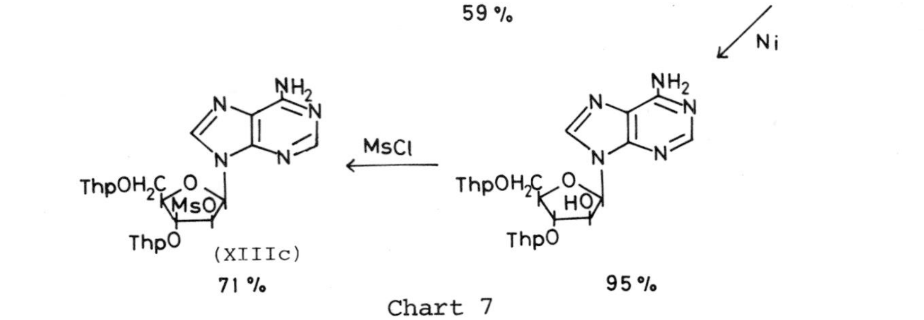

Chart 7

The cyclonucleoside was then protected with the nonparticipating tetrahydropyranyl group at 3'- and 5'-OH and derivatized
as before to 3',5'-di-O-tetrahydropyranyl-2'-O-mesylarabinosyla-
denine (XIIIc) in an overall yield of 39%. When compound XIIIc
was heated with sodium azide in DMF at 150°C for 7.5 hr, the 2'-
deoxy-2'-azido compound (XV) was obtained in a yield of 47%. Deblocking with acetic acid gave 2'-deoxy-2'-azidoadenosine (XVI)
in a yield of 57%. This compound had a melting point of 221-
222.5°C (13) and showed a correct analytical value. Ultraviolet
absorption properties similar to those of adenosine and an ir
band at 2110-2130 cm^{-1} suggested the structural assignment to
be correct. The ^{1}H-NMR spectrum (Fig. 2) of 2'-deoxy-2'-azido-
adenosine (SVI) showed that the signals of $H_{1'}$ and $H_{3'}$ were
shifted towards low field relative to adenosine, presumably due
to the magnetic anisotropy of the 2'-azido group. Furthermore,
in the 3',5'-diacetyl compound, the $H_{3'}$ and $H_{5'}$ signals were
shifted towards low field relative to the parent compound. In
the ^{13}C-NMR spectrum of XVI, the $C_{2'}$ signal appeared 9 ppm upfield relative to that of adenosine (14). These facts clearly
demonstrated that the azido group was in the $C_{2'}$ position.

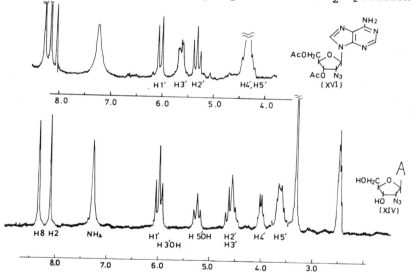

Figure 2. NMR spectra of 2'-deoxy-2'-2'-azido-3',5'-di-O-
acetyladenosine and 2'-deoxy-2'-azidoadenosine.

Catalytic hydrogenolysis over palladium charcoal of the
azido compound (XVI) gave 2'-deoxy-2'-aminoadenosine (XVII)
in a yield of 76%. This sample showed uv absorption properties
and color tests with ninhydrine and 10$_4^-$-benzidine spray identical to a sample provided by Dr. Mengel (15). The elemental an-

alyses also supported the structure of XVII as deoxyaminoadeno-
sine.   Thus, starting from adenosine, 2'-deoxy-2'-azido and 2'-
aminoadenosine was obtained in rather simple way.

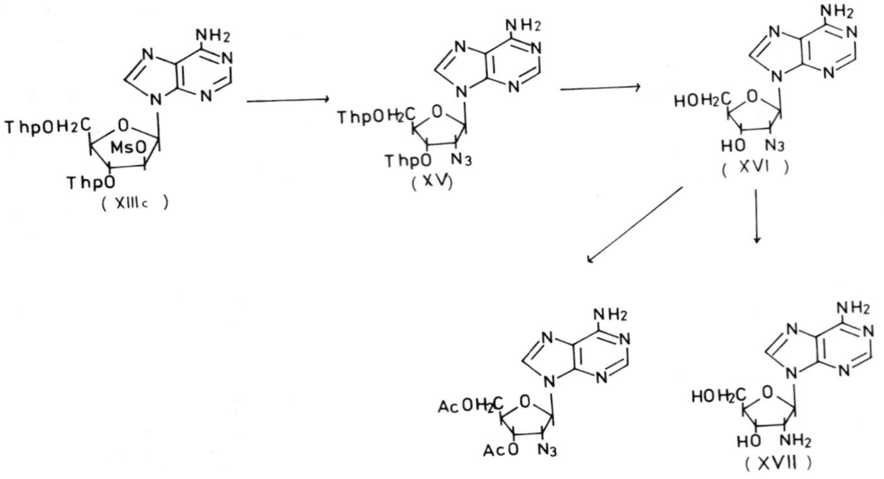

Chart 8

This method was next applied to guanosine to synthesize the
antibiotic 2'-deoxy-2'-aminoguanosine.   The 8,2'-anhydro-8-oxy-
9-β-D-arabinofuranosylguanine (XVIII) (8) was transformed, as
shown in the case of adenosine, to N^2,3',5'-triacetylarabino-
furanosylguanine (XIX) in an overall yield of 55%.   The triace-
tyl compound was mesylated and the protecting groups were changed
from acetyl to tetrahydropyranyl in an overall yield of 36%.

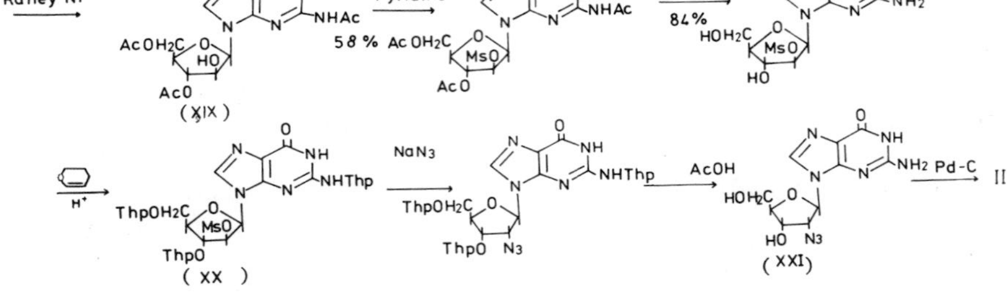

Chart 9

When N^2,3',5'-tri-tetrahydropyranyl-2'-mesylarabinofuranosyl-guanine (XX) was heated with sodium azide in DMF, none of the desired product was obtained. The reaction condition was then changed to acetamide at 210°C for 10 min and an azido compound was obtained in a yield of 18%. Deblocking of this compound with acetic acid gave 2'-deoxy-2'-azidoguanosine (XXI) as crystals with a melting point of 192°C (decomposition) in a yield of 40%. Ultraviolet absorption similar to guanosine and an ir band at 2100 cm^{-1} suggested that the structure was the required azido compound. Elemental analyses also supported the structure of XXI. Compound XXI was then hydrogenated over palladium charcoal to give 2'-deoxy-2'-aminoguanosine (II) as a powder with a melting point of 253-255°C (decomposition). Compound II, thus obtained, was compared with a sample from the *Aerobacter* species (3). As summarized in Table I, the two samples were completely identical by criteria of m.p., uv absorption properties, migration in paper chromatography, and color tests. Elemental analyses also suggested that the structure of the chemically synthesized 2'-deoxy-2'-aminoguanosine was correct.

TABLE I.   Properties of 2'-Deoxy-2'-Aminoguanosine

	Synthetic[d]	Natural[d]
Melting Point (°C)	250–252	252–254
u.v.		
H_2O	252, 275 (sh)	252, 275 (sh)
0.1N HCl	256, 280 (sh)	257, 280 (sh)
0.1N NaOH	256, 268	258, 267
PPC		
Solvent A[a]	0.25	0.26
Solvent B[b]	0.28	0.28
Solvent C[c]	0.51	0.51
Ninhydrine	Pink	Pink
10_4^--Benzidine	+	+

[a] *i*-PrOH-conc. NH_4OH-H_2O (7 : 1 : 2).

[b] *n*-BuOH-AoOH-H_2O (5 : 2 ; 3).

[c] EtOH-1 M NH_4OAc (75 : 30).

[d] Both samples showed correct elemental analysis values.

Thus the structure of the antibiotic 2'-deoxy-2'-aminoguan-
osine was confirmed. The present method for the synthesis of
2'-azido and 2'-aminopurine nucleosides may be suitable for
large-scale preparations and a variety of corresponding nucleotides and polynucleotides are being synthesized in our laboratory.

ACKNOWLEDGMENTS

We are greatly indebted to Dr. R. Mengel and Dr. T. Nakanishi for the generous gifts of samples. This research was
supported by a Grant-in-Aid for Scientific Research from the
Ministry of Education, to which authors' thanks are due.

REFERENCES

1. R. J. Suhadolnik, *Nucleoside Antibiotics*, Wiley Interscience,
    New York, 1970, p. 3.
2. B. R. Baker, R. E. Schaub, J. P. Joseph, and J. H. Williams,
    J. Am. Chem. Soc. 77, 12 (1955); B. R. Baker, R. E.
    Schaub, and H. M. Kissman, *J. Am. Chem. Soc. 77*, 5911
    (1955); H. M. Kissman and M. J. Weiss, *J. Am. Chem. Soc.
    80*, 2575 (1958).
3. T. Nakanishi, F. Tomita, and T. Suzuki, *Agric. Biol. Chem.
    38*, 2465 (1974).
4. M. L. Wolfrom and M. W. Winkley, *J. Org. Chem. 32*, 1823 (1967)
5. R. Mengel and H. Wiedner, *Chem. Ber. 109*, 433 (1976).
6. M. Ikehara, *Acc. Chem. Res. 2*, 47 (1969).
7. M. Ikehara, H. Tada and M. Kaneko, *Tetrahedron 24*, 3489
    (1968); M. Ikehara and K. Muneyama, *Chem. Pharm. Bull.,
    18*, 1196 (1970).
8. M. Ikehara and T. Maruyama, Tetrahedron, *31*, 1369 (1975).
9. M. Ikehara and T. Maruyama, *Chem. Pharm. Bull 24*, 565 (1974).
10. M. Ikehara and Y. Ogiso, *J. Carbohydr. Nucleosides Nucleo-
    tides, 1*, 401 (1974).
11. M. Ikehara and Y. Ogiso, *Tetrahedron 28*, 3695 (1972).
12. M. J. Robins, Y. Fouron and R. Mengel, *J. Org. Chem. 39*,
    1564 (1974).
13. The melting point reported by Mengel and Wiedner (5) was
    205°C.
14. Unpublished experiment by S. Uesugi and M. Ikehara.
15. Melting point of our sample was 197-198°C, which was the
    same as that reported by Wolfrom and Winkley (4) (195-
    197°C), but 6-7°C higher than that of Mengel and Wiedner

SYNTHESIS AND REACTION OF NUCLEOSIDES CONTAINING
SULFUR FUNCTIONS IN THE SUGAR PORTION*

*TOHRU UEDA, AKIRA MATSUDA, TAMOTSU ASANO,
AND HIDEO INOUE*

*Faculty of Pharmaceutical Sciences, Hokkaido University,
Sapporo, Japan*

ABSTRACT

Various nucleosides having sulfur functions in the sugar moiety
have been prepared and utilized for further transformations.
The present chapter outlines the synthesis of 2',3'-episulfides
derived from uridine and cytidine. The synthesis of C-cyclonu-
cleosides by the photoirradiation of 5'-deoxy-5'-phenylthio
nucleosides derived from adenosine and guanosine is described in
Section II.

I. SYNTHESIS OF 2',3'-EPISULFIDES DERIVED FROM URIDINE AND
    CYTIDINE

Although the synthesis of 2,3-ribo- and lyxo-episulfide derivatives of pentofuranosides has been achieved by Goodman and co-
workers (1-3) their nucleosides have not yet been prepared. S^2
,2'-Cyclo-2-thiouridine (1), previously prepared in our laboratory (4), was converted to the 5'-O-trityl-3'-O-methanesulfonyl
derivative (2, mp. 191-193°C) by the standard procedure. The
mesylate (2) was treated with methanolic hydrochloric acid at
room temperature and after the removal of the liberated trityl

*This work was supported, in part, by the Grant-in-Aid for
Scientific Research from the Ministry of Education, and Welfare
of Japan.

alcohol, the product was treated in refluxing 0.1 *N* hydrochloric
acid solution for 30 min. The product (3, 76%) isolated as a
gummy solid had lost the ultraviolet and infrared absorptions
characteristics of the *S*-cyclo-2-thiouridine system and the
mesyl function of 2. Treatment of 3 with acetic anhydride
in pyridine afforded a crystalline acetate (4, 85%, m.p.
214-216°C, mass, m/e: 284(M⁺), NMR (DMSO)δ: 7.59(*d*, H-6), 5.6(*d*,
H-5, $J_{5,6}$=7.8 Hz), 6.21(*d*, H-1', $J_{1',2'}$=1.5 Hz), 4.8-3.7(*m*, H-2',
3',4',5'), 2.02(*s*, AcO)). The analytical and spectrometric data
were in good agreement with the proposed structure, 1-(5-*O*-acetyl-
2,3-dideoxy-2,3-epithio-β-D-lyxofuranosyl)uracil (4).

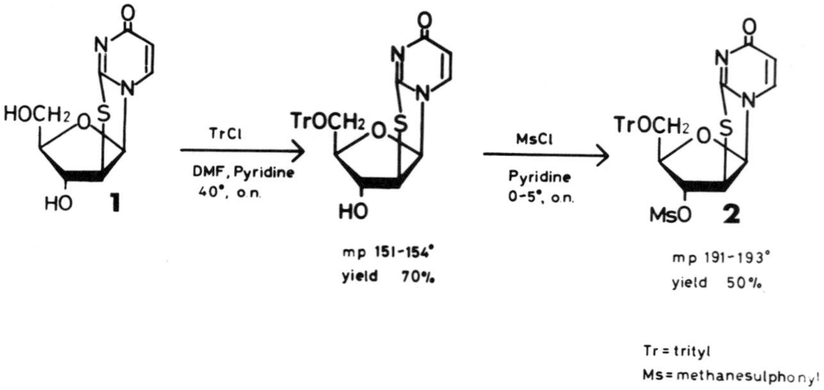

Chart 1

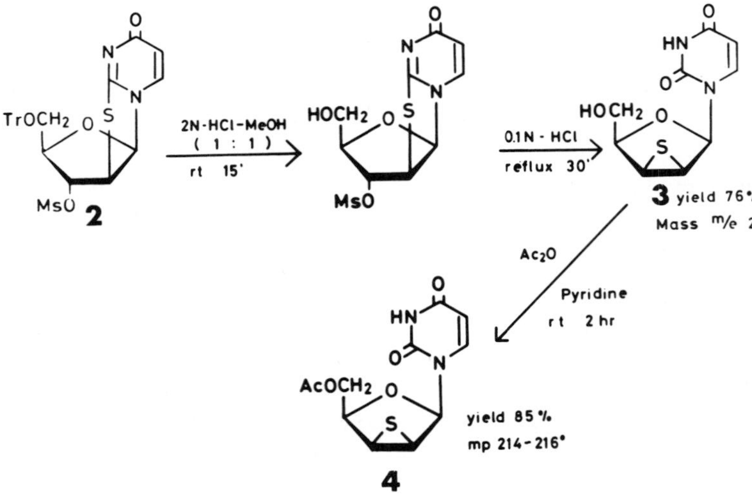

Chart 2

The epithio derivative with ribo configuration was prepared from 1-(5-O-benzoyl-2,3-anhydro-β-D-lyxofuranosyl)uracil (5) (5). Treatment of 5 with ammonium thiocyanate in boiling dioxane gave a mixture of the 3'- and 2'-thiocyanato derivative, the former being the main product. Mesylation of the mixture afforded the 2'-O-mesyl-3'-deoxy-3'-thiocyanato derivative (6, m.p. 174-175°C, 2160 cm^{-1} (SCN)). Treatment of 6 with potassium thioacetate in dimethylformamide in a refrigerator afforded the 2',3'-episulfide (7a, mp. 166-167°C, mass m/e: 346(M$^+$)) in almost quantitative yield.

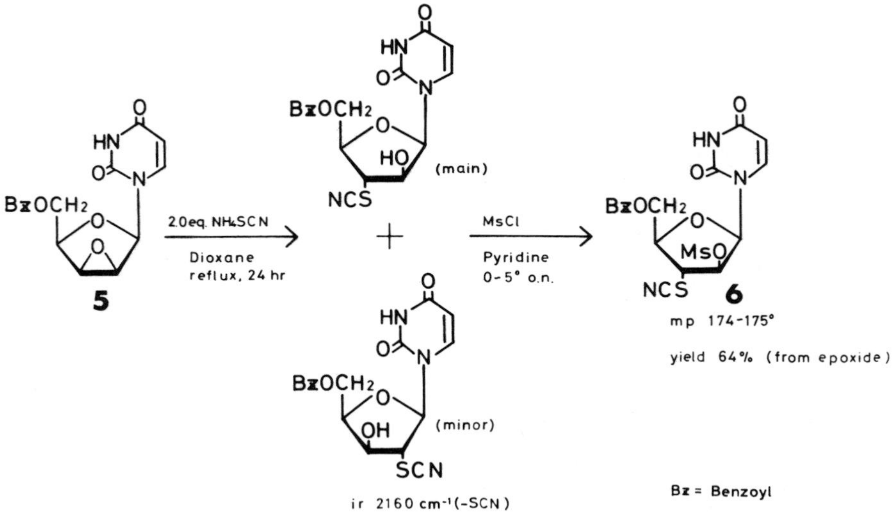

Chart 3

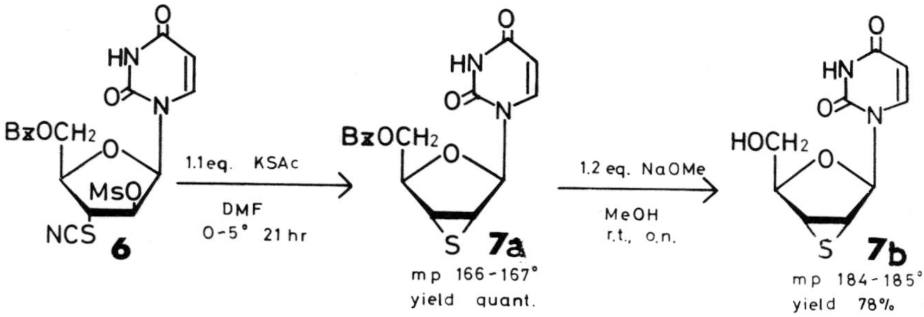

Chart 4

It is to be noted that the ribo-epoxide of pyrimidine nucleosides has been unknown although its intermediary formation was
postulated in many instances, especially in the case leading to
the 2,2'-cyclonucleoside formations (6). The ribo-episulfide
(7a), on the other hand, was fairly stable and was converted to
the free 2',3'-dideoxy-2',3'-epithiouridine (7b, m.p. 184-185°C,
mass m/e: 242 (M$^+$)) by treatment with sodium methoxide in methanol at room temperature. This was converted to the 5'-O-acetate
(7c, 96%, m.p. 171-173.5°C, NMR(DMSO) δ: 7.73 (d, H-6), 5.62 (d,
H-5, $J_{5,6}$=8.1 Hz), 5.94 (s, H-1'), 4.6-4.15 (m, H-4',5'), 4.10
(d, H-2'), 3.86 (d, H-3', $J_{2',3'}$=4.6 Hz), 2.02 (s, AcO)). The
CD spectra of the two 5'-O-acetyl epithionucleosides (Fig. 1) are
characteristic in that the sign of the band around 250 nm derived from the episulfide chromophore (7) is reversed, the lyxo-
episulfide being positive.

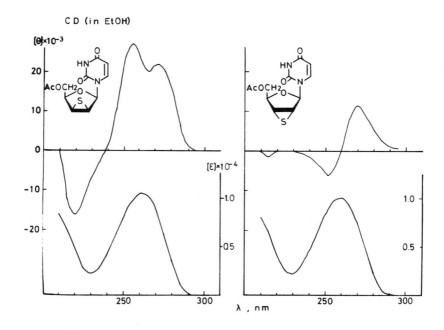

Figure 1

Both 4 and 7c were desulfurized by treatment with triphenylphosphine in boiling dioxane to give 1-(5-O-acetyl-2,3-dideoxy-
β-D-pent-2-enofuranosyl)uracil (8, 54 and 75%, respectively, m.p.
128-128.5°C), which was identical with reported data (8).

The ribo-episulfide derived from cytidine was prepared in
a similar manner as that described in the preparation of 2',3'-

epithiouridine. 1-(2,3-Anhydro-β-D-lyxofuranosyl)cytosine (9),
synthesized by Kanai (9) and Fox (10), was acetylated to give
the N^4,5'-O-diacetate(10, m.p. 216-216.5°C) and treated with
pyridinium thiocyanate in refluxing dioxane for 3 hr. The product, a mixture of the 3'- and 2'-thiocyanato derivative, was
mesylated in pyridine, and after purification through silica

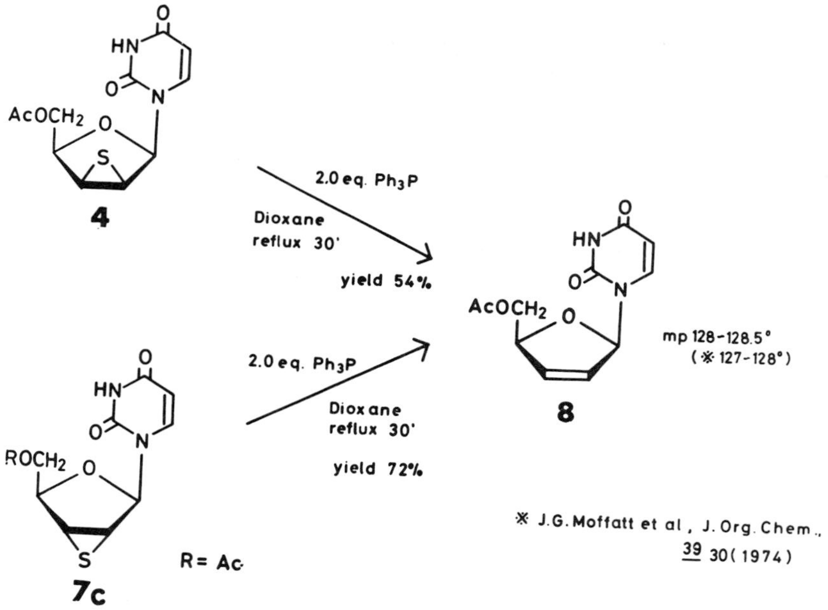

Chart 5

gel chromatography the 3'- and 2'-O-mesyl derivatives (11) were
obtained as a foam. These could be separated by thin-layer
chromatography in pure form. A mixture (11) was treated with
potassium thioacetate in a refrigerator overnight and the product was separated into two components by preparative thin-layer
chromatography on silica gel. The product showing higher rf
value was determined as the expected 2',3'-epithiocytidine diacetate (12, 39.5%, m.p. 200°C(dec), mass m/e: 325(M⁺)) by analytical and spectrometric investigations. The other product was
found to be 1-(5-O-acetyl-3-deoxy-3-acetylthio-2-O-mesyl-β-D-
arabinofuranosyl)-N^4-acetylcytosine(13, 20%, m.p. 168-169°C),
as confirmed by analytical and spectrometric data.
    Prolongation of the reaction period of the above reaction
resulted in a formation of additional by-product, which was
found to be derived from 12, and the structure was tentatively
assigned as the 2',3'-dideoxy-2',3'-diacetylthioxylosyl derivative of N^4-acetylcytosine (14). The CD spectra of diacetyl

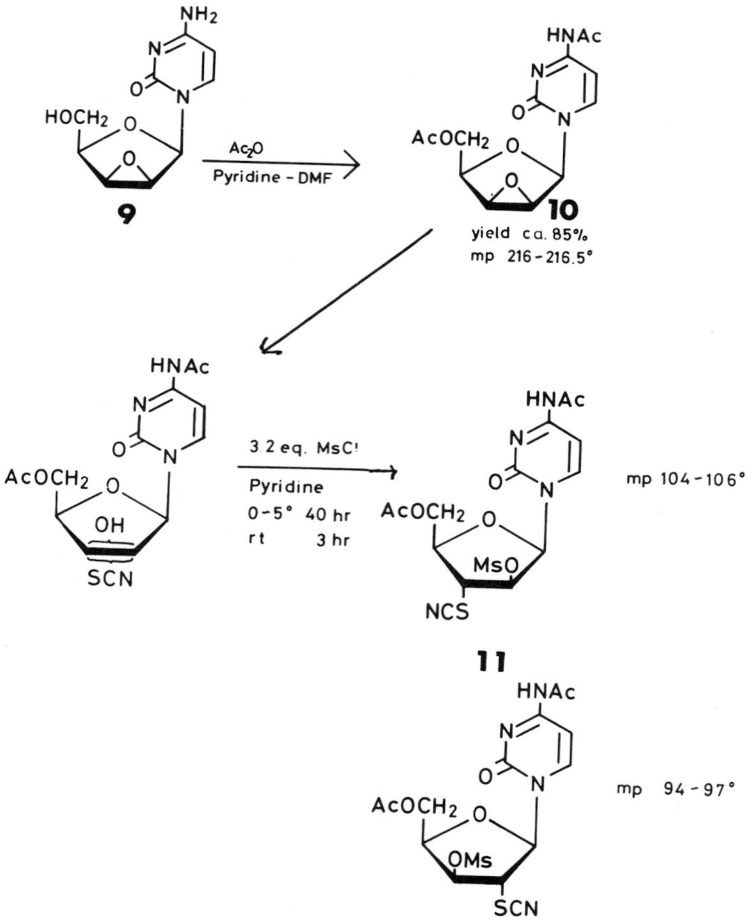

Chart 6

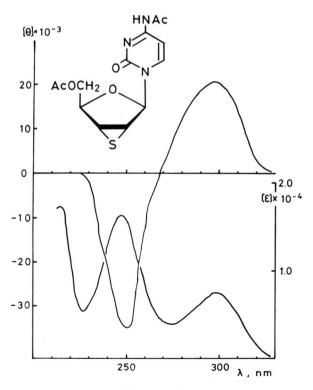

Chart 7

Mixture of mesylate **11** 1.5 eq. KSAc → DMF 0-5° 18 hr

13

NMR (DMSO-d₆)

δ 6.34 (d, 1, J₁',₂' = 5.3 Hz, C₁'-H)

δ 3.12 (s, 3, - SO₂CH₃)

δ 2.43 (s, 3, - SCOCH₃)

12

mass ᵐ/ₑ 325 M⁺

NMR (DMSO-d₆ + CDCl₃)

δ 5.95 (s, 1, C₁'-H)

δ 4.06 (d, 1, J₂',₃' = 4.5 Hz, C₂'-H)

δ 3.72 (d, 1, C₃'-H)

14

NMR (DMSO-d₆)

δ 6.26 (d, 1, J₁',₂' = 6.3 Hz, C₁'-H)

δ 2.38 (s, 3, -SCOCH₃)

δ 2.32 (s, 3, -SCOCH₃)

CD (in EtOH)

Figure 2

Chart 7

Mixture of mesylate **11** 1.5 eq. KSAc → DMF 0-5° 18 hr

13

NMR (DMSO-d_6)

δ 6.34 (d, 1, $J_{1',2'}$ = 5.3 Hz, $C_{1'}$-H)

δ 3.12 (s, 3, - SO_2CH_3)

δ 2.43 (s, 3, - $SCOCH_3$)

12

mass m/e 325 M⁺

NMR (DMSO-d_6 + $CDCl_3$)

δ 5.95 (s, 1, $C_{1'}$-H)

δ 4.06 (d, 1, $J_{2',3'}$ = 4.5 Hz, $C_{2'}$-H)

δ 3.72 (d, 1, $C_{3'}$-H)

14

NMR (DMSO-d_6)

δ 6.26 (d, 1, $J_{1',2'}$ = 6.3 Hz, $C_{1'}$-H)

δ 2.38 (s, 3, -$SCOCH_3$)

δ 2.32 (s, 3, -$SCOCH_3$)

CD (in EtOH)

Figure 2

epithiocytidine(12, Fig. 2) revealed a negative CD band around the 250-nm region with a positive at 300 nm, which is consistent with that of epithiouridine.

II. SYNTHESIS AND PROPERTIES OF 8,5'-C-CYCLOPURINE NUCLEOSIDES

In recent years several cyclonucleosides bearing a carbon bridge instead of oxygen, nitrogen, or sulfur have been prepared and utilized as the conformationally fixed analogs of nucleosides and nucleotides (11-17). The formation of 5'-deoxy-8,5'-cyclo-adenosine from 5'-deoxyadenosylcobalamin by illumination is particularly interesting (11). We have found that the photoirradiation of 5'-alkylthio- or arylthio-adenosine derivatives afforded the 5'-deoxy-8,5'-cycloadenosine derivatives in satisfactory yields (18).

For example, irradiation of 2',3'-O-isopropylidene-5'-deoxy-5'-phenylthioadenosine (15) with a 60-W low-pressure mercury vapor lamp in the presence of trimethyl phosphite under argon bubbling in acetonitrile for 1.5 hr afforded 2',3'-O-isopropyl-idene-5'-deoxy-8,5'-cycloadenosine (16, m.p. 230.5-231.5°C) in 70% yield. Other examples are summarized in Table I. Deaceto-nation of 16 with 0.1 N HCl at 85-90°C for 60 min gave 5'-deoxy-8,5'-cycloadenosine(17, m.p. 300°C, NMR shown in Fig. 3). Other 5'-deoxy-8,5-cyclopurine nucleosides have been prepared by sim-

R	(MeO)$_3$P	Time(hr)	Yield (%)
By	—	10 ^{a}	17.5
"	+	3 ^{a}	44.3
Ph	+	5 ^{a}	69.6
"	+	1.5 ^{b}	65.8

Reaction condition
  0.4 ~ 0.5 mM solution (CH$_3$CN)
  argon bubbling
Light source
  a : Ushio 100-W high-pressure mercury vapor lamp(Quartz filter)
  b : Eikōshya 60-W low-pressure mercury vapor lamp

Table 1

ilar approaches. Photolysis of 2',3'-*O*-isopropylidene-5'-deoxy-
5'-benzylthioinosine (<u>18</u>) gave 8,5'-cycloinosine (<u>19</u>), which
was also derived by the hydrolytic deamination of <u>16</u>. 2',3'-*O*-
Isopropylidene-5'-deoxy-5'-benzylthio-N^2-benzoylguanosine (<u>20</u>),
prepared from N^2-benzoylguanosine by 5'-*O*-tosylation and successive substitution with benzyl mercaptide in liquid ammonia,
was cyclized by irradiation to (<u>21</u>). The N^2-benzoyl group of
<u>21</u> was removed to give 2',3'-*O*-isopropylidene-5'-deoxy-8,5'-
cycloguanosine (<u>22</u>). The N^2-benzoyl group can be removed prior
to photocyclization.

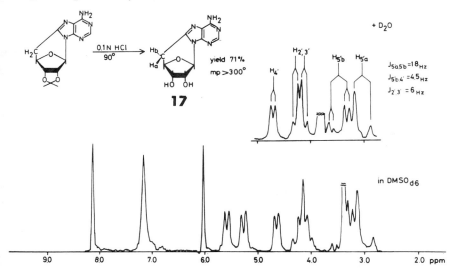

Figure 3

For the investigation of the structure-activity relationship
of purine-nucleoside 2',3'-cyclic phosphates to various cleases
the 2',3'-cyclic phosphates of these 8.5'-cyclopurine nucleosides.
    Adenosine and guanosine were directly transformed to the
respective 5'-deoxy-5'-phenylthio derivatives(<u>23</u>,<u>24</u>) by the procedure of Nakagawa and Hata (19), and were phosphorylated by the
method previously established in our laboratory (20) to the respective 2',3'-cyclic phosphate(<u>25</u>,<u>26</u>). Compound <u>25</u> was irradiated in aqueous acetonitrile with a 60-W low-pressure lamp for
2 hr to afford 5'-deoxy-8,5'-cycloadenosine 2',3'-cyclic phos-
phate(<u>27</u>). Alternately, <u>17</u> was phosphorylated to give <u>27</u>. Compound <u>26</u> was similarly irradiated to give 5'-deoxy-8,5'-cyclo-
guanosine 2',3'-cyclic phosphate (<u>28</u>). The nucleotides were purified through DEAE-cellulose column and isolated as triethyla-
mmomium salt.

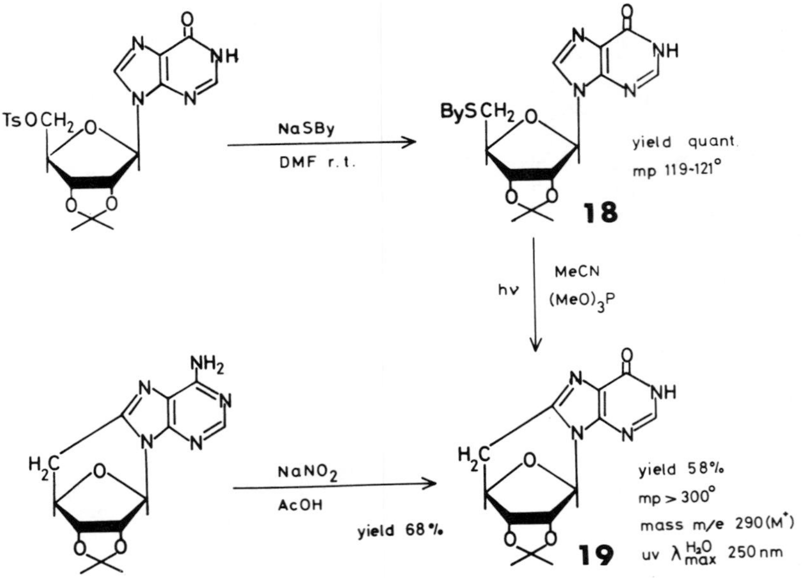

Chart 8

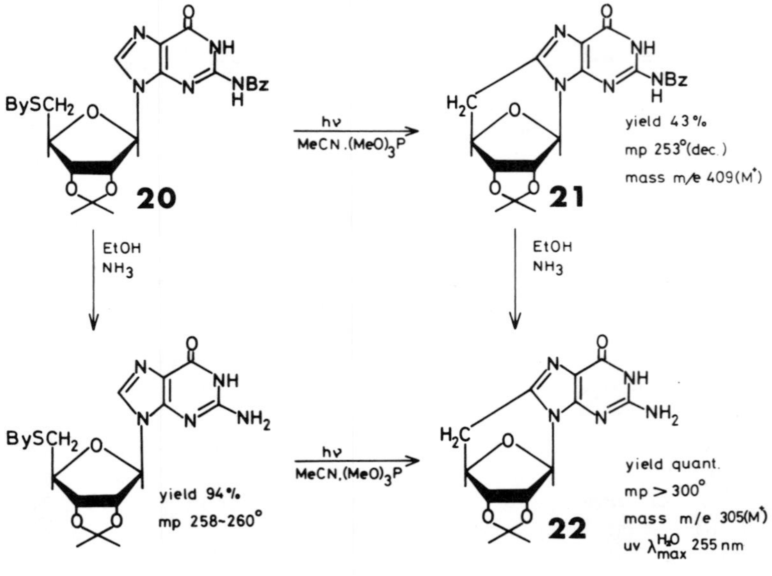

Chart 9

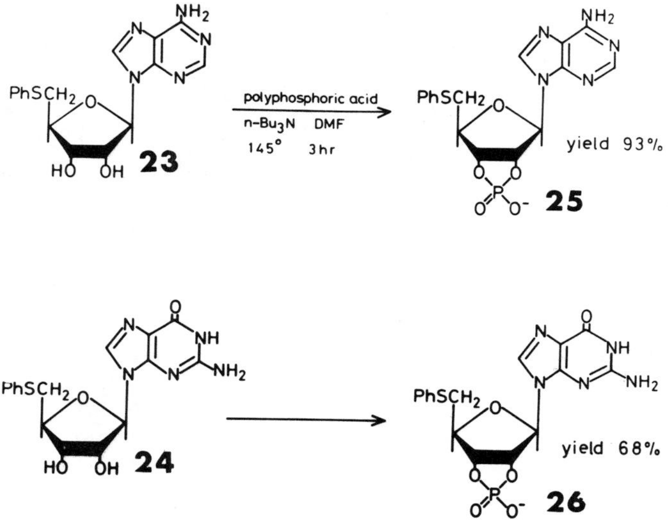

Chart 10

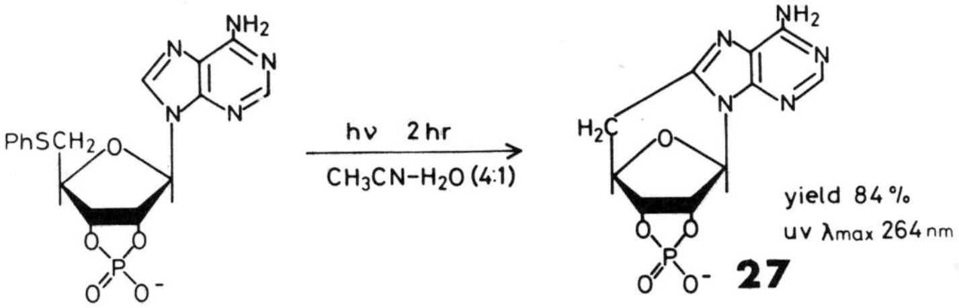

Chart 11

Compound 28 was found to be a substrate for RNase T_1 digestion giving the 3'-phosphate(29), though the rate of hydrolysis was very low(Fig. 4). Both cyclopurine nucleotides, 27 and 28, were good substrate for RNase T_2 digestion.

It is to be emphasized that the change of the pattern of CD spectra of 28 and 29(Fig. 5) between neutral and N^7-protonated form is essentially similar to that (21) of guanosine and guanosine phosphate. These phenomena of the reversal of the sign of CD band have been attributed to the reversal of the conformation of the guanine moiety about the glycosylic linkage from anti to syn form on N^7-protonation (21). This conformational change was further suggested in the interaction of guanosine 3'-phosphate with RNase T_1, based on the measurements of difference uv spectra (22). However, similar difference uv spectra were again observed between 29 and its complex with RNase T_1 (Fig. 6). The difference CD spectra of 29 and its complex with RNase T_1 are also

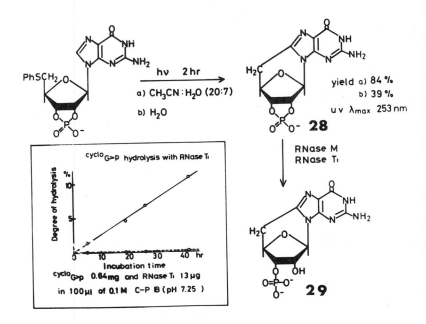

Figure 4

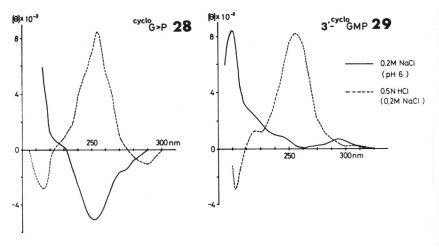

Figure 5

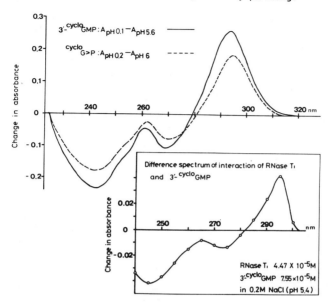

Figure 6

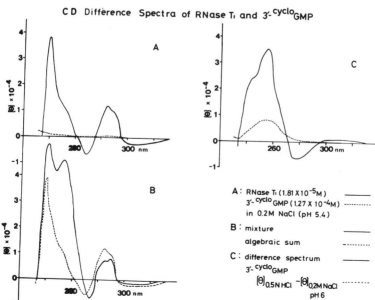

Figure 7

similar to those of guanylic acid (23) (Fig. 7). Since compound
28 and 29 are fixed as anti conformation yet exhibit similar
spectral changes to those of guanosine and guanylic acids, it
should be stated that the change of CD and uv characteristics
between neutral and protonated (or complexed with the enzyme)
guanosine derivatives cannot necessarily be attributed to the
change of the conformations about the glycosylic bond.

For better models of purine nucleosides having anti conformations it is necessary to introduce a hydroxyl function into
the 5'-position of 8,5'-cyclonucleosides, which was realized by
the following procedures. Treatment of 16 with an excess of
selenium dioxide in refluxing dioxane for 3 days afforded 2',3'-
O-isopropylidene-5'-keto-8,5'-cycloadenosine (30, m.p. 279°C(dec)
in 80% yield, which had been prepared by Harper and Hampton (16a)
via the cyclization of the 5'-carboxylate of adenosine derivative
with methyl-lithium in low yield (∿5%). Reduction of 30 with sodium borohydride in aqueous methanol for 30 min gave a single
product, 2',3'-O-isopropylidne-8,5'(S)-cycloadenosine (31, m.p.

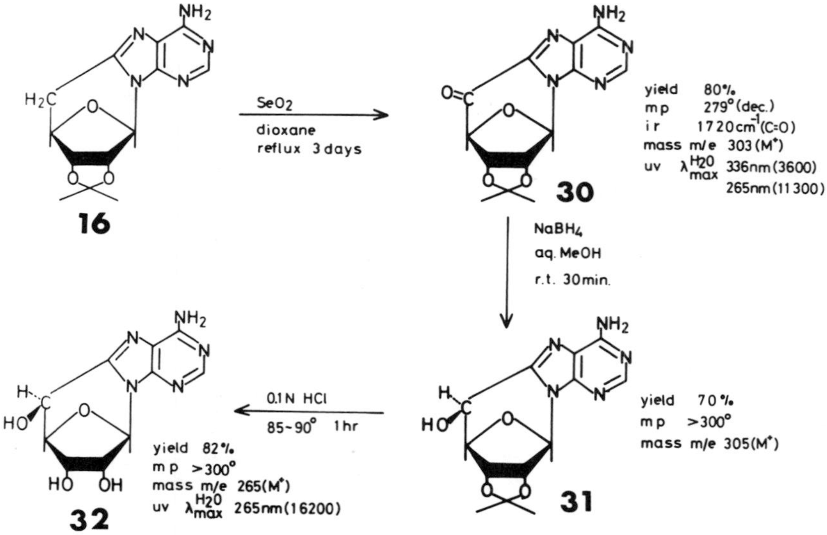

Chart 12

300°C, Mass m/e: 305 (M⁺)) and no formation of the (R)-epimer
was observed, which was in contrast with the preceding result
(16a). Deacetonation of 31 afforded free 8,5'(S)-cycloadeno-
sine(32, mp 300°C, Mass m/e: 265(M⁺)). The (R)-epimer of 32 was
derived from the (S)-epimer(31) by the 5'-O-mesylation to 33
followed by treatment with sodium acetate in dimethylformamide

at 80°C for 5 hr to give the (R)-epimer (34). On deacetonation with 0.1 N HCl 8,5' (R)-cycloadenosine (35, m.p. 266°C(dec), mass m/e:265 (M$^+$)) was obtained in high yield. The structure of (S)- and (R)- were clearly distinguished by NMR analysis of their 5'-O-acetate as shown in Fig. 8.

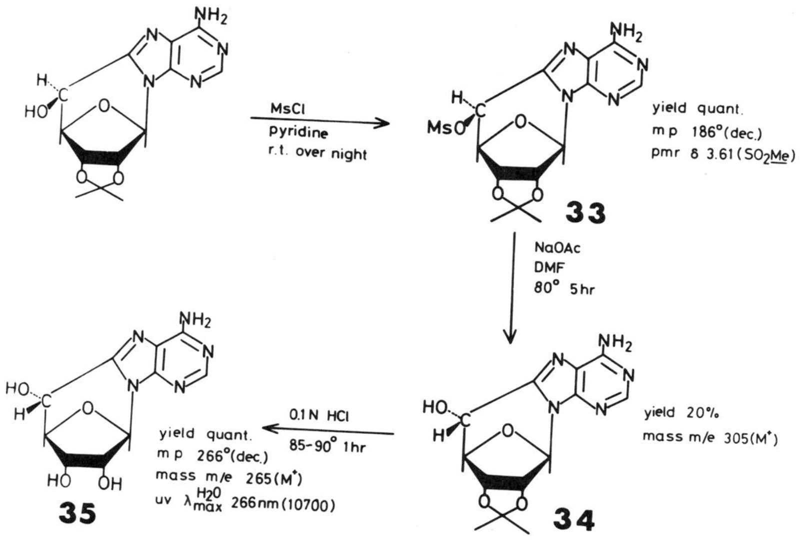

Chart 13

A different approach to prepare 32 and 35 was investigated. 2',3'-O-Isopropylidene-8-phenylthioadenosine (36), prepared from the respective 8-bromo derivative by substitution with sodium thiophenoxide in methanol at 50-60°C, was irradiated with a 400-W high-pressure mercury vapor lamp in the presence of trimethyl phosphite under argon bubbling with pyrex filter. The main product was 2',3'-O-isopropylideneadenosine(37) and the minor products were 31 and 34. The yields of the latter two compounds were improved by the addition of a peroxide in the photoreaction in place of trimethylphosphite, and the results were summarized in Table II.

The CD spectra of 32 and 35 are shown in Fig. 9. It is evident that 8,5-(R)-cycloadenosine (35), in spite of being fixed in anti conformation, has a positive band at the main absorption region. This means that the configuration of a newly introduced chiral carbon(5'-position) is also responsible for the sign of the CD spectra in addition to the anomeric carbon.

Studies of the synthesis of their 5'-nucleotides as well as those of guanosine derivatives are in progress and will be reported elsewhere.

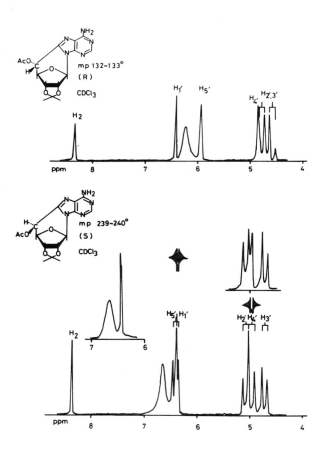

Figure 8

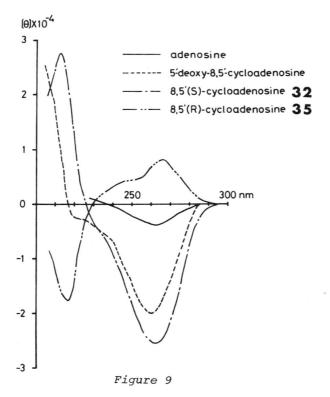

Solvent	Additions	Time(hr)	Yield (%)		
			isop A	cyclo(S)	cyclo(R)
CH₃CN	(CH₃O)₃P	3.25	41.8	13.0	8.1
(CH₃)₂CO	''	6.5	53.1	9.6	—
CH₃OH	''	3	63.5	4.2	6.6
CH₃CN	Ph₂CO	2	18.4	8.1	—
''	BHP[a]	4	6.2	19.1	10.1
''	DBP[b]	4.5	36.0	13.4	7.4
''	DCP[c]	3	20.5	20.2	13.1

o Reaction condition   o Light source   a: tert-Butyl hydroperoxide
  0.3 mM solution   Riko 400-W high-prssure   b: Di-tert-butyl peroxide
  argon bubbling   mercury vapor lamp(Pyrex filter)   c: Dicumyl peroxide

Table II

$[\theta] \times 10^{-4}$

— adenosine
----- 5'-deoxy-8,5'-cycloadenosine
—·— 8,5'(S)-cycloadenosine **32**
—··— 8,5'(R)-cycloadenosine **35**

250   300 nm

Figure 9

REFERENCES

1. L. Goodman in *Basic Principles in Nucleic Acid Chemistry*,
    Vol 1 (P.O.P. Ts'o, ed), Academic Press, New York, 1974
    p. 93.
2. L. Goodman, *Chem. Commun.* 219 (1968).
3. K. J. Ryan, E. M. Acton, and L. Goodman, *J. Org. Chem. 33*,
    3727 (1968).
4. T.Ueda, and S. Shibuya, *Chem. Pharm. Bull 18*, 1076(1970);
    T. Ueda and H. Tanaka, *Chem. Pharm. Bull 18*, 149(1970).
5. J. F. Codington, R. Fecher, and J. J. Fox, *J. Org. Chem.
    27*, 163(1962).
6. For example see: J. F Codington, R. Fecher, and J. J. Fox,
    J. Amer. Chem. Soc. 82, 2794(1960) and Ref. 4.
7. C. Djerassi, H. Wolf, D. A. Lightner, E. Bunnenberg, K.
    Takeda, T. Komeno, and K. K. Kuriyama, *Tetrahedron 19*,
    1547(1963).
8. T. C. Jain, I. D. Jenkins, A. F. Russell, J.P.H. Verheyden,
    and J. G. Moffatt, *J. Org. Chem. 39*, 30(1974).
9. T. Kanai, PhD thesis, Hokkaido University, 1973.
10. U. Reichman, D. H. Hollenberg, C. K. Chu, K. A. Watanabe,
    and J. J. Fox, *J. Org. Chem. 41*, 2042(1976).
11. H.P.C. Hogenkamp, *J. Biol. Chem. 238*, 477(1963).
12. A. W. Johnson, D. Oldfield, R. Rodrigo, and N. Shaw, *J.
    Chem. Soc.* 4080(1964).
13. A. W. Johnson, L. Mervyn, N. Shaw, and E. L. Smith, *J. Chem.
    Soc.* 4146(1963).
14. K. Keck, *Naturwiss. 53*, 1034 (1966).
15. T. Kunieda and B. Witkop, *J. Am. Chem. Soc. 91*, 7752(1969).
16. (a) P. J. Harper and A. Hampton, *J. Org. Chem. 37*, 795
    (1972); (b) A. Hampton and R. R. Chawla, *J. Carbohydr.
    Nucleosides Nucleotides 2*, 281(1975); (c) A. Hampton,
    P. J. Harper, and T. Sasaki, *Biochemistry 11*, 4736(1972);
    (d) A. Hampton, P. J. Harper, and T. Sasaki, *Biochemistry
    11*, 4965 (1972).
17. J. A. Rabi and J. J. Fox, *J. Org. Chem. 37*, 3898(1972).
18. For a preliminary result see: K. Muneyama, T. Nishida, and
    T. Ueda, Abstract Papers of Symposium on Nucleic Acid
    Chemistry, Osaka, 1973, p 7.
19. I. Nakagawa and T. Hata, *Tetrahedron Lett.* 1409(1975).
20. T. Ueda and I. Kawai, *Chem. Pharm. Bull 18*, 2303(1970).
21. (a) W. Gushlbauer and Y. Courtois, *FEBS Lett. 1*, 183(1968);
    (b) D. W. Miles, L. B. Townsend, M. J. Robins, R. K.
    Robins, W. H. Inskeep, and H. Eyring, *J. Am. Chem. Soc.
    93*, 1600(1971).
22. T. Oshima and K. Imahori, *J. Biochem. 69*, 987(1971); T.
    Oshima and K. Imahori *J. Biochem. 70*, 197(1971).
23. C. Sander and P.O.P. Ts'o, *Biochemistry 10*, 1953(1971).

NEW C-NUCLEOSIDE ISOSTERES OF SOME
NUCLEOSIDE ANTIBIOTICS*

*JACK J. FOX, KYOICHI A. WATANABE, ROBERT S.
KLEIN, CHUNG K. CHU, STEVE Y-K. TAM, URI REICHMAN,
KOSAKU HIROTA, J.-S. HWANG, FEDERICO G. DE LAS HERAS,
AND IRIS WEMPEN*

*Laboratory of Organic Chemistry, Memorial-Sloan
Kettering Cancer Center, Sloan-Kettering Institute,
Sloan Kettering Division of Graduate School of Medical
Sciences, Cornell University, New York*

Since the discovery (1,2) of pseudouridine in 1957, a host
of naturally occurring C-nucleosides have been reported (3)
(Fig. 1). All of these (except pseudouridine) are antibiotics
and several exhibit anticancer and antiviral activities. The
unique structural feature that distinguishes them from ordinary
nucleosides is the presence of a carbon to carbon linkage
(instead of a C-N linkage) between the aglycon and sugar
moieties. This unique feature renders these nucleosides
resistant to the action of such enzymes as nucleoside hydrolases
and phosporylases, which cleave normal C-N sugar-base linkages,
thus eliminating one aspect of potential catabolic deactivation
(4), while still allowing binding to and possible inhibition of
such enzymes. We report herein an overview of our recent
efforts in the syntheses and anticancer evaluation of certain
new C-nucleosides structurally related to pseudouridine and to
the formycins.

Our interest was initially directed toward the synthesis
of a 5-(B-D-ribofuranosyl) isocytosine [ψ-isocytidine], which
is isosteric to both cytidine and 5-azacytidine. 5-Azacytidine,
synthesized originally by Piskala and Sorm (5) in 1964, was

*These investigations were supported in part by funds
from the National Cancer Institute, U.S. Public Health Service,
DHEW (Grant No. CA-08748, CA-18601 and CA-18856).

later isolated from *S. ladakanus* as a nucleoside antibiotic.
Reviews on its biological activity have appeared (6,7). 5-
Azacytidine has shown particularly impressive activity (8)
against acute myelogenous leukemia resistant to ara-C. Objective remissions have been induced by 5-azacytidine in some
patients with breast cancer, melanoma, and colon cancer. In
almost all clinical cases it exhibited undesirable side effects
such as severe nausea and vomiting at therapeutic doses, although this can be diminished to some extent by continuous administration over several days (9). 5-Azacytidine is readily
deaminated by cytidine deaminase to 5-azauridine, which may com-
promise the effectiveness of this drug. Additionally, this antibiotic, like other 5-azapyrimidines, is generally unstable in
aqueous media forming ribosyl derivatives of *N*-amidinourea
or *N*-formylbiuret and biuret, and other degradation products.
Whether the adverse host toxicity is caused by the products of
catabolism of 5-azacytidine is not certain. It was hoped that
ψ-isocytidine would be more resistant to both enzymatic and
nonenzymatic degradation and would have a better therapeutic
effect with less undesirable side effects. Toward this end,
the synthesis of ψ-isocytidine was undertaken and achieved in
our laboratory (10,11).

Our earlier synthetic ventures (12) into C-nucleosides
(Fig. 2) utilized 2,3-*O*-isopropylidene-5-*O*-trityl-β-D-ribofur-
anosyl chloride (3), which is easily prepared (12,13) by tri-
tylation of 2,3-*O*-isopropylidene-D-ribose followed by chlorination with triphenylphosphine and CCl$_4$ in DMF. This halogenose
(3) was condensed with diethyl sodiomalonate to afford the anomeric mixture of the bifunctionalized *C*-glycosyl derivative (4),
which, after reaction with urea and sodium ethoxide in ethanol,
gave the 5-β-D-ribosylbarituric acid C-nucleoside derivative
(5). While the overall yields are quite good, the method has
the built-in limitation of affording only 6-hydroxy ψ-uridine
type nucleosides.

Attempts in our laboratory to prepare a bifunctionalized derivative of 4 containing the formylacetate rather than the malonate side chain at C-1 by condensation of halogenose 3 with
ethylformylacetate or its dimethyl acetal were not successful.
However, we found (10,11) that formylation of ethyl(5-*O*-trityl-
2,3-*O*-isopropylidene-D-ribosyl)acetate (6) (14) with ethyl formate and sodium hydride in a mixture of anhydrous ether and absolute ethanol gave the desired formylacetate derivative as the
sodium enolate (7). Though treatment of 7 with guanidine in
ethanol in the presence of sodium ethoxide did produce the protected ψ-isocytidines (9) (10), the yields were low and their
reproducibility was inconsistent; owing most probably to the
preponderance of the enolate form (rather than the aldehydic
form) of 7 in base (Fig. 3). Attack by the nitrogen nucleophile on the aldehydic (enolic) carbon atom would be electroni-

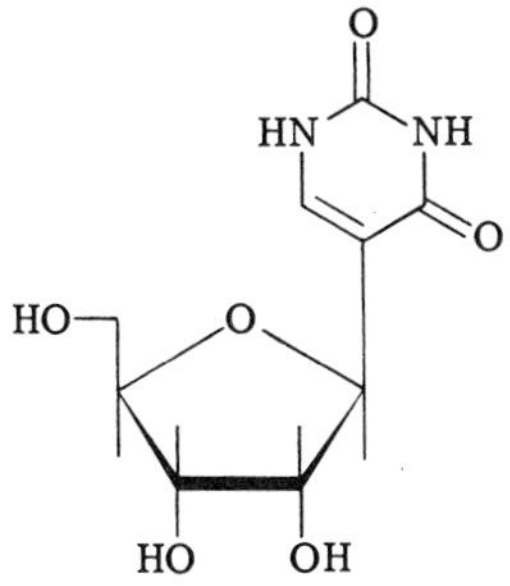

Figure 1

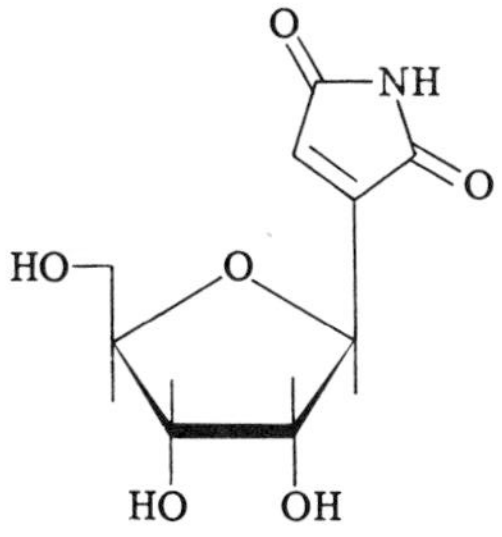

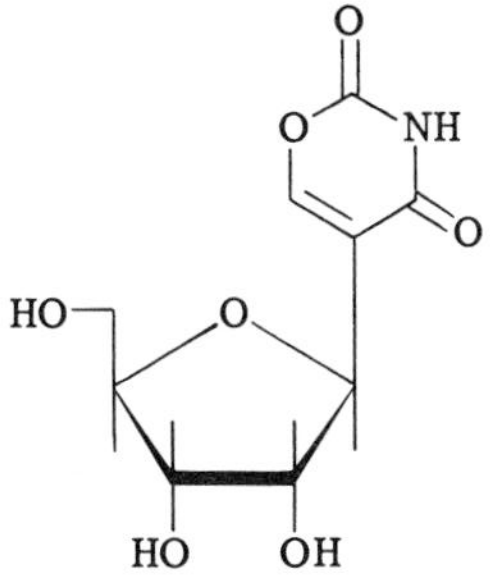

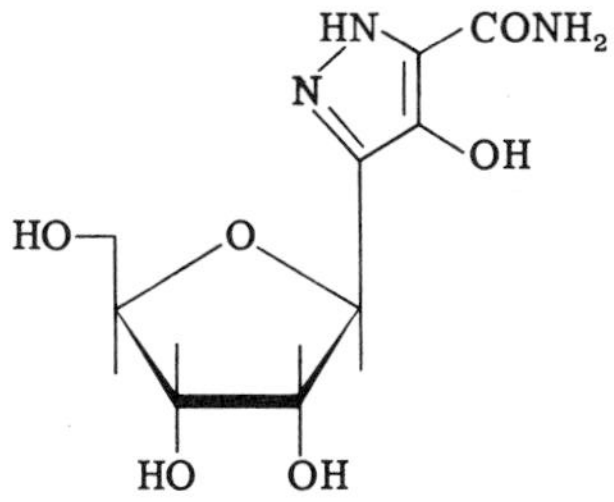

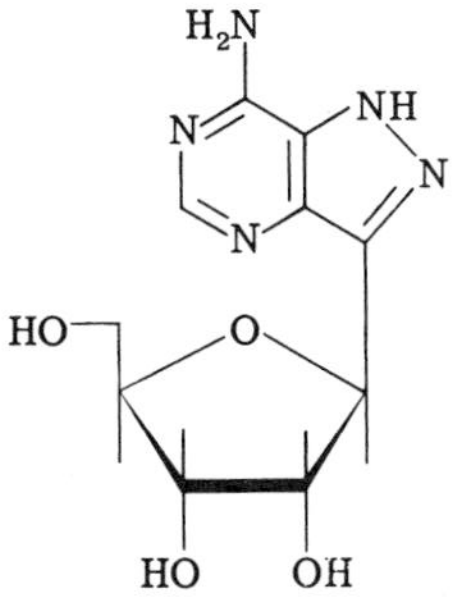

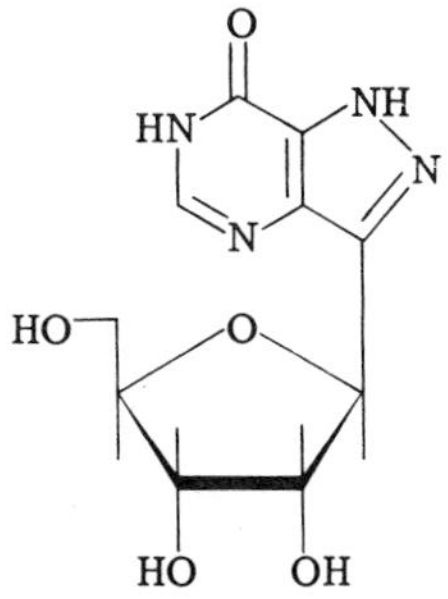

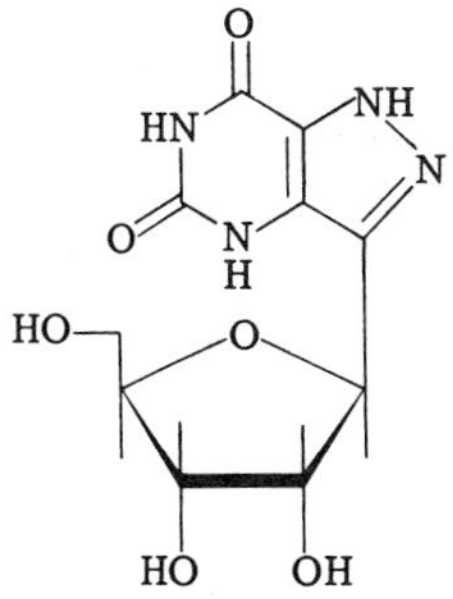

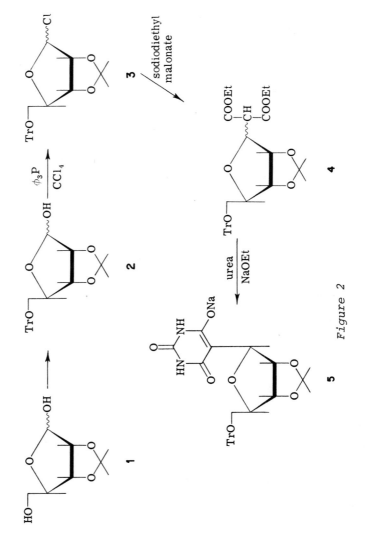

Figure 2

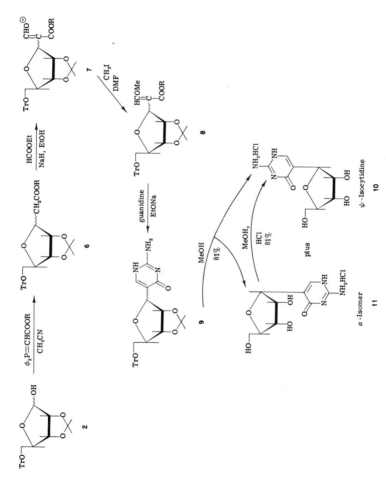

Figure 3

cally hindered by the adjacent negatively charged oxygen. In
order to remove the negative charge, crude enolate 7 was methylated with methyl iodide in DMF to afford the desired β-methoxyacrylate derivatives 8, one isomer of which was isolated in crystalline form after column chromatography in ∿ 25% overall yield
from 2 (11).

In large-scale runs, the isolation of crystalline 8 by column
chromatography was not practical. Cyclization of crude 8 with
guanidine afforded 9 in 15% overall yield from 2. Brief treatment of 9 with 10% methanolic hydrogen chloride gave predominantly the α-isomer 11, whereas prolonged treatment afforded
mainly the desired β-isomer (ψ-isocytidine) 10, which crystallized from the reaction mixture as its hydrochloride salt while
the α-isomer remained in solution. The yield of ψ-isocytidine
from 9 was thus readily raised to ∿ 80% (10,11). Pure α-isomer
11, when treated with dilute deuterium chloride, underwent epimerization at C-1' to ψ-isocytidine, as indicated by ^{1}H NMR.
At the equilibrium point, the α:β ratio was ∿ 1:4. The mechanism for the interconversion of 10⇌11 is probably akin to
that proposed (2,15) for the isomerization of pseudouridines involving protonation of the sugar ring oxygen followed by opening
of the furanoid ring.

Pseudoisocytidine (10) was shown to be chemically stable at
pH 7.4 for at least 6 days at 22°C and at least 3 days at 37°C,
during which time no evidence of epimerization was found. Enzymatic studies by W. Kreis at our Institute have shown that,
whereas cytidine, ara-C (1-β-D-arabinofuranosylcytosine), and
5-azacytidine are readily deaminated by cytidine deaminase from
mouse kidney, no deamination occured with ψ-isocytidine. These
data suggest that, in line with our original design and in contrast to ara-C and 5-azacytidine, ψ-isocytidine is stable against
both enzymatic and chemical catabolism.

Biological evaluation (16) of ψ-isocytidine was most encouraging. The ability of ψ-isocytidine to inhibit growth of mouse
and human leukemic cells in culture (Table I) was equal to or
perhaps better than that of 5-azacytidine. In lines of P815
resistant to ara-C there was a five- to ten-fold increase in
sensitivity to ψ-isocytidine as contrasted to 5-azacytidine and
also as contrasted to the parent ara-C-sensitive P815. The ara-C
resistant line of L5178Y (L5178Y/ara-C), on the other hand was
somewhat less sensitive to both ψ-isocytidine and 5-azacytidine
than the parent line. ψ-Isocytidine was somewhat more active
than 5-azacytidine against the SKL 7 line of human acute leukemic cells.

The inhibitory effect of ψ-isocytidine in concentrations as
high as 10μg/ml against L5178Y *in vitro* was completely blocked
(Fig. 4) by uridine or cytidine but not by 2'-deoxycytidine or
thymidine. Similarly, uridine and cytidine (but not thymidine

TABLE I. *Comparison of the 50% Inhibitory Concentrations of ψ-Isocytidine and 5-Azacytidine in Various ara-C-Sensitive and Resistant Lines of Mouse and Human Leukemic Cells in Vitro*

Compound	Concentration inhibitory to 50% of cells (μg/ml)					
	L5178Y	P815	L1210	L5178Y/ara-C	P815/ara-C	SKL 7
ψ-Isocytidine	0.8	0.5	0.5	3.8	0.04	0.9
5-Azacytidine	1.0	1.9	0.6	3.6	0.5	2.1

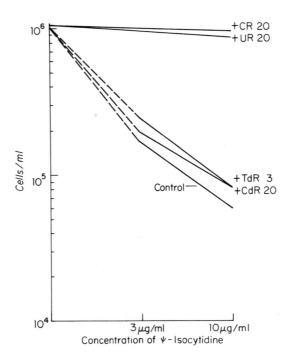

FIGURE 4.   Blocking of ψ-isocytidine and uridine, but
not by deoxycytidine or thymidine in cells of mouse leukemia
L51787.

or deoxycytidine) blocked the inhibitory effect of 5-azacytidine,
which suggests that its mechanism(s) of action and that of ψ-
isocytidine may be somewhat alike.

In vivo studies (16) were also quite encouraging (Figs. 5
and 6). A dose of 60 mg/kg intraperitoneally administered
daily for 5 days had an approximately equal antileukemic effect
in mouse leukemia L1210 or P815 as did 5-azacytidine. Against
the P815 ara-C resistant mouse leukemia (Table II), ψ-isocytidine gave a better effect than did comparably toxic doses of
5-azacytidine. To our best knowledge, ψ-isocytidine is the
first synthetic pyrimidine C-nucleoside for which anticancer
effects have been demonstrated. Thus, since ψ-isocytidine (a)
has antileukemic effects in mouse leukemia in vitro and in vivo
when administered intraperitoneally (i.p.) or orally (p.o.)
equal to or better than 5-azacytidine, (b) is metabolically
more stable, and (c) shows no cross-resistance with ara-C, it
might be useful in patients with acute myeloblastic leukemia,
whose disease has developed resistance to ara-C or its analogs.

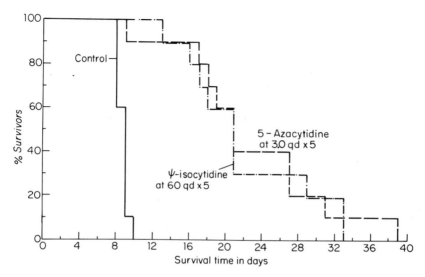

FIGURE 5.   The effects of 5-azacytidine and ψ-isocytidine
against L1210/0 leukemia.

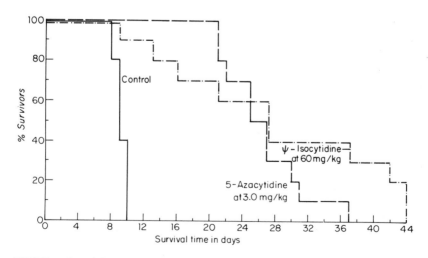

FIGURE 6.   The effects of 5-azacytidine and ψ-isocytidine
against P815.

TABLE II. Comparative Survival Times of Mice Treated with ψ-Isocytidine and 5-Azacytidine in ara-C-Resistant Mouse Leukemia

Line	Compound	Dose (mg/kg)	Schedule	Survival time (days)	% increase in life span
P815/ara C	Controls			9.3 ± 1.0	
	ψ-Isocytidine	60	Daily for 4 doses i.p.	28.8 ± 4.4	145
	5-Azacytidine	5	Daily for 4 doses i.p.	19.3 ± 5.3	106
P815/ara C	Controls			10.4 ± 1.6	
	ψ-Isocytidine	150	Every 4th day for 3 doses i.p.	20.8 ± 4.7	100
	5-Azacytidine	10	Every 4th day for 3 doses i.p.	15.3 ± 3.5	47
P815/ara-C	Controls			9.7 ± 0.8	
	ψ-Isocytidine	100	Daily for 10 doses p.o.	15.2 ± 2.6	57
	5-Azacytidine	10	Daily for 10 doses p.o.	12.0 ± 1.1	24
	ψ-Isocytidine	50	Daily for 10 doses i.p.	19.5 ± 1.7	97
	5-Azacytidine	3	Daily for 10 doses i.p.	13.9 ± 1.8	43
P815/ara-C (inoculated s.c.)	Controls			10.5 ± 0.7	
	ψ-Isocytidine	80	Daily for 4 doses i.p.	16.2 ± 4.6	54
	5-Azacytidine	5	Daily for 4 doses i.p.	14.6 ± 1.9	39

Preclinical pharmacology on ψ-isocytidine leading to phase 1 clinical trials is underway at our Institute. For this purpose, (^{14}C) labeled ψ-isocytidine was synthesized. Cyclization of *crystalline* 8 (Fig. 3) with ^{14}C guanidine followed by removal of protecting groups afforded 2-^{14}C-ψ-isocytidine in ∿ 80% overall yield from 8.

These biological data obviously warrant further efforts into the syntheses of C-nucleoside analogs of pyrimidine nucleosides occurring in the nucleic acids. For example, 4-thiouridine, a component of tRNA, has been shown by Bloch *et al.* (17) to have activity against Ehrlich ascites and S-180 in mice. The synthesis of the isosteric ψ-2-thiouridine (Fig. 7) was achieved (11) from the versatile intermediates 7 and 8, previously employed for the synthesis of ψ-isocytidine. Thus treatment of 7 or 8 with thiourea and sodium ethoxide afforded the protected ψ-2-thiouridine. Brief treatment of this intermediate with methanolic HCl afforded ψ-2-thiouridine (12). The overall yield of 12 was better when the β-methoxyacrylate 8 rather than 7 was cyclized with thiourea. A total synthesis of pseudouridine was also achieved (11) from the common β-methoxyacrylate intermediate 8 by treatment with urea snd sodium ethoxide in ethanol. After separation from the α-isomer by preparative thin-layer Chromatography, the protected ψ-uridine was deblocked with methanolic hydrogen chloride to afford a crystalline nucleoside identical with ψ-uridine.

The synthesis of pyrimidine C-nucleosides bearing an amino function in the 4-position (cytosine types) was also achieved (Fig. 8) by use of an α,β mixture of the acrylonitrile C-ribosyl derivative 14 (18,19), which is readily prepared from the known (14) acetonitrile derivative 13 (also an α,β mixture). The separation of isomers of 14 has been accomplished (19). However, from a practical viewpoint this is not necessary since *either* isomer gave in the subsequent step an α,β mixture of the protected 5-ribofuranosyl-4-aminopyrimidines, which were separated by column chromatography. Treatment of each blocked nucleoside isomer with methanolic hydrogen chloride afforded the corresponding C-nucleoside, e.g., 2,4-diamino-5-(β-D-ribofuranosyl)pyrimidine (15) or the α-isomer (19).

Cyclization of 14 with an excess of thiourea with sodium ethoxide in ethanol gave an isomeric mixture of the protected ψ-2-thiocytidines, which were separated by fractional crystallization and deprotected with methanolic hydrogen chloride to ψ-2-thiocytidine (16) and to its α-isomer (19). Ring closure of 14 with urea afforded the α,β mixture of protected ψ-cytidines, which was also separated by fractional crystallization. However, treatment of either isomer with methanolic hydrogen chloride under a variety of conditions gave an intractable mixture, which from PMR analysis indicated that "anomerization" as

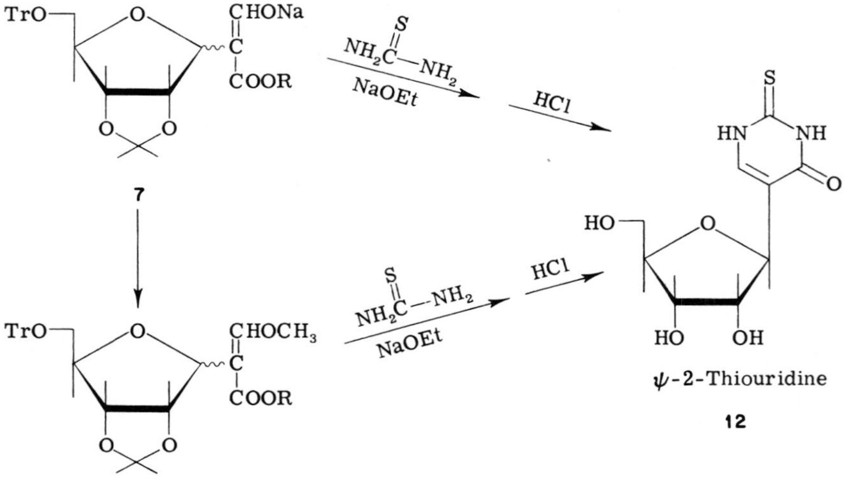

Figure 7

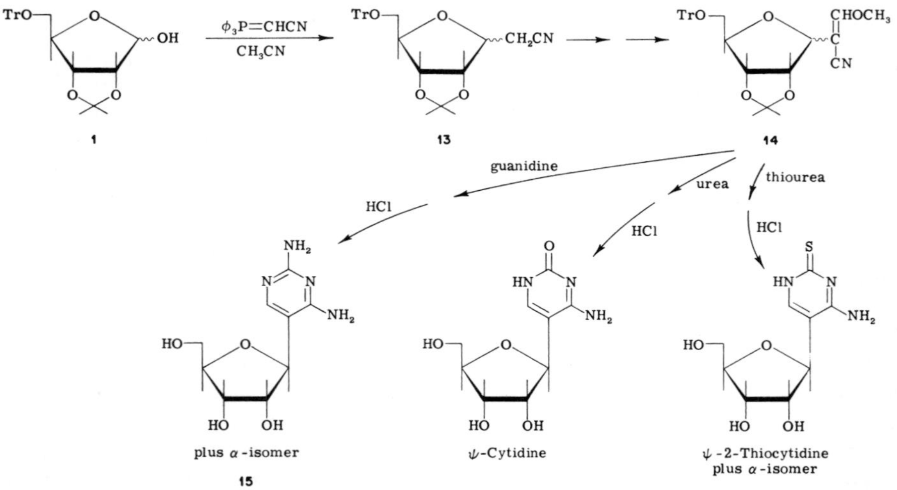

Figure 8

well as ring-size isomerization (furanosyl→pyranosyl) had occurred along with the formation of other impurities. ψ-Cytidine (5-β-D-ribofuranosylcytosine) had been previously obtained by David and Lubineau (20) by a different route and was purified by chromatography on Dowex 50 (H⁺) using 0.1 *M* sulfuric acid as eluent. Using these conditions we were unable to elute ψ-cytidine from this Dowex 50 column even using higher acid concentrations as eluents. The synthesis of ψ-cytidine by a different approach is therefore under investigation in our laboratory.

Of interest would be the synthesis of a 5-(β-D-arabinofuranosyl)isocytosine, which may be viewed as an isostere of the very effective antileukemic agent, ara-C. Two approaches have been investigated to achieve the syntheses of these C-nucleosides. The first (21) involves the synthesis of the arabinosyl-β-methoxyacrylates (18, Fig. 9) from commercially available tri-0-benzyl-D-arabinofuranose using procedures previously employed for the synthesis of the ribosyl analog (8, Fig. 3). Ring closure of 18 with guanidine afforded the tri-0-benzylated 5-arabinosyl-isocytosine (19) which, after deblocking with boron trichloride in methylene chloride, afforded only the *alpha* isomer (20) in fair yields.

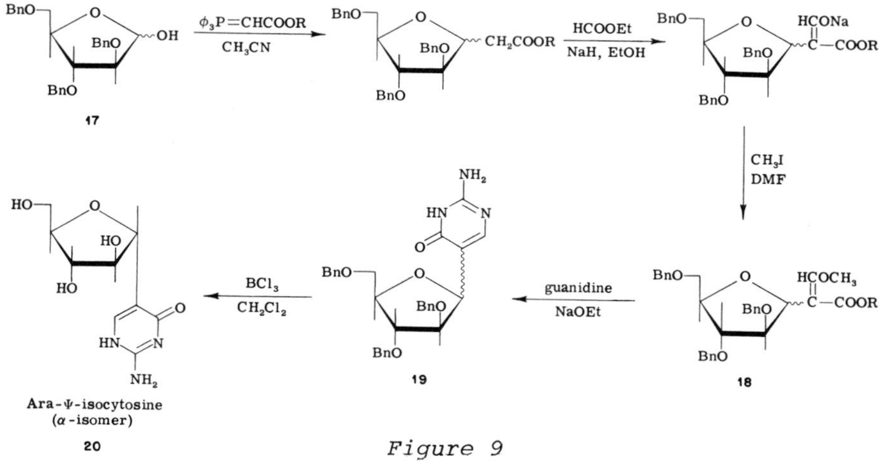

Figure 9

The desired β-nucleoside (22, Fig. 10) was readily obtained
(22) from ψ-isocytidine by use of the anhydro-nucleoside
approach (23). Thus, treatment of ψ-isocytidine (10) with α-
acetoxyisobutyryl chloride (24) in acetonitrile under reflux
gave the crystalline 4,2'-anhydronucleoside (21, R = 2,3,5-tri-
methyldioxolanon-2-yl) in good yield. Similarly, treatment of
(10) with salicyloyl chloride (25) afforded crystalline 21
(R = acetyl) in equally good yields. Deacylation of 21 with
methanolic hydrogen chloride afforded the crystalline 4,2'-an-
hydro nucleoside as the hydrochloride salt in high yield. The
4,2'-anhydro linkage of 21 (R = H or acetyl) was found to be
much more stable to base than is 2,2'-anhydro-1-(β-D-arabinosyl)
cytosine (26) and stringent conditions (10% sodium hydroxide at
reflux for 0.5 hr) were required to cleave this linkage with
the exclusive formation of the *beta* isomer, ara-ψ-isocytosine
[(22), 5-(β-D-arabinofuranosyl)-isocytosine] in high yield.
Compound 22 underwent epimerization at C-1' in dilute acid solution. The PMR studies of this solution showed the initial
formation of the α-isomer (20). At equilibrium, the major product was the 5-(α-D-arabinopyranosyl)isocytosine along with
small amounts of 20,22, and the β-pyranosyl isomer (22).

Reaction of ψ-uridine with acetoxyacyl chlorides in acetonitrile proceeded rather differently (22) from that with ψ-isocytidine (Fig. 11). Even under varied conditions (time, temperature, etc.), mixtures of variously protected anhydronucleo-
sides (23) were obtained along with 2'-chloro-2'-deoxy-ψ-uri-
dines (24) and other isomers (25) and (27). (Fraction 25 may
contain α-isomers of 24 along with other uncharacterized components.) Shorter reaction time favored formation of anhy-
dronucleosides 23 as the major products. Treatment of 23 (R =
acetyl, R' = 2,3,5-trimethyldioxolanon-2-yl) with 0.5 *M* sodium
methoxide gave the deprotected anhydronucleoside 26. The same

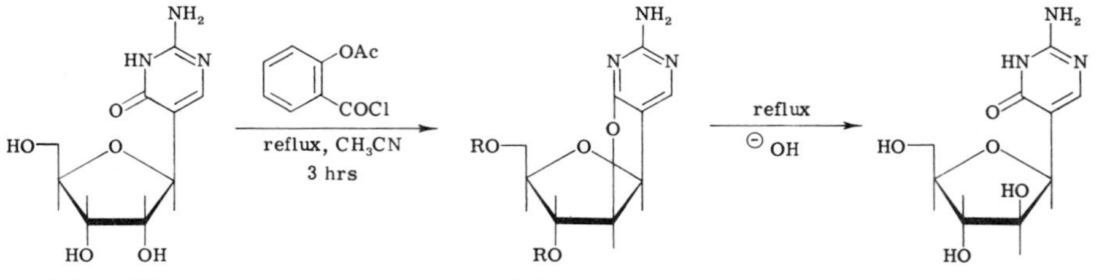

Ψ-Isocytidine
10

4,2'-Anhydro-
ara-Ψ-isocytosine
21

Ara-Ψ-Isocytosine
22

Figure 10

compound is also obtained by treatment of the 2^l-chloro-β-D-ribofuranosyl derivatives (24) with sodium methoxide. For a practical synthesis of crystalline 26, isolation of intermediates was not necessary. The crude product from the reaction of ψ-uridine with α-acetoxyisobutyryl chloride or with salicyloyl and chloro derivatives (24) converted into 26. The 4,2 -anhydro linkage of 26 was very labile to acid. Treatment of 26 with Dowex 50 (H+) in water for 10 min afforded crystalline 5-(β-D-arabinofuranosyl) uracil (27) in high yield. Prolonged Dowex 50 (H+) treatment slowly isomerized the β-isomer (27) to its α-counterpart (22).

A recent report by Argoudelis and Mizsak (28) described the isolation of 1-N-methyl-ψ-uridine from the culture filtrate of *Streptomyces platensis* var. *clarensis* (which produces an antibacterial and antiviral antibiotic U-44590). The synthesis (29) of 1-N-methyl-ψ-uridine was readily achieved by exhaustive trimethylsilylation of ψ-uridine followed by methylation of the syrupy product 28 with methyl iodide in acetonitrile to 29 (Fig. 12). Hydrolytic removal of the trimethylsilyl groups gave the desired product 30 in good yields. The conversion of 28 into 30 probably proceeded via a mechanism (30) closely related to the Hilbert-Johnson reaction that is known to occur

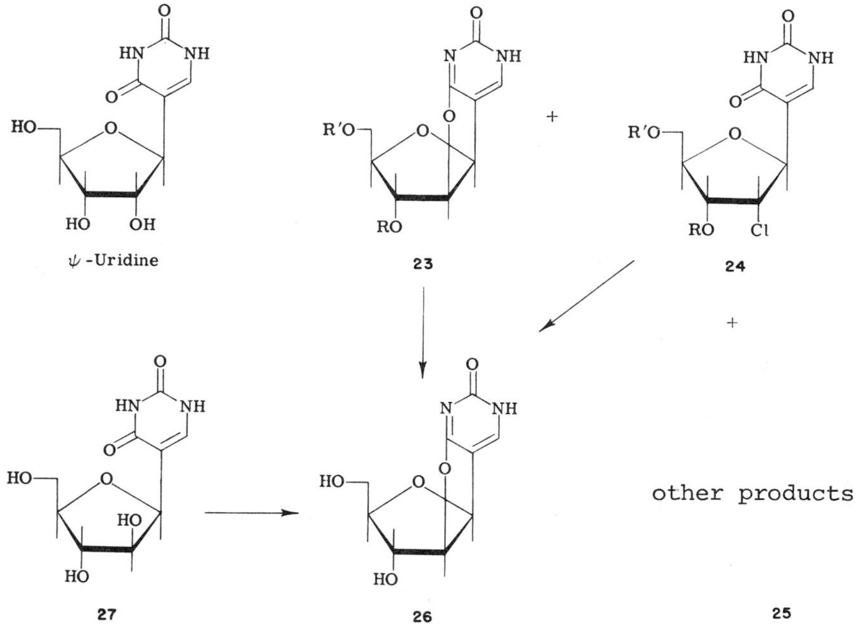

Figure 11

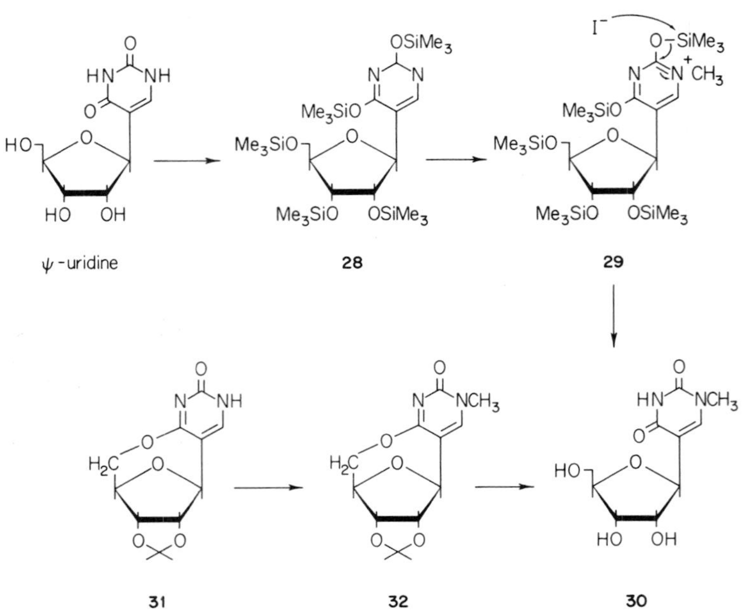

Figure 12

preferentially at N-1 of 2,4-dialkoxy (or *bis*-trimethylsilyloxy)·
pyrimidines. The same product <u>30</u> was obtained (29) from the
known (31) 4,5'-anhydronucleoside (<u>31</u>) by methylation of the
latter with dimethylformamide dimethyl acetal followed by hydrolytic removal of protecting groups and cleavage of the anhydro
linkage with 80% formic acid. Crystalline <u>30</u> thus obtained
was identical with that produced via <u>28→29</u> (29).

As mentioned previously (Fig. 2,<u>3</u>), 2,3-*O*-isopropylidene-5-
O-trityl-β-D-ribofuranosyl chloride is easily prepared by tri-
tylation and chlorination of 2,3-*O*-isopropylidene-D-ribose.
This intermediate (<u>3</u>) was converted by reaction with the silver
acetylide of ethyl propiolate into an isomeric mixture of ethyl
3-(2,3-*O*-isopropylidene-5-*O*-trityl-D-ribofuranosyl)propiolates
(32), from which the pure β-isomer (<u>33</u>, Fig. 13) was readily obtained by chromatographic separation. From intermediate <u>33</u>,
entry was gained into a new class of pyrimidine C-nucleosides,
namely the 6-ribosylated derivatives (32). Thus treatment of
<u>33</u> with guanidine afforded the 6-ribosyl-isocytosine nucleoside
<u>34</u>. Reaction with pyrrolidine afforded the enamine <u>35</u>, which
after hydrolysis with Dowex 50 (H⁺) afforded the β-ketoester
<u>36</u> in quantitative yield. The latter, after treatment with
thiourea and sodium methoxide followed by removal of the protecting groups with methanolic hydrogen chloride, gave 6-ribo-
furanosyl-2-thiouracil (<u>37</u>). The corresponding uracil analog

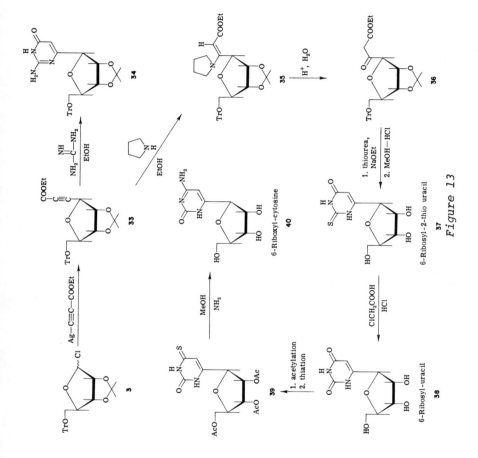

Figure 13

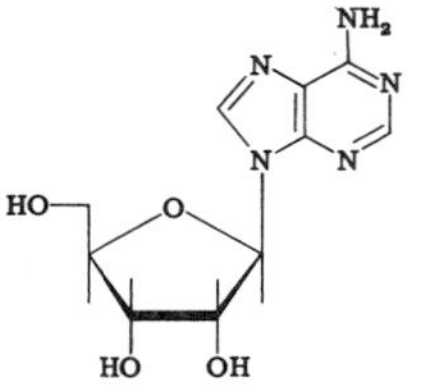

Adenosine

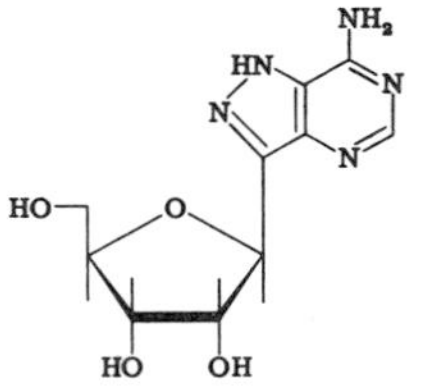

Formycin

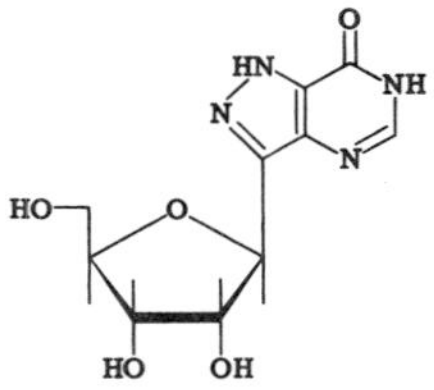

Formycin B

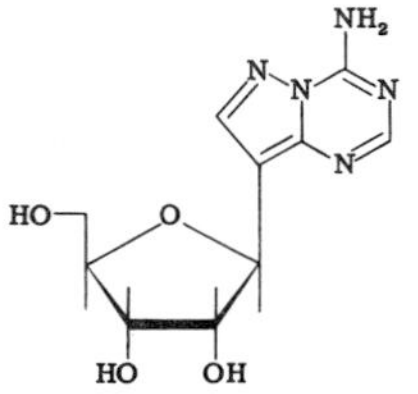

4-Amino-8-[β-D-ribofuranosyl]-pyrazolo [1,5-a]-1,3,5-triazine (APTR)

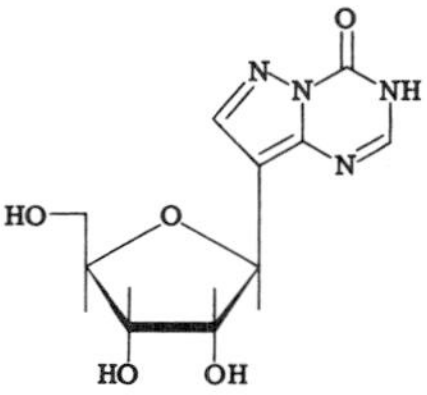

4-Oxo-3H-8-[β-D-ribofuranosyl]-pyrazolo [1,5-a]-1,3,5-triazine (OPTR)

Figure 14

38 was readily obtained by treatment of 37 with chloracetic acid. Conversion of 6-ribofuranosyluracil into the corresponding cytosine analog 40 was accomplished (32) by standard thiation procedures to 39 followed by reaction with methanolic ammonia. It is noted that these 6-ribosylated pyrimidine nucleosides do not undergo epimerization at C-1' in acidic media, whereas the 5-ribosylated analogs (ψ-uridines, ψ-isocytidines, etc.) isomerize under these conditions. This phenomenon is readily explained by the structure of the 6-ribosylated nucleosides, in which the lone pair of electrons on the nitrogen atoms cannot be delocalized to assist in the lactol ring opening.

We come now to our studies on C-nucleosides related to purine nucleosides, as well as to the formycins. Figure 14 clearly illustrates the close structural relationships that exist between adenosine, formycin and formycin B, and two of our target nucleosides belonging to a hitherto unknown pyrazolo-*s*-triazine class of C-nucleosides (viz. APTR and OPTR). Two approaches (18) were employed to obtain the key intermediate, aminopyrazole C-nucleoside 44 (Fig. 15). The first begins with the ribosyl propiolate derivative 41, which was prepared by the same procedure used for the synthesis of 33. Compound 41 underwent a 1,3-dipolar cycloaddition with diazomethane to ester

<u>42</u> (33). The latter was converted to its hydrazide and then to the acylazide <u>43</u> by brief treatment with nitrous acid. The azide <u>43</u> underwent a Curtius rearrangement in boiling ethanol and the carbamate thus produced was hydrolyzed with sodium hydroxide to the aminopyrazole C-nucleosides <u>44</u>. This alkaline hydrolysis step was accompanied by epimerization at C-1' (α/β ratio, 2:1). These isomers could be separated by silica gel column chromatography. It can be seen therefore that no advantage is gained by use of the pure beta isomer of propiolate <u>41</u> (18). An alternate procedure to a tri-O-benzylated derivative of <u>44</u> has recently been described by Gupta *et al.* (34).

A somewhat shorter approach to the aminopyrazoles <u>44</u> was achieved (18) from our previously employed methoxyacrylonitrile C-ribosyl derivative <u>14</u> by treatment with anhydrous hydrazine and sodium ethoxide.

Conversion of the β-isomer of <u>44</u> with N-carbethoxyisothiocyanate (Fig. 16) in acetonitrile gave intermediate <u>45</u>, which was ring closed by N sodium hydroxide in methanol to the 4-oxo-2-thio-8-(β-D-ribofuranosyl)pyrazolo(1,5-a)-1,3,5-triazine derivative <u>46</u> in high yield (18). A 2,4-dioxo derivative of <u>48</u> has been recently synthesized (34) by a somewhat analogous procedure. Treatment of the β-isomer of <u>44</u> with ethyl N-cyanoformimidate in benzene gave directly the 4-aminopyrazolo-triazine nucleoside, which after very brief treatment with methanolic hydrogen chloride afforded the beta 4-amino-pyrazolo-s-triazine C-nucleoside ("APTR" - <u>48</u>) in good yield (35).

Preliminary biological evaluation of APTR (<u>48</u>) by J. H. Burchenal at this Institute showed that this compound (which is isosteric to both adenosine and formycin) is approximately 10 times as active as formycin against mouse leukemic cells in culture with ID_{50} ranging between 0.01 and 1.0 μg/ml. APTR is roughly equivalent to thioguanine, is 50-100 fold more active than 6-mercaptopurine, and is active in cell lines of leukemia resistant to ara-C and ψ-isocytidine. As with formycin, the inhibitory effects of APTR are blocked by adenosine.

With the small quantities of APTR available preliminary *in vivo* studies have shown that APTR at 300 mg/kg i.p. qd x 10 produced an increase in life span (ILS) of 80% against mouse leukemia L1210 and a 58% ILS against a leukemia line made resistant to vincristine and cross-resistant to ara-C and adriamycin. Further *in vivo* studies against sensitive and resistant leukemias are under way.

The aminopyrazole C-nucleoside (<u>44</u>) also served as a precursor for the synthesis of the 4-oxo-pyrazolo-s-triazine nucleoside (<u>50</u>, Fig. 17) (35). Treatment of <u>44</u> with N-carbethoxy-formimidate in ethanol followed by cyclization with potassium

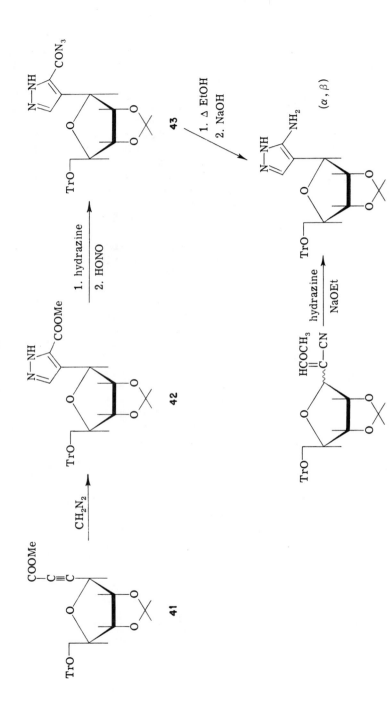

Figure 15

434

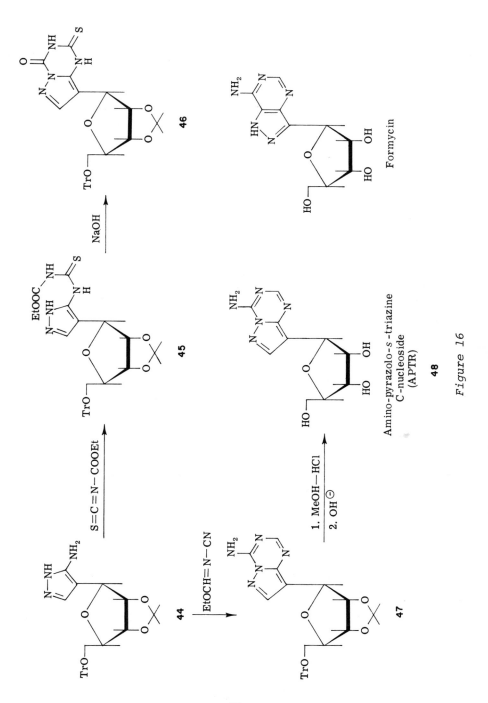

Figure 16

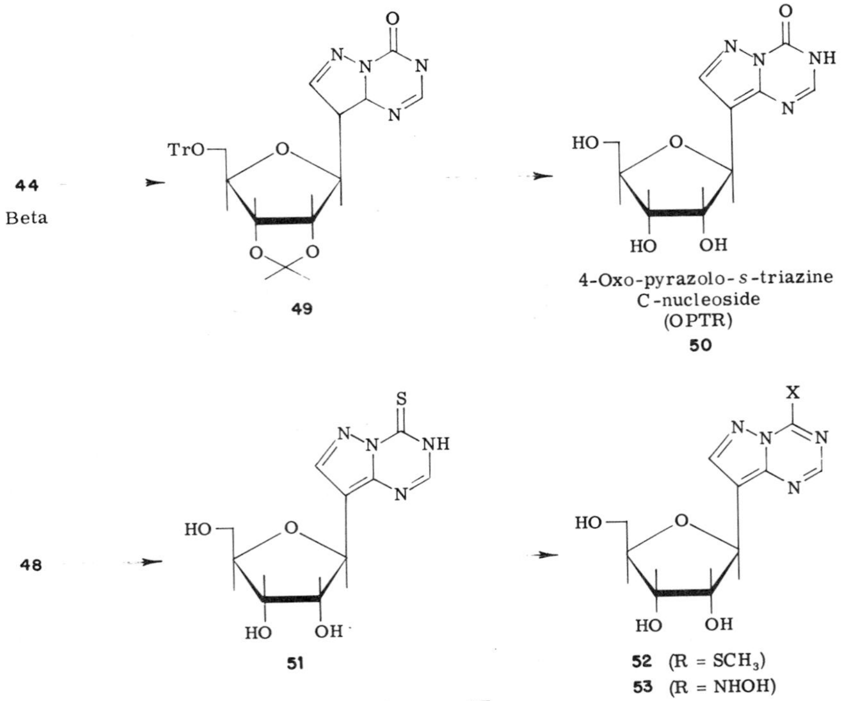

Figure 17

carbonate and neutralization afforded the C-nucleoside 49.
Again, deblocking was achieved by *brief* treatment of 49 with
methanolic hydrogen chloride to give only the pure beta C-nucleoside 50 ("OPTR"). The α-isomers of 50 and of 48 were obtained from the α-isomer of 44. Preliminary tissue culture
studies showed that OPTR is ∿ tenfold more active (ID_{50} = 0.5
than formycin B (ID_{50} = 5.3 μg/ml) against mouse leukemia
cells.

W. Kreis at this Institute found that, using an L1210 cell
homogenate as a source of adenosine deaminase, APTR is deaminated about half the rate of formycin. It is hoped, of course,
that unlike formycin (which is enzymatically deaminated to the
less active formycin B) APTR will be less susceptible to catabolic deamination by adenosine deaminase in man.

APTR served as a precursor for the syntheses of variously 4-substituted pyrazolo-*s*-triazine C-nucleosides (36). When treated with hydrogen sulfide-pyridine-water (37), APTR was smoothly converted into the 4-thioxo analog 51 (36). Methylation of 51 with methyl iodide and sodium methoxide afforded the 4-methylthio analog 52, which, after treatment with hydroxylamine, gave the 4-hydroxylamino analog 53. Isomerization at C-1' was not observed in the reactions of 48→53.

In preliminary tissue culture studies, the methylthio derivative 52 showed activity against L1210 and P815 (ID_{50} of 0.09 and 0.07 µg/ml, respectively), whereas the free base, 4-methylthio-pyrazolo(1,5-*a*)-1,3,5-triazine (36), was relatively inactive ($ID_{50} > 10$ µg/ml). In these tissue culture systems (L1210 and P815), 6-methylthio-9-β-D-ribofuranosylpurine (38) gave ID_{50} of 0.3 and 0.09 µg/ml, respectively.

This concludes the report of our recent studies in the area of C-nucleosides. As stated at the outset, some of the naturally occurring C-nucleosides have exhibited interesting antiviral and/or anticancer properties. It is reasonable to expect that *synthetic* C-nucleoside analogs will become an ever increasing source of biologically and biochemically useful agents.

REFERENCES

1. W. E. Cohn, *Biochim. Biophys. Acta 32*, 569 (1959); *J. Biol. Chem.*, *235*, 1488 (1960).
2. W. E. Cohn, *Fed. Proc. 16*, 166 (1957).
3. For a review, see R. J. Suhadolnik, *Nucleoside Antibiotics* Wiley-Interscience, New York, 1970.
4. F. F. Snyder and J. F. Henderson, *J. Biol. Chem. 248*, 5899 (1973).
5. A. Piskala and F. Sorm,*Coll. Czech. Chem. Commun. 29*, 2060 (1964); L. J. Hanka, J. S. Evans, D. J. Mason, and A. Dietz, *Antimicrob. Agents Chemother.*, 619 (1966).
6. R. J. Suhadolnik, *Nucleoside Antibiotics*, Wiley-Interscience, New York, 1970, p. 271.
7. C. Heidelberger, in *Cancer Medicine* (J. F. Holland and E. Frei, eds), Lea and Febiger, Inc., Philadelphia, 1973, p. 768.
8. M. Karon, K. Sieger, S. Leimbrock, J. Finklestein, M. Nesbit, and J. Swaney, *Blood 43*, 359 (1973); K. B. McCredie, G. P. Bodey, M. A. Burgess, V. Rodriguez, M. Sullivan, and E. J. Freireich, *Blood, 40*, 975 (1972).
9. W. R. Vogler, D. Miller, and J. W. Keller, *Proc. Am. Assoc. Cancer Res. 16*, 155 (1975).

10. C. K. Chu, K. A. Watanabe, and J. J. Fox, *J. Heterocycl. Chem. 12*, 817 (1975).

11. C. K. Chu, I. Wempen, K. A. Watanabe, and J. J. Fox, *J. Org. Chem. 41*, 2793 (1976).

12. H. Ohrui and J. J. Fox, *Tetrahedron Lett. 22*, 1951 (1973).

13. R. S. Klein, H. Ohrui, and J. J. Fox, *J. Carbohydr. Nucleosides Nucleotides 1*, 265 (1974).

14. H. Ohrui, G. H. Jones, J. G. Moffatt, M. L. Maddox, A. T. Christensen, and S. K. Byram, *J. Am. Chem. Soc. 97*, 4602 (1975).

15. R. W. Chambers, V. Kurkov, and R. Shapiro, *Biochemistry 2*, 1192 (1963).

16. J. H. Burchenal, K. Ciovacco, K. Kalaher, T. O'Toole, R. Kiefner, M. D. Dowling, C. K. Chu, K. A. Watanabe, I. Wempen, and J. J. Fox, *Cancer Res. 36*, 1520 (1976).

17. A Bloch, R. K. Robins, M. W. Winkley, and J. T. Witkowski, Abstract presented to the American Association for Cancer Research, 60th Meeting, San Francisco, March 1969, p. 7.

18. F. G. de las Heras, C. K. Chu, S. Y-K. Tam, R. S. Klein, K. A. Watanabe, and J. J. Fox, *J. Heterocycl. Chem. 13*, 175 (1976).

19. C. K. Chu, U. Reichman, K. A. Watanabe, and J. J. Fox, *J. Org. Chem. 42*, 711 (1977).

20. S. David and A. Lubineau, *Carbohydr. Res. 29*, 15 (1973).

21. C. K. Chu, K. A. Watanabe, and J. J. Fox, unpublished.

22. U. Reichman, C. K. Chu, I. Wempen, K. A. Watanabe, and J. J. Fox, *J. Heterocycl. Chem. 13*, 933 (1976).

23. For a review of the chemistry of pyrimidine anhydronucleoside transformations see J. J. Fox, *Pure Appl. Chem. 18*, 223 (1969).

24. S. Greenberg and J. G. Moffatt, *J. Am. Chem. Soc. 95*, 4016 (1973).

25. U. Reichman, C. K. Chu, D. H. Hollenberg, K. A. Watanabe, and J. J. Fox, *Synthesis*, 533 (1976); E. K. Hamamura, M. Prystasz, J. P. Verheyden, J. G. Moffatt, K. Yamguchi, N. Uchida, K. Sato, A. Nomura, O. Shiratori, S. Takase, and K. Katagiri, *J. Med. Chem. 19*, 654 (1976).

26. I. L. Doerr and J. J. Fox, *J. Org. Chem. 32*, 1462 (1967).

27. U. Reichman, K. A. Watanabe, and J. J. Fox, unpublished.

28. A. D. Argoudelis and S. A. Mizsak, *J. Antibiot. 29*, 818 (1976).

29. U. Reichman, K. Hirota, C. K. Chu, K. A. Watanabe, and J. J. Fox, *J. Antibiotics, 30*, 129 (1977).

30. For a discussion of the mechanism of the Hilbert-Johnson reaction see K. A. Watanabe, D. H. Hollenberg, and J. J. Fox, *J. Carbohyd. Nucleosides Nucleotides 1*, 1, (1974).

31. A. M. Michelson and W. E. Cohn, *Biochemistry 1,* 490 (1962).
32. S. Y-K. Tam, F. G. de las Heras, R. S. Klein, and J. J. Fox, *Tetrahedron Lett. 38,* 3271 (1975) and unpublished work.
33. F. G. de las Heras, S. Y-K. Tam, R. S. Klein, and J. J. Fox, *J. Org. Chem. 41,* 84 (1976).
34. C. M. Gupta, G. H. Jones, and J. G. Moffatt, *J. Org. Chem. 41,* 3000 (1976).
35. S. Y-K Tam, J-S. Hwang, F. G. de las Heras, R. S. Klein, and J. J. Fox, *J. Heterocycl. Chem. 13,* 1305 (1976).
36. S. Y-K. Tam, R. S. Klein, and J. J. Fox, unpublished.
37. T. Ueda, K. Miura, M. Imazawa, and K. Odajima, *Chem. Pharm. Bull. 22,* 2377 (1974).
38. J. J. Fox, I. Wempen, A. Hampton, and I. L. Doerr, *J. Am. Chem. Soc., 80,* 1669 (1958); A. Hampton, J. J. Biesele, A. E. Moore, and G. B. Brown, *J. Am. Chem. Soc. 78,* 5695 (1956).

Index

A

9-(3-Acetamido- 3-deoxy- α-D-arabinofuranosyl)- 2-chloroadenine, 40
9-(3-Acetamido- 3-deoxy- β-D-arabinofuranosyl)- 2-chloroadenine, 40
9-(2-Acetamido- 3-α- D-ribofuranosyl)- 2-chloroadenine, 40
9-(3-Acetamido- 3-deoxy- β-D-ribofuranosyl)- 2-chloroadenine, 40
5'-Acetamidomethyl-AMP, 92
3-Acetamido- 1,2,5-(tri-O-acetyl- α-D-arabinofuranose), 39
3-Acetamido- 1,2,5-(tri-O-acetyl- β-D-ribofuranose), 39
Acetonitrile, 27, 29, 31, 34, 37, 41, 43–45, 47–48, 51–52
   as eluant in high pressure liquid chromatography, 41–52
   solvent of choice in condensations to form nucleosides, 221
1-Acetoxy-pentofuranosides, reaction with silylated heterocycle, 244, 247
2-Acetoxyisobutyryl chloride, 118
Acetyl, blocking group for 3'-hydroxyls, 23, 26
5'-O-Acetyl-2'-3'-O-1-adamantylphosphonates, isomers, 332
Acetyl-2'-chloro-2'-deoxypseudouridine, 118
N⁴-Acetylcytosine, derivative of 2',3'-dideoxy-2',3'-diacetylthioxylosyl, 401
1-(5-O-Acetyl-3-deoxy-3-acetylthio- 2- O-mesyl- β-D-arabinofuranosyl)-N⁴-acetylcytosine,
   preparation of, 401
1-(5-O-Acetyl-2,3-dideoxy- 2,3-epithio- β-D-lyxofuranosyl)- uracil, 398
1-(5-O-Acetyl-2,3-dideoxy-β-D-pent- 2-enofuranosyl) uracil, 400
1-O-Acetyl-3,5-di-O-p-toluoyl- 2-deoxy- D-erythro-pentofuranose, 275
5'-O-Acetyl epithionucleosides, CD spectra, 400
5'-O-Acetyl-2',3'-O-isopropylidene- 5-bromouridine, 331
Acetyl phosphate, hydrolysis, 87
N₁-Acetyl-N₃-(tribenzoyl-β-D-ribofuranosyl)-5,6-dihydro-5-methyl-s-trazine-2,4-
   (1H,3H) -dione,222
2-Acetoxyisobutyryl chloride, reaction with pseudouridine, 118
Acid-catalysis of fusion procedure, 122
Ackermann–Potter plot of adenosine deaminase preincubated with coformycin, 181
Activating agents, energy-rich phosphate, 22
Acyclic sugar 4-thiazolyl nucleoside analogs, 381, 385–386
Acyl, protecting groups, 32
5'-C-Acylaminomethyl AMP, derivatives, 89, 91
Acylated 1-halosugars, 252
Acylated sugar cation, from 4-thio sugar, 253
1,2-Acyloxonium ions, 252–254, 262
5'-O-Acyluridine 2'-and 3'-O-methylphosphonate formation, 340
5'-and 3'-O-(1-Adamantyl) chlorophosphonates, 329
5'-O-1-Adamantylchlorophosphonate of thymidine, 343
2',3'-O-Adamantylphosphonate of 5-azauridine, 5'-acetate, CD spectra of, 329
Adamantylphosphonates, 329, 333–334

Adamantylphosphonates, 5-membered, stability, 340
Adamantylphosphonates, 6-membered, 344
2'(3')-1-Adamantylphosphonates of nucleosides, split from cyclo forms, 339
2',3'-*O*-1-Adamantylphosphonates of pyrimidine nucleosides
   configurations and conformations, 336
   of purine nucleosides, 338
Adamantylphosphonic acid (APA), 335
1-Adamantylphosphonodichloride (APD), interactions with, uridine, 5-bromouridine,
   5-fluorouridine, 6-azauridine, 5'-fluoro-5'-deoxyuridine, adenosine, or 9-β-D-ribofu-
   ranosyl-6-methyl-mercapto purine, 329
1-Adamantylphosphonodichloride (APD), 340
   interaction with thymidine, 340–345
   PMR spectra of derivatives, 341–342
Adenine, 39–42, 69, 159–160, 185–187, 232–233, 387–388, 390
Adenine nucleosides, bridged N^6-$N^{6'}$ chemistry, 69
Adenine nucleotides, 55, 85–95
Adenine phosphoribosyltransferase, 123
Adenosine, 38, 69, 92, 193, 314, 390, 392, 394, 411–413
Adenosine and 5'-*O*-acetyladenosine 2',3'-*O*-1-adamantylphosphonates, PMR spectra, 342
Adenosine and analogs, 181
Adenosine analogs, 132, 160, 185
Adenosine concentration, 182
Adenosine cytotoxicity, 183
Adenosine deaminase, 49, 50
Adenosine deaminase deficiency with (SCID) Severe Combined Immunodeficiency Disease,
   183
Adenosine deaminase deficient erythrocytes of children, 183
Adenosine deaminase enzyme, 160–181, 190
   application of theory of tight-binding inhibitors, 171–180
   4-ADP riboside as substrate, 122
   binding class determination by ribose moiety, 180
   effect on cellular incorporation of 8-aza-adenosine, 190
   effect of ADA inhibition on cellular adenosine analogs, 181–193
   kinetic constants (Km + Ka) for inhibitors and substrates, 180
   protection of cells against ATP level surges, 183
   site of activity, 180
   theoretical studies of tight-binding enzymes, 160–170
Adenosine deaminase, human erythrocytic, 173, 176–178
Adenosine deaminase, inhibition, 122, 171–193
   coformycin, 159
   Covid-arabine®, 159
   deoxycoformycin, 159
   immunosuppressive agents, 195
   in intact human erythrocytes, 159
   mechanisms, 159, 181
Adenosine deaminase, intraerythrocytic, 183
Adenosine deaminase protein, 194
Adenosine deaminase, reactions, 171
Adenosine deaminase, spleen or mouse kidney, 435
Adenosine kinase, 48–49, 122
Adenosine, n^6-(ω-alkyl), 69
Adenosine nucleoside derivatives, 122

Adenosine 5'-triphosphate (ATP), effectors of many enzymes, 85–95
Adenylate kinase, rabbit muscle, 89
Adenylate lyase, 55
ADP, N^6-aroyl-derivatives, 90
Aglycon, trimethysilyl derivative, glycosylation, 287
AIV, 5-iodo-5'-dideoxyuridine, 85–95
Aldehydes, reversion from cyanohydrins, 362
Alkoxide, as nucleophile, 114
N^6-(ω-Alkyl) adenosines, 69
Alkylating agents for pseudouridines, 113
Alkylation of 5-mercapto groups, 60
Alkyldiphenylchlorosilane, protector of hydroxyls in nucleotides, 33
Alkylmercaptides, as nucleophile, 114
Alkylphosphonodichlorides, from five-membered cyclophosphonates, 340
5-Alkyluracils, condensations with protected 2-deoxyribofuranosyl chlorides, 221
Allopurinol riboside
    acetylation (pyridine/acetic anhydride), 123
    structure, 124–125
Alloses, 361–364, 373
D-Altro-ethnylpentitol derivative, 375
Amber mutations, 18–19
Amidine, ring closing, derivative of, 1,9-diribofuranosyladenine, 231
Amidoxime, from chloroxime, 367
Amines, methyl-,benzyl-,furfuryl-, and aniline-, 124
Aminoacyl functions, substitution of phenylalanyl or leucyl on 2'-or 3'-hydroxygroups, 80
Aminoacyl tRNA, 70
2'-Aminoadenosine, synthesized, 392
N^6-Aminoalkyladenosine 5'-phosphate, iodoacetylation with N-iodoacetoxy succinimide, 92
5'-Amino analogs, 101
5'-Amino analog of Id-Urd, 101, 103
4-Amino-7-(2,3-anhydro-β-D-lyxofuranosyl)-pyrrolo-(2,3-d) pyrimidine, 314
7-Amino-1-(β-D-arabinofuranosyl) imidazo-(1,2-c) pyrimidin-5-one, 295
5'-Amino and 5'-azidopyrimidine, 85
2-Amino-4-chloro-5-(β-D-ribofuranosyl) pyrimidine, 116
5(4)-Amino-4(5)-cyanoimidazole, 232
5-Amino-4-cyano-1-β-D-ribofuranosyl-imidazole, 229–231
5'-Amino-5'-deoxythymidine, 97–99, 101
5'-Amino-5'-deoxythymidine, 5'-N'-triphosphate, 105
5'-Amino-2',5'-dideoxyuridine, iodination, 100
5-Amino-2,4-dimethoxypyrimidine, pKa value, 256
N^6-(2-Aminoethyl) $a$denosine, reaction with methyl orthoacetate, 69
Amino groups of oligonucleotides, deacylation by ammonia, 23
Aminomethyl nucleoside, 91
5-Amino-3-nitro-1-β-D-ribofuranosyl-1,2,4-triazole, 274
2'-Amino nucleosides, in ribo configuration, 388
5'-Amino nucleosides, 98, 103
Aminoxazoline, 200, 203
Aminopyrazole-C-nucleosides, 431–432
4-Aminopyrazolo(3,4-d)pyrimidine(4-APP), 122–123
4-Amino-pyrazolo-s-triazine C nucleoside (APTR), 432, 435, 436
7-Amino-1-(β-D-ribofuranosyl)imidazo(1,2-a), pyrimidin-5-one, 296
7-Amino-1-(β-D-ribofuranasyl)imidazo(1,2-c)pyrimidine-5-one, 293–294
3-Amino-4-ribofuranosyl-pyrazole, 359

4-Amino-1-(β-D-ribofuranosyl)pyrazolo(3,4-*d*)pyrimidine(4-APP riboside), 122
4-Amino-1-(β-D-ribofuranosyl)pyrazolo(3,4-*d*)pyrimidine-3-carboximidate, 121–127
7-Amino-2-β-D-ribofuranosyl-*v*-triazolo(4,5-*d*)pyrimidine, 305, 306
5-Amino-3-(β-D-ribofuranosyl)-1,2,4-triazolo(1,5-*a*) pyrimidin-7-one, 290
2'-Amino ribose structure of natural nucleosides, synthesis attempts, 388
6-Amino-substituted purine ribonucleosides, general procedure for preparation, 71
5'-Amino sugar nucleosides, 98, 103
3-Amino-1,2,4-triazole, 274, 275
7-Amino-*v*-triazolo(4,5-*d*)pyrimidine, 305
3(5)-Amino-4(2,3,5-tri-*O*-benzyl-β-D-ribofuranosyl)pyrazole, 376
AMP amino-hydrolase, 55
AMP, carboxylic-phosphoric anhydride isosters, 86
AMP derivatives, 55, 90–92
AMP kinases, muscles (rabbit, pig, carp), 55, 88–90, 94–95
AMP-utilizing enzymes, studies, 91
Analog nucleotides of arabinosyl adenine, 159
2',3',-Anhydroadenosine, exceptions to J'$_{1-2}$' $\leqslant$ H$_2$ coupling limit, 314

2,5-Anhydro-D-alloses, 361, 362, 364, 367
2,5-Anhydro-D-allose derivatives, 361
2,5-Anhydro-D-allose, tribenzoate, 364
2,2'-Anhydro-dihydro-5-azathymidine, 204
3,6-Anhydro-D-glucose, furanose form, 372
N-2,2'-Anhydro isomer of dihydro-5-azathymidine, 209
9-(2,3-Anhydro-β-D-lyxofuranosyl)adenine, 314
1-(2,3-Anhydro-β-D-lyxofuranosyl)cytosine, acetylation to N^4, 5'-*O*-diacetate, 401
4,2'-Anhydro nucleosides, stability to bases, 428
8,2'-Anhydro-8-oxy-9-β-D-arabinofuranosyladenine, 387–388
2,2'-Anhydro synthon, 200
2,2'-Anhydro triazine nucleoside, 203
2,5-Anhydro-3,4,6-tri-*O*-benzyl-D-allose, 361, 373
2',3'-Anhydrotubercidin, exceptions to J'$_{1-2}$' $\leqslant$ H$_Z$ coupling limit, 314
Anistrophy effects, 317
Anion-exchange chromotography, 216
Annulation reagent, 1,3-electrophilic (C-N-C), 199–210
Anomeric configuration determination for ribofuranosyl compounds, 311–324
Anomeric configuration of nucleosides, 312
Anomeric 2',3'-*O*-isopropylidene-ribofuranosyl nucleosides, 316
Anomeric proton resonance of 3',5'-cyclic monophosphates, 324
Anomeric ribofuranosyl nucleoside, 313
Anomeric triols, 313
Antibacterials, deoxyuridine analog, 136, 137, 138, 211
Antibiotic, nucleosides, 123, 200, 211, 259, 387–395, 415–436
Anticancer agents, 38, 415, 436
Antifols, 38
Antifungal agents, 362
Antileukemic activity in mice, of 4-hydroxylaminopurine riboside (HAPR), 133
Antimetabolites, macromolecular, 56
Antineoplastic agents and nucleoside combinations, 149–157
Antitemplate, activities of S-methyl MPC (S-alkylated partially thiolated polynucleotides), 66
Antitumor, activity of, 56, 65, 121–133, 137, 149, 193–195
Antiviral agents, 98–101, 160, 185, 200, 225, 267, 436
APD interaction with nucleosides, 339

Aplastic anemia in dogs, from dihydro-5-azathymidine, 199
4-APP ribonucleotide, intracellular inhibitor, 123
4-APP riboside, 122, 123, 132
α-Ara-8-Aza A and related compounds, 51
9-(α-D-Arabinofuranosyl)-8-azaadenine, 47, 52
9-β-D-Arabinofuranosyl)-8-azaadenine(3-α-D-arabinofuranosyl-3 H-(1,2,3),triazolo(4,5-
    d)pyrimidin-7-amine, 47–48
Ara-C(1-β-D-arabinofuranosylcytosine), 420, 427
Ara-C resistant leukemia, 118
5-(β-D-Arabinofuranosyl)isocytosine, 427
β-D-Arabinofuranosyl nucleoside-3′,5′-cyclic phosphates, 322
β-Ara-nucleotides, cyclic, 322
1-β-D-Arabinofuranosyl-1,2,4-triazole-3-carboxamide-5′-phosphate, synthesis, 276
3-α-D-Arabinofuranosyl-3H-(1,2,3)triazolo(4,5-d)pyrimidin-7-amine, radiolabeled assays,
    48–50
Arabinonucleosides, α and β anomers, 275
Arabinosyl adenine, (Ara-A), 159–160, 185–187
Arabinosyl cytosine, 189
Arabinosyl-β-methoxyacrylates, 427
"Aromatic" nuclesides, configuration, 323
N⁶-Aroyl derivatives of AMP, ADP, and ATP, 89, 90
Ascites cells (Ehrlich), 65
ATP anhydride, preparation reaction, 87
ATP, carboxylic-phosphoric mixed anhydride isosteres, 86
ATP derivatives, 55, 85, 92–94
ATP site of action, 90
ATP-site-directed exosite enzyme reagents, 88, 92–95
ATP-utilizing enzymes, 88
Azaadenine, 52
8-Azaadenosine, 159, 187, 190
5-Azacytidine, 415–416, 420, 424
8-Aza-6-(methythio) purine, 48
5-Aza-nucleoside, 199–210
5-Azapyrimidines, 416
8-Aza-6-thioguanosine(5-amino-3,6-dihydro-3-β-D-ribofuranosyl-7H-(1,2,3) triazolo (4,5-d)
    pyrimidinone-7-thione, 42
8-Aza-6-thioinosine (3,6-dihydro-3-β-D-ribofuranosyl-7H-(1,2,3)-triazolo-(4,5-d)
    pyrimidine-7-thione, 42, 45
Azathymidine, 199, 211
6-Azauracil, 140–141
6-Azauridine, 140, 141, 312, 416
Azido, 101
5′-Azido-5′-deoxythymidine, 101
3′-Azidoxylofuranosyladenine, 387
Azole heterocycles, 267
Azole nucleosides, 267–184
6-Azatoyocamycin, 129

B

Bacteria, 211
Bacterial gene, total synthesis, 1–36
Baker's "trans-rule," 312

Barbituric acid β-C-glycoside, 372

Base pair nucleotides of DNA, 5, 21–22

Base stacking, 71, 75–77, 79

N-Benzoyl-3′-*O*-acetyl-cytidine, silylation, 34

1-(5-*O*-Benzoyl-2,3-anhydro-β-D-lyxofuranosyl) uracil, 399

Benzyolation of 1,2-dihydro-2-oxo-5-methylpyrazine, 135

N-Benzoyl-cytidine, 348

N-Benzoyl-guanosine, silylation reactions, 348

N-Benzoyl-guanosines series 5′-*O*-silyl groups, deblocking, 351

5-*O*-Benzoyl-2,3-*O*-isopropylidene-β-D-ribofuranosylcyanide, 361

7-(Benzylthio)-3-β-D-ribofuranosyl-3H-(1,2,3) triazolo (4,5-*d*)pyrimidine-5-amine, 42

Bihelical complexes, formation using T_4 polynucleotide ligase, 3–4

1-(3,5-*bis*-*O*(4-chlorobenzoyl)-2-*O*-methyl-α-D-arabinofuranosyl)-5-fluoro-
    4-methoxy-2-(1H)-pyrimidinone, 41, 43

1-(3,5-*bis*-*O*(4-chlorobenzoyl)-2-*O*-methyl-β-D-arabino-furanosyl)-5-fluoro-4-methoxy
    -2-(1H)-pyrimidione, 40, 41, 43

1-(3,5-*bis*-*O*-(4-chlorobenzoyl)-2-*O*-methyl-α-D-arabinofuranosyl)-5-fluorouracil, 41

1-(3,5-*bis*-*O*-(4-chlorobenzoyl)-2-*O*-methyl-β-D-arabinofuranosyl)-5-fluorouracil, 41

p-N,N-*bis*-(2-chloroethyl) aminobenzaldehyde, 125, 127

Bis (p-nitrophenyl) phosphate, catalyst in fusion reactions, 269

N_1,N_3-*Bis*-riboside, 262–263

Bis-silyl-dihydro-*s*-triazine base, 221

Bis-silyl-5-ethyluracil, 216

Bis-trimethyl silyl base, condensation with 2,3,5-tri-*O*-acetyl-D-ribofuranosyl bromide, 224

Bis-trimethylsilyl-5-bromouracil, metalation, 139

Bis-trimethylsilytrifluoroacetamide, 213

Bleomycin, unspecific tumor inhibition, 151

Bleomycin-inosine combinations, 151, 156

Blocking groups for amino groups, 23–26

μ Bondapak C_{18}, nonpolar packing material, 38

Bondapak C_{181} Porasil β, preparative column, 45

Bredninin, 259, 280

Bridged adenine nucleosides, N^6-N^6, chemistry of, 69

Bridged nucleosides, 74–82

Bridge formation, 70–74, 80–84

Bridgehead carbons of heterocyclic compounds, 305, 307

Bridgehead nitrogen atom containing nucleosides, nucleotides, and derivatives, 287–298

Bridgehead nitrogen atom, 287, 292

8-Bromoguanosine, reversed sign of Cotton effect, 312

5-Bromo-2′-deoxyuridine (AIV or Ald Urd), 101

3-Bromo-1,2,4-triazole nucleosides, 273

5-Bromouracil derivatives, reaction with sulfur nucleophiles, 57

5-Bromouridine, CD spectra, 332

5-Bromouridine 2′, 3′-*O*-1-adamantyl phosphonates, mixture of diastereoisomers, 330

n-Butyldiphenylsilyl (nBDPS), 33–35

n-Butyldiphenylchlorosilane, protecting group for hydroxyls in oligonucleotide synthesis, 33

t-Butyldimethylsilyl group, 200

C

Calf intestinal ADA and coformycin, 179

Cancer calls, 56, 65

Carbamoylmethylenetriphenylphosphorane, 362–363

Carbon–carbon linkage, 415

Carbon-13 chemical shift(s), 289–290, 301–309
C–C linkage, 415
^{13}C NMR spectroscopy, 262, 289, 301–309, 318
Carboximidate group modifications on nucleosides, 121
3(5)-Carboxamido-4-(β-D-ribofuranosyl)-pyrazole, synthesis, 363
Carrier conjugation with purine and pyrimidine analogs inhibitory effect, 145–147
Catabolic enzymes, 85–95, 122
Catalysts
   bis(p-nitrophenyl) phosphate in fusion procedure, 269
   mercuric bromide, 221
   palladium on charcoal, 99
   PtO$_2$, 99
   in "ribosylation reaction," 239–240
   SnCl$_4$, 247
   toluene sulfonic acid, 240
Catalytic dehalogenation of imidazopyrimidines, with palladium on carbon, 294
8,5'-C-cyclopurine nucleosides, 404–413
CD maxima of nucleosides, 73–75
CD spectra, 69, 312, 400, 411, 413
Cell growth, inhibition of 4-APP and 4-APP riboside, 122–123
C-glycoside(s), 359–377
C-glycosyl malonates, 370
C-glycosyl nucleosides, 377, 384
C-glycosyl nucleosides, blocked, 382, 384
C-glycosyl nucleotides, advances in synthesis methods, 359–377
4-C-glycosyl thiazole nucleosides, synthesis, 381–386
Chemical shifts of, 315–316, 351
Chemotherapeutic, activity, 65, 159
Chemotherapeutic, agents, antitumor, 56, 65, 149, 189, 193–195, 415, 436
Chlorinating agents, dimethylchloromethyleneammonium chloride, 123, 125
1-Chloroacetonylidenetriphenylphosphorane, 364
Chloroadenine, separation of mixed, 39–42
2-Chloroadenine nucleosides, separation by HPLC, 39
1-(5-*O*-(4-Chlorobenzyl)-2-*O*-methyl-β-D-arabinofuranosyl)-5-fluoro-4-methoxy-2-(1H)-pyrimidinone,
   41–43
2'-Chloro-2'-deoxypseudouridine, 118
5'-Chloro-5'-deoxythymidine, 329
5'-Chloro-5'-deoxythymidine 3'-*O*-1-adamantylchlorophosphonate, structure determination, 340–341
5-Chloro-2'-deoxyuridine, 5'-amino analogs, 101
2-Chloro-4-ethoxy-5-(β-D-ribofuranosyl)pyrimidine, 115
4-Chloro group, relative reactivity with various amines, 123–124
7-Chloroimidazo-(1,2-a)pyrimidin-5-one, 295
5-Chloro-4-nitro-1-methylimidazole, 123, 124
Chloromercuri procedure, 122
Chloronucleoside, 81
Chloronucleoside orthoester, 81
m-Chloroperbenzoic, cyclizing agent , 204
6-Chloropurine ribonucleosides, 71
6-Chloropurine riboside and n^6-(*w*-aminoalkyl) adenosine substitution with orthoester, 80
7-Chloro-1-(α or β-D-ribofuranosyl)imidazo(1,2-C)pyrimidin-5-one, 292–293
6-Chloro-9-β-D-ribofuranosylpurine, 69
5-Chloro-1-β-D-ribofuranosyl-1,2,4-triazole, 274

5-Chloro-3-(β-D-ribofuranosyl)- 1,2,4- triazolo(1,5-*a*)pyrimidin-7-one, 100
N-Chlorosuccinimide, oxidation of TPTE, 26
7-Chloro-1-(2,3,5-tri-*O*-acetyl-β-D-ribofuranosyl)imidazo (1,2-*a*) pyrimidin-5-one, 296
7-Chloro-1-(2,3,5-tri-*O*-acetyl-α-D-ribofuranosyl)imidazo (1,2-*c*) pyrimidin-5-one, 292
7-Chloro-1-(2,3,5-tri-*O*-acetyl-β-D-ribofuranosyl)imidazo (1,2-*c*) pyrimidin-5-one, 292
3-Chloro-1,2,4-triazole nucleosides
    synthesis using acid-catalyzed fusion procedure, 173
5-Chloro-1,2,4-triazolo(1,5-*a*)pyrimidin- 7-one
    condensation of TMS derivative with 2,3,5-tri-*O*-acetyl-α-D-ribofuranosyl-bromide, 290
7-Chloro-1-(2,3,5-tri-*O*-benzyl-α - β-D-arabinofuranosyl)-imidazo (1,2,-*e*) pyrimidine-5-one,
    293, 294
α-Chlorovinyl ketone, 364
Chromatography, liquid, in polynucleotide synthesis, 26–35
N^6-(CH$_2$)$_n$NHCOCH$_2$I derivative of ATP, 85
N^6-(CH$_2$)$_n$NHCOCH$_2$R derivative of ATP, 93
(CH$_3$)$_3$SiClO$_4$, 256, 260
(CH$_3$)$_3$SiO$_3$CF$_3$, 256, 262
(CH$_3$)$_3$SiO$_3$SF, 256
Citrovorum factor, 38, 151
C-nucleoside(s)
    aminopyrazole, 431–432
    antibiotics, 415–436
    natural, 436
    related to formycin, 431
    related to pruine nucleosides, 431
    pyrazolo-*s*-triazine class, 431
    synthetic, antibiotic properties, 436
Coformycin, 159–197
Complementary bases, 70
Condensations of nucleotides, 23, 26
Condensation reaction, chlorosugar and silylated aglycone, 247
Condensing agents for oligonucleotide synthesis, 23, 24
Configuration determination of, 232, 311–318
Conformation, Cs envelope, limiting dihedral angle, 314
Corey–Pauling–Koltun, models, 70, 72, 78
Cotton effects(s) of ORD, of purine and pyrimidine nucleosides, 312–313
Cotton effect, reversed sign, 312
Coupling efficiencies of nucleoside pairs, 80
Covidarabine®, deoxycoformycin, 159–197
2-C-pseudoisocytidine, synthesis, 425
Cyanohydrins, reversion to aldehydes, 362
3-Cyano-1,2,4-triazole, 269, 270, 275
Cyclic adamantylphosphonate nucleosides, 333
3′,5′-Cyclic nucleoside monophosphates, 311, 318–321
3′,5′-Cyclic nucleotides, 311, 318–321
Cyclic phosphate of a "ribofuranosyl-acetonitrile" derivative, 323
Cyclization of intermediate olefins, 371–372
Cyclo-1-adamantylphosphonates, 5-or 6-membered preparation, 329
Cyclo-1-adamantylphosphonates of nucleosides, 339
(R)-Cycloadenosine, preparation of, 411
8,5′-(R)-Cycloadenosine, 411, 413
8,5′-(S)-Cycloadenosine, 5′-nucleotides, 410, 413
8,5-(R)-Cycloguanosine, 5′-nucleotides, 411

Cycloguanylic acid, uv spectra, 409
C-Cyclonucleosides, 397
Cyclonucleoside bonds, cleavage, 388
Cyclonucleoside formation, 292
Cyclophosphamide, 149–153
Cyclophosphonates, 329, 340
4,O$^{2'}$-Cyclopseudouridine, 118
Cyclopurine nucleosides, 5'-deoxy-8-5-, 404
8,5'-Cyclopurine nucleosides-2',3'cyclic phosphates, 405
S^2,2'-Cyclo-2-thiouridine, conversion to 5'-O-trityl-3'-O-methane sulfonyl derivative, 397
Cytidine, 348, 400–404, 415, 420
Cytidine deaminase, 420
Cytidine 3'-phosphate, in bridged nucleoside studies, 77–80
Cytidine or guanosine, 149
Cytidylyl-(3'-5')cytidine (CpC), 101, 189, 256, 355, 401
Cytosine, 1-(2,3-anhydro-β-D-lyxofuranosyl)-, 401
    arabinosyl-, 101, 189
    1-β-D-arabinofuranosyl-, 420
    dinucleoside phosphate, 355
        synthesis with TBDMSi, 355
    silylated, 256
Cytotoxic compounds, 42, 43, 48

D

DEAE-cellulose, 26
Deazaguanine, 3- and 7-β-D-ribofuranosyl -3-, 304
Debenzylation of pyrazoles, using sodium in liquid ammonia, 364
Deblocking nucleosides, 112–113, 116–118, 121
Desulfurization methods, with Ra, 202
Deoxyadenosine, 193
2'-Deoxy-2'-aminoadenosine, chemical synthesis, 394
2'-Deoxy-2'-aminoadenosine, chemical synthesis, 388
3'-Deoxy-3'-aminoadenosine, 388
2'-Deoxy-2'-aminoguanosine, 387–388, 392–395
2'-Deoxy-2'-azidoadenosine, 387, 392, 394
2'-Deoxy-2'-azidoguanosine, 387, 395
2'-Deoxy-2'-azidoinosine, 387
Deoxycoformycin, 159–197
5'-Deoxy-8,5'-cycloadenosine, 404, 413
5'-Deoxy-8,5'-cycloadenosine2',3'-cyclic phosphate, 405, 407
5'-Deoxy-8,5'-cycloguanosine2',3'-cyclic phosphate, 405, 408
5'-Deoxy-8,5-cyclopurine nucleosides, 404–405
(R)-3-(2-Deoxy-β-D-erythro-pentofuranosyl)- 3,6,7,8- tetrahydroimidazo (4,5-*d*)
        (1,3)diazepin-8-ol, *d*-conformycin or Covidarabine®, 159–197
1-(2-Deoxy-β-D-erythro-pentofuranosyl)-1,2,4-triazole- 3-carboxamide 5'-phosphate, 276
5'-Deoxy-5'-fluoropseudouridine, deblocking of nucleoside, 112
5-(5-Deoxy-5-iodo-2,3-O-isopropylidene- β-D-ribofuranosyl)uracil, 112
2'-Deoxynucleosides, 275
5'-Deoxy and 5'-substituted nucleosides, 275
5'-Deoxypseudouridine, synthesis, 113
2-Deoxyribofuranosyl chlorides, in condensation of 5-alkyluracils, 221

1-(2-Deoxy-β-D-ribofuranosyl)5-methyl-5,6-dihydro-*S*-triazine-2,4(1H,3H)-dione, (5,6-dihydro-5-azathymidine, DHAdT)

  PMR spectra for α-and β-anomers, 220

  synthesis mechanism, 216

1-(2-Deoxy-β-D-ribofuranosyl)- 5,6-dihydro-5-methyl-*S*-triazin-2,4 (1H,3H) dione, 211–227

Deoxyribonuclease, pancreatic, 7

Deoxyribonucleoside, dihydro-5- azathymidine, 199

2'-Deoxyribonucleoside-5'-monophosphates, 4 common types, 22–23

Deoxyribopolynucleotide segments, 6–7

2'-Deoxy-2'-substituted purine nucleosides, 387–395

DHAdT, 3'esters, 221

DHAdT, 1-(2-Deoxy-β-D-ribofuranosyl) 5-methyl- 5,6-dihydro- *s*-triazine- 2,4(1H)-dione, 212–227

N^4,5-*O*-Diacetate, synthesis, 401

3',5'-Di-*O*-acetyl-9-β-D-arabinofuranosyladenine, 387, 390

Di-*O*-acetyl-2'-deoxy-2'-azidoinosine, 390

1,2-Di(adenin-N^6-yl)ethane, 69

1,4-Di(adenosin-N^6-yl)butane, synthesis, 69

1,2-Di(adenosine-N^6-yl)ethane, 69, 71

Di(adenosine-N^6-yl)methane, 71

1,4-Diaminobutane, 69

1,2-Diaminoethane , 69, 71

2,6-Diaminopurine ribonucleoside (DAPR), 159, 185–188

2,4-Diamino-5-β-D-ribofuranosylpyrimidine (or α-isomer), 425

3,5-Diamino-1-β-D-ribofuranosyl-1,2,4-triazole, 274

3,5-Diamino-1,2,4-triazole, anticancer activity, 274

Dibromovinyluracil, silylated, 15,7-Dichloroimidazo-(1,2-a)pyrimidine, 295, 296

2,6-Dichloropurine, 39

6,8-Dichloropurine riboside, 387

2,4-Dichloro-5-(2,3,5-tri-*O*-acetyl- β-D-ribofuranosyl)pyrimidine, 109, 114, 116

1-(2',3'-Dideoxy-2',3',-epithio- β-D-lyxofuranosyl)uracil, 400

2',3'-Dideoxy-2',3',epithiouridine, 400

Diethylaniline hydrochloride, 114

Dihydro-5-azathymidine, 199–208

5,6-Dihydro-5-azathymidine, 211–212

1,2-Dihydro-1-(2-deoxy -β-D-erythro-pentofuranosyl)- 2-oxo- 5-methyl-pyrazine-4-oxide, 138

1,2-Dihydro-2-oxo-methylpyrazine- 4-oxide, 135–136, 138

1,2-Dihydro-2-oxo-5-methylpyrazine, 135–138

1,2-Dihydro-2-oxopyrazine- 4-oxide, 136–138

Dihydrotriazine ring system, 201

Dihydro-*s*-triazinone, 213, 216

3,6-Dihydro-3-β-D-ribofuranosyl-7H- 1,2,3-triazolo (4,5-d)pyrimidine-7-thione, 42–45

5,7-Dihydroxyimidazo-(1,2-*a*)pyrimidine preparation from 2-aminoimidazolium sulfate, 295, 296

2,3,5,6-Di-*O*-isopropylidene-D-allose, 370

2,3,5,6-Di-*O*-isopropylidene-D-mannose, 370

6,8-Dimercaptopurine riboside, synthesis from 6,8- dichloropurine riboside, 390

4,5-Dimethoxycarbonylisoxazoles, 365

2,4-Dimethoxy pyrimidine, 256

2,4-Dimethoxy-5-(β-D-ribofuranosyl)pyrimidine, 115

1,9-Dimethyladenine, 233

Dimethylchloromethyleeneammonium chloride, 123, 125

N^6-Dimethyl-2'-deoxy-2'-aminoadenosine, 387, 390

N,N-Dimethylformamide, column wash, 38

6,8-Di(methylmercapto)-9-β-(2′-deoxy-2′-azido-D-ribofuranosyl)purine, 393
6,8-Di(methylmercapto)-9-β-D-ribofuranosyl-purine, 393
1,2-Dimethylpyrazol-3-one, 297
1,5-Dimethylpyrazolo(1,5-*a*)pyrimidin-7-one, 297
4,5-Dimethylpyrazolo(1,5-*a*)pyrimidin-7-one, 297
Di(methylthio)derivative, 116
Dimroth transposition of thermodynamic nucleoside, 246
Dimroth type rearrangement, 232, 235, 236
3′,5′,-Diol grouping, selective protection, 200
Dioxolane ring and other 5-membered rings, 234
2,4-Dioxo-8-(2,3,5-tri-*O*-benzyl-β-D-ribofuranosyl)-1H,3H-pyrazolo(1,5-*a*)-1,3,5-triazine, 376
6,8-Dioxy-9-β-(3′,5′-di-*O*-acetyl-2′-azido-D-ribofuranosyl)purine, 388
N,N-′-Diphenylethylenediamine, 361
Diphenylsilyl groups, acetyl-, n-butyl-, phenyl-, 21
5′-Diphosphates, 61
5′-Diphosphate of 5-methylmercaptouridine(sm₅UDP), 61
3′- and 5′-*O*-Di-P₂-(1-adamantyl)-P₂-chloropyrophosphonates, 329
3′,5′-*O*-Di-P₁,P₂-(1-adamantyl)pyrophosphonate, eight-membered, 329
1,3-Dipolar cycloaddition of ethyl diazoacetate to acrylate, 363
Dipyridamole, nucleoside transport inhibitor, 189
Diribofuranosyladenine, 232
1,9-Diribofuranosyl adenine, 229–232
1,9-Diribosyl purine, synthesis approach, 229–237
Disilyl dihydrotriazine, 215, 216
3,5-Disubstituted-1,2,4-triazole-nucleosides, 272
2′,5′-Di-*O*-tert-butyldimethylsilyl-*N*-benzoylcytidine, 355
2′,5′-Di-tert-butyldimethysilyl-*N*-benzoylcytidine, 351
3′,5′-Di-*O*-tetrahydropyranyl-2′-*O*-mesylarabinosyladenine, 394
2,4-Dithione derivative of 2,4-dichloro-5-(2,3,5-tri-*O*-acetyl-β-D-ribofuranosyl)pyrimidine, 116–117
2,4-Dithiopseudouridine, 117
Dithiothreitol (DTT), 57, 59
3,5-Di-*O*-toluoyl-D-ribofuranosyl chloride, 215
3,5-Di-*O*-toluoyl-2-deoxy-ribofuranosyl halide, 216
DNA
   analysis by enzymic digestion, 104
   directed by DNA polymerases, 64
   directed by RNA polymerases, 64
   duplex, 2, 10–15, 67
   nucleosides, 98
   polymerase α, 64
   polymerase I of *E. coli*, 64
   polymer, 98, 104
   and RNA, strands, base-base interaction (stacking), 70
   single-stranded segments, preparation, 8
   synthetic, 16
   viral and host cells, 105, 211
Duplexes of DNA for synthesis of gene, 6–7, 10–15

E

E. coli, 1–36, 87, 97, 105
EcoR1 restriction sequences, 4, 18–19

Ehrlich ascites cells, 65
Ehrlich-Carcinoma, implanted in Swiss-Mice, 153
β-Elimination, 26
Enzymatic studies with synthetic nucleosides and nucleotides, 267
Enzymatic synthesis of ribonucleotides, 60–63
Enzyme(s)
  acylation at AMP sites, 87
  adenine phosphoribosyltransferase, 122–123
  adenosine deaminase, 122, 160–181, 190, 194, 435
  adenosine kinase, 48–49, 122
  AMP amino hydrolase, 55
  AMP-utilizing, studies, 91
  ATP-utilizing tests, 88
  deoxyribonuclease, 7
  dTMP kinase, 98–99
  of *E. coli*, 16
  exosite enzyme reagent, 86
  hexokinase, 89
  nuclease(s), 7, 56, 64
  nucleoside hydrolase, 415
  nucleoside phosphorylases, 415
  nucleotides which inactivate enzymes, synthesis of, 85–95
  in oligonucleotide synthesis phosphorylation, 25
  phosphorylation at AMP sites, 87
  ribonuclease, 69
  RNase A, 64
  semi-tight binding kinetics, 169
  species or tissue-selective inactivators, 86
  substrate-identical inactivators, 86
  thymidine kinase, 97, 99, 105, 141, 200
  thymidylate kinase, 98
  thymidylate synthetase, 141
  tight-binding, kinetics, 169
  utilizing adenine nucleotides, 85–95
Enzyme-inhibitor complex dissociation
  constant (Ki), 89
  kinetics, 193
Enzyme inhibitor theory, 159–181
Enzymic adenine nucleotide sites, 85
Enzymic reactions, 77, 162
Episulfide chromphore, CD spectra, 400
2′,3′-Episulfides, derivatives of uridine or cytidine, 397
2′,3′-Epithiocytidine diacetate, preparation of, 401
Erthrocytes, human, 132, 159–195
Escherichia coli, see E. coli
5′Esters of DHAdT, 225
Estradiol, 145–147
5-Ethoxycarbonyl-3-β-D-ribofuranosylisoxazole, 365
Ethyl-*N*-benzyloxycarbonylorthoglycinate, 80
Ethyl iodide, as alkylating agent, 113
Ethyl-3-(2,3-*O*-isopropylidene)D-ribofuranosyl, propiolates, 429
Ethyl orthoformate, 69, 77, 231
Ethyl 2-(*D*-ribofuranosyl)-2-formyl acetate, 110

Ethyl(5-*O*-trityl-2,3-*O*-isopropylidene)D-ribosyl-acetate, 416
5-Ethynyl-2'-deoxyuridine, 139–143
5-Ethynyl 2'-deoxyuridine and 5-fluoro-2'-deoxyuridine, 143
5-Ethynyluracil, synthesis from 5-formyluracil, 137
5-Ethynyluracil ribonucleoside, synthesis, 139

F

Fibroblast cell culture, human, 69
Fluoride ions, removal of blocking groups (TBDMSi) on nucleosides, 348
Fluorine, 88
Fluorobenzamide residues, 88
N⁶-*o*-Fluorobenzoyladenosine 5'-triphosphate and *p*-fluoro analog, 88
N⁶-*o*-and *p*-Fluorobenzoyl-ADP, lack of inhibition of tested enzymes, 89
N⁶-*o*-and *p*-Fluorobenzoyl, AMP derivatives, 88, 89
N⁶-*o*-and *p*-Fluorobenzoyl, ATP derivatives, 89–90
5-Fluoro-2'-deoxyuridine, 5'-amino analogs, 101
5'-Fluoro-5'-deoxyuridine, with ADP in pyridine, 331
5-Fluoro-2,4-dimethoxypyrimidine, 40–41
5-Fluoro-1-(2-*O*-methyl-α and β-D-arabinofuranosyl)uracil, 40–41
5'-Fluorothymidine, 98
3-Fluoro-1,2,4-triazole nucleoside, 273
5-Fluoro-1,2,4-triazole nucleoside, 273
5-Fluorouracil, 1-(3,5-*bis*-*O*-(4-chlorobenzoyl)-2-*O*-methyl-α and β-D-arabinofuranosyl), 41
5-Fluorouracil (5-FU)

  chemotherapeutic effects, 149
  combination with various nucleotides, 151, 154
  synergistic potentiation by cytidine or guanosine, 149
Formycin, di-,mono- and triphosphate, 183
Formycin β, synthesis, 372–373
Formycin derivative, 364
Fourier transform carbon-13 NMR spectroscopy, 301–309
Friedel–Crafts catalysts, 251, 256–257
Friedel–Crafts catalyzed glycosylation, 293
Friend-virus-infected mouse spleen
  DNA directed RNA polymerases, 64
Furanose C-glycosides, 372
Furanose-pyranose rearrangement, 42–47
Fused five-membered rings, 371
Fusion method, silyl, 123
Fusion procedure(s), 122, 269, 271–275, 282
Fusion reactions, 39, 239–249

G

Gene
  bacterial, total synthesis, 1–20
  precursor, 4
  promoter region sequences for tyrosine, 2, 4
  segments, joining, 9
  synthesis, 1–35
  synthetic, 1–36
  structure–function studies, 19
Genetic defects, 18–19

"Geometry-only" 'H-NMR method, 311, 324
Glycolysis, allosteric inhibition, 183
C-Glycosides, 236
N-Glycosides of pyrazolo-(1,5-*a*)pyrimidines and pyrazolo(1,5-*a*)-1,3,5-triazines, 287
Glycosylation, 239–250, 268, 269, 293, 296, 301–309
Glycosyl azide, 280
Glycosyl bond (CN), 239
3-Glycosylpyrazoles, formation of, 363–364
1-Glycosyl-4-substituted 1,2,3-triazoles, 279
2-Glycosyl-4-substituted 1,2,3-triazoles, 279
Glycosyl-1,2,4-triazole-3-carboxamides, 276
1-Glycosyl-1,2,3-triazole, 278
Guanine, 28–29, 187
Guanine, deaza-, 3- and 7-β-D-ribofuranosyl-3-, 304
Guanosine, 42, 348, 387–395, 405, 411
Guanosine or cytidine, 149
Guanosine and inosine analogs, 287–288
Guanozole, ribonucleoside of, 274

H

Halogen, deactivation of neighboring nitrogen in glycosylation reaction, 289
Halogen atom, influence on site of glycosylation, 292
3-Halogen-1,2,4-triazole nucleosides, 274
5'-Halogenated analogs of thymidine, 98
Halosugar, 244, 247, 248
"Head-to-tail" joining, of DNA segments, 5
Hemolytic agent in man, 133
Herpes, 103
Herpes simplex virus, 85, 97, 103–104, 199, 211
Herpetic keratitis in rabbits, 103
Heteroaromatic compounds, 301–309
Heteroaromatic systems, N-ribosylation, 301
Heterocycles, 244, 289
Heterocyclic analogs of purine nucleotides and nucleosides, 287–298
Heterocyclic base(s), 23, 239, 301–309
Heterocyclic C-ribofuranosides, 367
Heterocyclic moiety of pseudouridine, transformation, 114
Heterocyclic system, six-membered ribosylation site assignment by ^{13}C NMR, 305
Hexokinase, yeast, 89
High-pressure liquid chromatography, 30–35, 37–52, 201
Hilbert–Johnson condensation, 211
Hilbert–Johnson reaction, 37, 39–42, 251
H NMR method for determination of anomeric configuration of ribofuranosyl compounds, 318–324
H NMR spectra, 294, 368
Hoard-OH procedure, 92
Homopolymer, 5-methylmercaptouridylic acid, 62
Hormonal nucleoside-3',5'-cyclic monophosphates, biologically active, 318
H-2' of ribofuranosyl nucleoside-3',5'-cyclic monophosphates, 311, 322–324
HRS-reticulum-cell-sarcoma, implanted in Swiss-Mice, 156
HSV-1 virus in mice, 212, 221, 222, 225
HSV-2 virus, 212

HSV-1 infected cells, 63
Hudson's rules of isoratation, 312–313
Human fibroblast cell culture, 69
Hydrogen bonding, 70
Hydroxylamine, 189, 191
Hydroxylaminoformycin, host toxicity, 132
N^6-Hydroxyaminopurine ribonucleoside (HAPR), 133, 159, 189, 191–193
4-Hydroxylamino..1-(β-D-ribofuranosyl)pyrazolo(3,4-d)pyrimidine
   host toxicity, 132
   structure, 125
p-Hydroxymercuribenzoate (HMB), 58
6-Hydroxypseudouridine type nucleosides, 416
7-Hydroxypyrazolo(1,5-a)pyrimidine, 296
4-Hydroxy-1,2,3-triazole-5-carboxamide, 280
Hypochromism (H), 69, 73–75

I

Imbach's rule with α and β anomers, 317
Imidazole anion,4-and 5-substituted, 304
Imidazole nucleosides, 259, 304
Imidazolidine derivatives, 361
Imidazo (4,5-d)pyridazine -4,7 (5H, 6H)-dione, ionized base, 308
Imidazo(1,2-a)pyrimidine, ring system
   glycosylation studies, 295
Imidazo(1,2-a)pyrimidin-5-one, 7-chloro-1-(2,3,5-tri-O-acetyl-β-D-ribofuranosyl), 296
Imidazo(1,2-c)pyrimidin-5-one
   7-chloro-1-(β-D-ribofuranosyl)-, and α-anomer, 293
   7-chloro-1-(2,3,5-tri-O-benzyl-β-D-arabinofuranosyl)-, and α-anomer, 294
6-Imino 1,9-diribosyl purines and related compounds, synthesis, 229–230
Immunosuppressive agents, 195
Inhibitors of (ADA) adenosine deaminase, 159–195
Inosine, 42, 146, 151, 387
Inosine analog, 288–289
Inosine-bleomycin combination, 151, 156
Inosine (or thymidine) with cyclophosphamide, 153
Inosine and guanosine analogs, 287–288, 294
Inosine or thymidine, 149
Interferon induction, 56, 65
Iodoacetylamino-n-alkyl, 92
Iodoacetylation of nucleosides, 92
5-Iodo-5'-amino-2',5'-dideoxycytidine (A1C), 85–95, 97, 101, 102
5-Iodo-5'-amino-2',5'-dideoxyuridine (AIU), 85–95, 101, 103–106
5-Iodo-5'-azido2,5'-dideoxyuridine, 101
5-Iodo-2'-deoxycytidine (A1C), 101
5'-Iodo-5'-deoxynucleoside, 275
5'-Iodo-2'-deoxyuridine (IdYUrd), 99
   5'-amino-analog(s), 99, 101
     antiviral effect IdUrd, 101
     inhibitor of thymidine kinase, 101
   5'-azido-analog, 99, 101
     reduction, 101
   (^{125}I)-IdUrd-DNA, 104
   mutagenicity, 103

3-Iodo-1,2,4-triazole nucleosides, 273
Isocytosine,5-(β-D-arabinosyl)-, 427
Isomerization of TBDMS, blocked nucleosides, 355
2',3'-O-Isopropylidene-8,5'(S)-cycloadenosine, 410
2',3'-O-Isopropylidene-5'-deoxy-5'-benzylthio-N²-benzoyl-guanosine, 405
2',3'-O-Isopropylidene-5'-deoxy-8,5'-cycloadenosine, 404
2',3'-O-Isopropylidene-5'-deoxy-8,5'-cycloguanosine, by photocyclization, 405
2',3'-O-Isopropylidene-5'-deoxy-5'-phenylthioadenosine, 404
2',3'-O-Isopropylidene derivatives of nucleosides, 294
2',3'-O-Isopropylidene-5,6-dihydrouridines, 317
Isopropylidene function, role in cyclization, 371–372
Isopropylidene group(s) of ribose ring, 232, 234, 236, 359
2',3'-O-Isopropylidene-5'-keto-8,5'-cycloadenosine, 410
2',3'-O-Isopropylidene-8-phenylthioadenosine, 411
O-Isopropylidene ribofuranose series, 236
2,3-O-Isopropylidene-β-D-ribofuranosyl cyanide, 385
2,3-O-Isopropylidene-D-ribofuranosylamine, 231, 234
5-(2,3-O-Isopropylidene-β-D-ribofuranosyl)uracil, 111–112
2,3-O-Isopropylidene-D-ribose, 416
2,3-O-Isopropylidene sugars, 359
2,3-O-Isopropylidene-5-O-trityl-β-D-ribofuranosyl chloride, 370, 416, 429
2,3-O-Isopropylidene-5-O-trityl-D-ribose, reaction with phenylphosphoranes to yield
    C-glycosides, 368
Isoxazole(s), 359, 364–366

K

Karplus relation ~3.5–8 Hz, 314
α-Keto sulfone derivative of dihydro-5-azathymidine, 204
Kinase(s), 5, 25, 55, 90
Kinetic nucleoside, 240, 246–249
Kinetic product, α anomer of DHAdT, 219
Kinetic studies of ADA, 162–165
Ki values, 91

L

Lanthanide-induced shift studies, 202–203, 205–206
Leukemia, 42–43, 103, 118–119, 130, 137–142, 195, 416, 420–436
Ligase, enzymatic joinings by, 5
Lipophilic region of rabbit enzymes, 92
Lipophilicity, 27, 30, 32, 34, 35, 91, 200
Liquid chromatography of nucleosides, high pressure, 37–52
Liquid chromatography systems, 26–52
Lyxo-episulfide, 397, 400

M

"Macromolecular antimetabolites," 56
Maleimides, synthesis, 359, 362
Malonates, C-gylcosides, 370
Mammary carcinoma, 149
D-Mannose-2,3-cyclic carbonate, furanose form, 372
Mechanism of action of synthetic nucleosides and nucleotides, 267
Mechanism of catalysis, 85–97
5-Mercaptocytosine, 61

5-Mercaptocytosine nucleosides and nucleotides, 56
5-Mercaptopurine, 189
6-Mercaptopurine riboside isomer, 123, 124
5-Mercaptouracil, 56
5'-Mercaptouridine, 61
Mercuric bromide, as catalyst in condensation reactions, 221
Mercuri procedure (classical) of Davoll and Lowry, 47
2'-Mesylarabinosyladenine, preparation of, 387, 391
1-(2'-O-Mesyl-3'-deoxy-3'-thiocyanato-β-D-arabinosyl)uracil, 399
Methanolic ammonia, deacetylation, 290
Methemoglobin formation in dogs, 133
Methotrexate, 149
Methotrexate stomatitis, 151
Methotrexate (MTX) and nucleoside combination, 155
Methotrexate/thymidine combination, 151
Methotrexate/tetrahydrofolic acid/thymidine/bleomycin/inosine combination, 151–157
4-Methoxy-2-methylthio derivative, 116
2-Methoxypyridine, pKa value, 256
4-Methoxypyridine, pKa value, 256
5'-(Methoxytrityl) groups, 26, 29
1-Methyladenine, 233
N⁶-Methyladenine, 235
N⁶-Methyladenosine, 233
Methyl 3,6-anhydro-D-glucopyranoside, 372
1-(2-O-Methyl-α-D-arabinofuranosyl)-5-fluorouracil, 44
1-(2-O-Methyl-β-D-arabinofuranosyl)-5-fluorouracil, 44
Methylated polymer, 60
N-Methylation of pyrazole and pyrazolo (1,5-a) pyrimidine, 297
N-Methylbiuret, ring closure with ethyl formate, 212
3-Methyl-2-butenylamine or ethylenimine, 15, 126
Methyl-4-chlorobenzoate, 41
Methyl 5(4)-cyanomethylimidazole-4(5)-carboxylate, 304
6-Methylcytosine, 136
Methyl 6-diazo-2,3-O-isopropylidene-β-D-ribo-hexofuran-5-ulose, 381
5-Methyl-5,6-dihydro-S-triazine-2,4-(1H,3H)-dione, 213–215
6-Methyl- or 6-ribosyl-imidazo (1,2-c)pyrimidin-5-one, 294
Methyl iodide, 113
Methyl 3-(2,3-O-isopropylidene-5-O-trityl-β-D-ribofuranosyl)propiolate, 374
Methyl isothiocyanate, 201
5-Methylmercapto-2'-deoxyuridine, 63
S-Methyl-5-mercaptodeoxyuridine-5'-triphosphate(sm⁵ d UTP), 63
5-Methylmercapto derivatives, 63
5-Methylmercaptouridylic acid, 62
Methyl-3-methyl-1,2,4-triazole-5-carboxylate, 272
5-Methylpyrazolo (1,5-a)pyrimidin-7-one, 296
3-Methyl-5-β-D-ribofuranosylisoxazole, 366
5-Methyl-4-(β-D-ribofuranosyl)pyrazolo(1,5-a)pyrimidin-7-one, 296–299
5-Methyl-3-(β-D-ribofuranosyl)1,2,4-triazole(1,5-a)pyrimidin-7-one, 290
Methylthio derivative of APTR, 436
4-Methylthio group of 6-mercaptopurine ribose derivatives, nucleophilic displacements, 123
2-Methylthiopyrazolo (1,5-a)-s-triazin-4-one, 305
6-Methylthio-9-β-D-ribofuranosylpurine, 436

7-(Methylthio)-3-(2,3,5-tri-*O*-acetyl-β-D-ribofuranosyl)-3H-(1,2,3) triazolo(4,5-*d*) pyridine, 42

N-Methyl-*s*-triazine, reduction, 212–213

Methyl 1,2,4-triazole-3-carboxylate, 268–269, 272

1-Methyl-*s*-triazolo (4,5-*c*) pyridine, 302

3-Methyl-1,2,4-triazolo (1,5-*a*)pyrimidin-7-one, 289

4-Methyl-1,2,4-triazolo (1,5-*a*)pyrimidin-7-one, 289

Methyl-3-(2,3,5-tri-*O*-benzyl-β-D-ribofuranosyl) propiolate, 374

6-Methyluracil, 260

Micrococcal nuclease, 7

M. Luteus, 64

5'-Modified nucleosides in the purine series, 277–278

Molecular sieve, AW-500, 48, 221

5'-*O*-Monomethoxytrityl-*N*-benzoylcytidine, 351

5'-Monomethoxy trityl group, 23, 26, 29

5'-*O*-Monomethoxytrityl-and 5'-*O*-*tert*-butyldimethylsilyl-*N*-benzoyldeoxy-adenosine model substances for oligonucleotide synthesis, 355

3'-Mononucleotides, 7

5'-Mononucleotides, 7

5-Monophosphate derivative of 3-carboxamide of 4-APP riboside, 132

Monophosphates of 5-mercaptocytosine, 61

Monophosphates of 5-mercaptouridine, 61

Monosylyl dihydrotriazinone, 213, 215

Murine leukemia L1210, 69, 72–73

Murine tumors, 149, 189

Mutaganicity, 103

Mutations, amber, 18

N

Nebularine, analog, 123, 125

Nitrile oxides, 365

2-Nitroestradiol, 147

5'-*D*-Nitromethyl-2',3'-*O*-isopropylideneadenosine, 91

5-Nitrouridine, derivative of, 254

N-Nonanoyl adenine, 48

N-1-ribosylation, percentage, 240

N-2-ribosylation, percentage, 240

Nuclear magnetic resonace, methods for determining anomeric configuration, 314–318

Nuclease(s), 7, 56, 64

Nucleic acids, 56, 60, 69–84

Nucleophilies, 114

Nucleoside, 2', 3'-*O*-(1-adamantyl) phosphonates, synthesis, 329

Nucleoside 5'-aldehydes, 361

Nucleoside alkylphosphonates, investigations of, 329–345

Nucleoside analogs, 135–147

Nucleoside antibiotics, 127, 211–227, 361, 415–436

Nucleosides, 5'-amino, 98, 103

Nucleosides, antineoplastic agents, 149–157

Nucleosides, antitumor, 135–147

Nucleoside, β-D-*O*-arabinofuranose derivatives, biological properties, 294

Nucleosides, blocked with methanolic ammonia, 48

Nucleoside(s) configuration, 126

Nucleosides containing sulfur functions in sugar portion, 397–413 α-Nucleoside-3′,5′-cyclic phosphates, 321

Nucleoside 3′, 5′-O-cyclophosphates, 344

Nucleoside, deblocking with trifluoroacetic acid in methanol, 112

Nucleoside derivatives, 287–298

Nucleoside(s), esters, 221

Nucleoside, five-membered ring, 246–247

α-Nucleosides, in formation of β-nucleosides, 253–254

β-Nucleoside, formation, 219, 253

N_1-Nucleoside; formation, 224, 257–258, 262

N_3-Nucleoside, formation, 224

Nucleoside(s) in fusion reactions, 247

Nucleoside of halogen derivative, 98, 101, 273

Nucleoside hydrolases, 415

Nucleosides, intact penetration into the cell, 277

Nucleoside metabolism, 48–51

Nucleoside metabolites from cell extracts, separation and identification by microparticulate reversed-phase packing in HPLC, 37, 46–52

Nucleoside(s) with modified glycosyl moieties, synthesis, 275–276

α-Nucleoside-5′-monophosphate, cyclization to α-cyclic nucleotides using N,N-dicyclohexyl-carbodiimide (DCC), 321

Nucleoside-5′-phosphate, synthesized from phosphorychloride, 275–276

Nucleoside phosphorylases, 415

Nucleoside protection in nucleoside interactions, 80

β-Nucleoside of pseudoisocytidine, 428

Nucleosides, purine and pyrimidine, separation by reversed phase HPLC, 37–52

Nucleoside, pyrrolopyrimidine antibiotics, 127

Nucleoside, ribosylation reaction, kinetic and thermodynamic products, 245–246

Nucleosides, regioselective synthesis, 288

Nucleoside-related compounds with furyl moiety, 384

C-Nucleosides related to pseudouridine, 109–120

Nucleosides, silylated, 251

Nucleosides, stereospecific syntheses, 288

Nucleoside, synthesis, 211, 251–264, 295

Nucleosides, synthetic, 267

Nucleosides of 4-and 5-substituted imidazole anion, 304

Nucleoside, thiation, 123

Nucleosides, transposition, 245–248

Nucleotide, building blocks, di-, tri-, and tetra-, 25

Nucleotide derivatives, 83, 85–95, 287–298

Nucleotide sequence, 3–4

Nucleotides, synthetic, 267, 276

Nucleotides of synthesized gene, 3

O

Oligonucleotide(s), 21–35, 347–357

    amion groups, 23

    chain, extraction by chromatography, 26–35

    cleavage by phosphodiesterase, 80

    Corey-Pauling Koltun model of stacking, 79

    derived from a bridged nucleoside, 77

    proton addition, second, 75

    segments, 39 chemically synthesized, 5

synthesis, chemically
  basic principles, 21
  building blocks, 25
  condensing agents, 23, 24
  diagram of chemical steps, 25
    duplex formation, 22
    duplex joining, 22
    enzymatic joining of thirty-nine segments, 22
    high-pressure liquid chromatography, 21
  methodology of organic chemistry, 22
  protective group(s), 21–24, 347
    tert-butyldimethylislyl group, 347
  sequence specific synthesis, 22
  typical flow chart, 24
Ovarian carcinoma, 149
Overadditive chemotherapeutic action of combination, 154–156
Overhauser effect, internal nuclear, (NOE) determination of anomer c configuration, 317
Oxidation, 362
Oxadiazoles, synthesis, 359
1,2,4-)xadiazole C-glycoside, 367
Oxidized thiol (disulfide) group determination, 59
Oxoformycin, synthesis, 372–373
2-Oxo-5-methyl-pyrazine 4-oxide trimethylsilyl derivative, 135
4-Oxo-pyrazolo-*s*-triazine-nucleosides (OPTR), 432, 435
4-Oxo-2-thio-8-(β-D-ribofuranosyl) pyrazolo (1,5-*a*)-1,3,5-triazine, 432

P

Palindromes, 7
Palladium on charcoal catalyst, 116
Pancreatic RNaseA, 64
P-^{32}ATP, 5
Peptide bond, synthesis, 70
Peptidyl tRNA, 70
λ phage, 18–19
Phenylurethane phosphorylation, 91
Phosphodiester, 32
Phosphodiesterase, 7, 77–78, 80, 357
Phosphorylation in gene synthesis, 5, 8–10
Phosphate-activating agents, energy-rich, 22
Phosphomonoester, conversion to phosphodiester, 22, 27
Phosphoranes, 361–370
Phosphorylating agents, 355
Phosphorylation, 103–104, 355
Phosphoryl chloride, 114
Photocyclization, 405
Photoirradiation of, 404
Photolysis, 405
PKa values of, 255
PMR spectroscopy, 39–40, 46, 343
^{31}P-NMR spectroscopy, 340–345
Polycytidylic acid (MPC), 65
Polydeoxyribonucleotide(s), 5, 63

Polydeoxyribonucleotidyl transferase, 63
Polymerases, 56, 66
Polymerization, 62–63
Polynucleotides, 21–35, 55–68
Polyoxin, 362
Porasil, permanently bonded with octadecyltrichlorosilane, 38
Promoter duplex, 2–4, 7–10, 19, 22
Protecting groups and strategy, 23–24, 32–35, 347–357
Proteins, incomplete, nonfunctional, 18
Protein, synthesis, 69–70
Proton coupling constant, changes, 297
Proton magnetic resonance spectra, 234
PRPP synthetase, 183
P_2S_5-dioxane, 123, 125
Pseudoarabinoside, 118
Pseudoisocytidine, 416, 420–425, 428, 431
Pseudorabies virus, 199, 212
Pseudo-2-thiouridine, 425
Pseudouridine, 109–120, 376, 425, 428, 431
Pseudouridine, acetyl-2'-chloro-2'-deoxy, 188
Pseudouridine, 2'-chloro2'-deoxy-, 118
Pseudouridine, 4,O$^{2'}$-cyclo-, 118
Pseudouridine, 5'-deoxy-, 118
Pseudouridine, 5'-deoxy-5'-fluro-, 112
Pseudouridine, 2,4-dithio-, 117
Pseudouridine, 5-β-D-ribofuranosyl, 312
1-β-D-Psicofuranosyl-1,2,4-triazole-3-carboxiamide, 275
Pteridines, 38
Purine nucleosides, 27–28, 101, 122
Purine nucleoside 2',3'-O-1-adamantylphosphonates, 338–339
Purine nucleoside analogs, 288, 372–377
Purine nucleoside 2',3',-cyclic phosphates, 405, 410
Purine nucleoside ring, 229, 236
Purine nucleoside series, 277–278
Purine and pyrimidine nucleosides, 135–147, 313
Purine ribonucleoside, 2,6-diamino-, 159, 185–187
Purine skeleton containing a bridgehead nitrogen atom, synthesis, 288–298
Puromycin, analogs of purine nucleosides related to, 388
Pyrazine analog(s) of natural pyrimidine, 138

Pyrazine analog(s) of thymidine, 136–138
Pyrazofurin (Pyrazomycin), 268, 280, 363–364
    biological activities, 363
    preparation of 2',3'-O-isopropylidene derivative, 372
    pyrazole C-glycoside, 363
    3-β-D-ribofuranosylpyrazole system, 364
    similarity to 1,2,3-triazole nucleoside, 280
    structures, 268
Pyrazole, 3-amino-4-ribofuranosyl-, 359
Pyrazole, 3(5)-amino-4-(2,3,5-tri-O-benzyl-β-D-ribofuranosyl)-, 376
Pyrazole, 3(5)-carboxamido-4-(β-D-ribofuranosyl)-, 363
Pyrazole C-glycosides, 359
Pyrazole(s), functionalized, 363

Pyrazole and pyrazolo (1,5-α) pyrimidine, changes in proton coupling constants as a result of N-methylation, 297

Pyrazole(s), synthesis, 359, 364

Pyrazoline, 363

Pyrazolo (1,5-α) pyrimidine(s), 125, 128

Pyrazolo (3,4-*d*) pyrimidine, 4-acetamido-3-cyano-1-(2,3,5-tri-*O*-acetyl-β-D-ribofuranosyl)-, 128

Pyrazolo (3,4-*d*) pyrimidine, 4-anilino-1-(2,3,5-tri-*O*-acetyl-β-D-ribofuranosyl-, 124, 126

Pyrazolo (3,4-*d*) pyrimidine, 4-chloro-(2,3,5-tri-*O*-acetyl-β-D-ribofuranosyl-, 123, 125

Pyrazolo (3,4-*d*) pyrimidine-3-carboximidate-4-amino-1-(β-D-ribofuranosyl), 121

Pyrazolo (3,4-*d*) pyrimidine nucleosides, 121–134

Pyrazolo (1,5-α) pyrimidin-7-one, 305

Pyrazolo-s-triazine C-nucleosides, 431, 435

Pyrazomycin (pyrazofurin), 268, 280, 363, 364, 372

D,L-Pyrazomycin, 363

Pyridines, O-methyl, 255

2-Pyridone, 256

4-Pyridone, 256

Pyrimidin-7-amine, 3-α-D-arabinofuranosyl-3H-(1,2,3)-triazolo(4,5-*d*)-, 48–50

Pyrimidine, 109, 115, 122, 129–133, 256, 295–296, 305, 425

Pyrimidines, analogs of natural, 138–146

Pyrimidines, basic, 255

Pyrimidine deoxynucleosides, 221

Pyrimidine C-nucleosides, 425

Pyrimidines in obgonucleotide synthesis, 27

Pyrimidine(s), O-methyl, 255

Pyrimidine nucleosides, 97–107, 109, 112, 121–134, 142, 330–338

Pyrimidines and purines, 135–147

Pyrimidine ring, 247

Pyrimidin-7-one, 5-amino-3-(β-D-ribofuranosyl)-1,2,4-triazolo (1,5-α)-, 290

2 (1H)-pyrimidione, 40–43

Pyrazole nucleoside, 282

Pyrazolo-(3,4-*d*) pyrimidin-4-yl-hydrazone,β-D-ribofuranosyl), 127

Pyrrolopyrimidine nucleoside, 127

Pyruvate kinase, 86, 88, 89

R

Rabbit muscle, 87–90

Radioactivity, 59

Rate saturation effect, 91

Relazation times (T,), 318

Reese's Rules, 351

Resistance to ara-C and pseudoisocytidine in leukemia, 432

Restriction sequences, EcoB!, 4

Retention behavior in liquid chromatography, 28–30

Retention times in HPLC, 38

Reversed phase HPLC of nucleosides, 27, 30–35, 37–52

Ribavirin, 267, 278, 302–303

2,3-Ribo-and lyxo-episulfide derivatives of pentofuranosides, 397

Ribofuranoses, 37

C-Ribofuranosides, heterocyclic, 367

Ribofuranosyl-acetonitrile derivative, cyclic phosphate of, 323

1-β-D-Ribofuranosyladenine, 229, 232

1-β-D-Ribofuranosylbenzotriazole, 316
D-Ribofuranosyl chloride, 113
Ribofuranosyl compounds, 311–324
Ribofuranosylcyanide, 361, 367
1-(α-D-Ribofuranosyl) imidazo (1,2-C) pyrimidin-5-one 5'-monophosphate, 293–294
1-β-D-Ribofuranosylimidazo (4,5-d) pyridazine-4,7-(5H, 6H)-dione, 308
1-β-D-Ribofuranosyl) imidazo (1,2-a) pyrimidin-5-one, 296
1-(β-D-Ribofuranosyl) imidazo (1,2-C) pyrimidin-5-one, 293–294
5-(β-D-Ribofuranosyl)-isocytosine, 110
5-(β-D-Ribofuranosyl) isocytosine (ψ-isocytidine), 415
9-β-D-Ribofuranosyl-6-methyl-mercaptopurine, 339
2-β-D-Ribofuranosylnaphtho (2,3-d) triazole-4,9-dione, 307
Ribofuranosyl nucleosides, anomeric, 311–318
1-β-D-Ribofuranosyl-2-oxopurine, 229
1-Ribofuranosylpurine, 229, 230
4-β-D-Ribofuranosyl pyrazole-3,5-diester, 363
3,1-(β-D-Ribofuranosyl) pyrazolo (3,4-d) pyrimidin-4-one, 123–125
1-(β-D-Ribofuranosyl) pyrazolo (3,4-d) pyrimidin-4-thione, 123–124
3-β-D-Ribofuranosylpyrazole system, 364
4-(β-D-Ribofuranosyl) pyrazolo (1,5-a) pyrimidin-7-one, 296
1-(β-D-Ribofuranosyl) pyrazolo (3,4-d) pyrimidine, 123, 125
5-(β-D-Ribofuranosyl) pyrimidin-2,4-dithione, 117
5-β-D-Ribofuranosylpyrimidine, 116
Ribofuranosyl sugar, substituted in glycosyl bond, 239
N-α-D-Ribofuranosyl-1,2,3-thiadiazolo-(5,4-d)-pyrimidin-7-amine, 47, 49
1-β-D-Ribofuranosyl-1,2,4-triazole-3-carboxamide, 275
1-β-D-Ribofuranosyl-1,2,4-triazole-3-carboxamide phosphate, 275, 277
1-β-D-Ribofuranosyl-1,2,4-triazole-5-carboxylic acid, 270
1-β-D-Ribofuranosyl-1,2,4-triazole-3-carboxylic acid 5-phosphate, 276
1-β-Ribofuranosyl-1,2,4-triazole-3-thiocarboxamide 5'-phosphate, 276
1-β-Ribofuranosyl-1,2,4-triazoles, 3- and 5- substituted, 30
1-β-D-Ribofuranosyl-s-triazolo (4,5-c) pyridine, 302
3-(β-D-Ribofuranosyl)-1,2,4-triazolo (1,5-a) pyrimidin-7-one, 288
3-β-D-Ribofuranosyl-6,7,8-trihydroimidazo (3,4-d)(1,3 diazepin-8-(R) ol or coformycin, 159–195
6-Ribofuranosyluracil, 431
5-β-D-Ribofuranosyluracil, 312
9-(β-D-Ribofuranosyluronic acid) adenine, 86
Ribonuclease, pancreatic, 78
Ribonucleosides, 42, 274–275
Ribopyranosylamine, 234
N-β-D-Ribopyranosyl-(1,2,3)-thiadiazolo-(5,4-d)-pyrimidin-7-amine, 47, 49
N-α-D-Ribopyranosyl-(1,2,3)-thiadiazolo-(5,4-d)-pyrimidin-7-amine, 45, 49
Ribose ring, 236
Ribose structure, 2'-amino-, 388
Riboside, 80, 123, 233, 235, 262–263, 390
N-6 Ribosides, 9α- and 9-β anomers, 232–235
α-N-1 Riboside derivative of 8-Δ-, 235
β-N-1 Riboside derivative of 8-β-, 235
N_1 and N_3 Ribosides of DHAdT, 224
Ribosomes, 70
6-Ribosylaminopurine, 232
6-Ribosylated pyrimidine nucleosides, 431

Ribosylation in nitrogen heterocycles, site assignment, 301–309
N-1 Ribosylation, percentage, 240
N-2 Ribosylation, percentage, 240
Ribosylation reaction, 239–250
Ribosylation site
  compared with uv-spectral data of model methyl compounds, 289
  determinations on nucleosides, 122
  on purine analog nucleosides, 289
N-Ribosyl formininoether derivative, 231
6-Ribosyl-or 6-methyl-imidazo-(1,2-c)-pyrimidin-5-one, 294
5-Ribosylisoxazoles, synthesis scheme, 366
5-β-Ribosyluracil, 110
Ribotetranucleotide, C-C-C-G, 16
Ring-nitrogen, 203
Ring-size isomerization (furanosyl → pyranosyl), 427
RNA, transfer, 1–20, 70
RNase
  pancreatic, 16, 77
  T_1, 16
  T_1 digestion, 407–409
    cyclopurine nucleotide as substrate, 407
    interaction with guanosine 3'-phosphate, 407
  T_2 digestion, 407–409
    cyclopurine nucleotide as substrate, 407
RNA viruses, 199
RS-reticulum-cell sarcoma, implanted intramuscularly in Swiss-Mice, 154

S

S-acyl groups of polynucleotides, 60
"Salk intermediate," 200
S-alkyl groups, 60
Sangivamycin, 127
Sarcoma "180," implanted intramuscularly in female Swiss-Mice, 155
  ACID–ADA deficiency, 181–184
Selective protection for 3',5'-diol group, 200
Selectivity, 137
Severe combined immunodeficiency disease (SCID)-adenosine deaminase deficiency, 181–
    184
S-100 extract, 18
5-SH group, 65
Sh⁵UDP, 62
Sh⁵UDP and sh⁵CDP, 61–62
Sh⁵UDP and UDP, 62
Sm⁵UDP (5-methylmercaptouridylic acid), 63
Showdomycin and derivatives, 361–363
Silica gel, support material for HPLC stationary phase, 38
Silanes, sterically hindered, 33–35
Silyl group, 32–35, 353–354
Silyl triazine, 216–217
Silylated bases, 251, 255–256, 262
Silylated N-benzoylcytidine, 348
Silylated N-benzoylcytidine derivatives, 350, 352
Silylated N-benzoylguanosine derivatives, 348, 352

Silylated cytosines, use in nucleoside formation, 264
Silylated 5-ethynyluracil, 139
Silylated 6-methyl-5-nitro-uracil, steric factor increase by 5-nitro group, 259
Silylated 5-methoxyuracil, 261
Silylated 2-pyridone, 261
Silylated ribonucleotides, 348, 349
Silylation reactions, 113, 281, 348
Silyl fusion method, 123
Snake venom, 49, 77–78, 357
SnCl$_4$, 216, 224, 244, 247–249, 251–261, 264, 281
Sodium iodide in acetone, 112
Sodium methyl mercaptide in methanol, 115–118
Solvent in ribosylation reaction, CH$_3$-DN, 244
SPA-tumor, 152
Species-selective inactivation of enzymes, 95
Spectral properties of 1-glycosyl-3-substituted-1,2,4-triazole, 268
Spectral properties of 1-glycosyl-5-substituted-1,2,4-triazole, 268
Spectroscopy, UV, 26–27
Spin–spin coupling parameters, 314
Spleen phosphodiesterases, 357
Stacking form in oligonucleotides, 79
Stereospecific and regioselective synthesis of, 287
Streptomyces platensis (var. clarensis), 199
Structural gene, 3–5
Structure activity relationships, 137
S-Substituted partially thiolated polynucleotides, 66
5'-Substituted pseudouridine derivatives, 109
5-and 6- Substituted pyrimidines, 141
Sugar cation, 254, 257, 262
Sugar moiety, 301–309
Suppressor, 2
Suppressor gene of *E. coli* tyrosine tRNA, 1–36
Synthesis of nucleotides, 85–95
Synthesis (total) of tyrosine transfer, RNA gene, 3–5
Synthetic gene(s), 16–19
Synthetic nucleosides and nucleotides, 267
Synthetic nucleotides, 276
Synthetic tyrosine transfer RNA gene, 3, 5, 17, 18
Synthon, 2,2'-anhydro-, 200

T

T and B lymphotic functions, 181
Terminator functions, 19
Terminator region of gene, 4, 22
5'-*tert*-butyldimethylsilyl (TBDMSi), 221
Tert-butyldimethylsilyl blocked nucleosides, 355
Tert-butyldimethylsilyl (TBDMSi) group, 347–357
2'- and 3'-*O-tert*-butyldimethylsilyl ribonucleosides, 351
1,2,3,5-Tetra-*O*-acetyl-β-D-ribofuranose (2 β) plus indazole, 239
Tetrahydrofolic acid (citrovorum factor), 151
Tetrazole, 267, 282–283
Thermal neutron flux activation, 59
Thermodynamic nucleoside, conditions for production, 246–249

Thermodynamic product, β-anomer of DHAdT, 219
Thiadiazolopyrimidine, chromatogram, 43
Thiazole derivatives, 381–386
4-Thiazolyl nucleoside analogs, acyclic sugar, synthesis, 381, 385–386
Thioamides, 381
2-Thiocarboxamide-5-benzoyl-methylfuran, 381
6-Thioinosine, 146
Thiolated (10) polycytidylic acid (MPC), 60
Thiolated polynucleotides, 60
5-Thiolation of uracil and cytosine residues of polynucleotides, 57–59
Thiol (free) compounds, polymerization by enzyme in presence of UDP and CDP, 62
4-Thiouridine, 425
Thymidine, 29–30, 63, 137, 149, 329, 340–345
Thymidine adamantylphosphonates plus pyrophosphonates, PMR spectral data, 345
Thymidine derivatives, 97–99, 101, 105, 200, 204, 329, 341–345
Thymidine or inosine, 149–152
Thymidine or inosine with cyclophosphamide, 149, 152–153
Thymidine combined with tetrahydrofolic (citrovorum factor), 151
Thymidine kinase, 97, 99, 105, 141, 200
Thymidine, uridine, N^4-benzoylcytidine (nucleosides), 101
Thymidylate kinase, 98
Thymidylate synthetase, 141
Thymine, *bis* silyl ether, 219
Thymine and its nucleosides, 137
Thymidine and vinblastine, 149
$TiCl_4$, 256
Tin-dihydro-*S*-triazine complexes, 218–219
TLC on silica gel G, 39
TMP carboxylic-phosphoric mixed anhydride analogs, 87
dTMP, 98
dTMP kinase, 98–99
Toluene, 48
Toluene sulfonic acid, 239
p-Toluenesulfonic acid, 77
Tosylation, 112
5'-*O*-Tosyl derivative, 112
5'-*O*-Tosyl-2,3-*O*-isopropylidinepseudouridine, starting material, 109
Toxicity, histological, 103
Toyocamycin, structure, 127, 129
Transcription, 2–5
Transcription of synthetic gene, 16, 18–19
Transfer ribonucleic acids (see tRNs), 70
Transglycosidylation of nucleosides, 260
N^2, 3', 5'-Triacetylarabinofuranosylguanine, 392
N^4, 2', 5'-*O*-Triacetylcytidine 3'-phosphate, 77
2,3,5-*Tri-O*-acetylpseudouridine, 113
2', 3', 5-*Tri-O*-acetylpseudouridine, 114
2,3,5-*Tri-O*-acetyl-D-ribofuranosyl bromide, 224
2,3,5-*Tri-O*-acetyl-β-D-ribofuranosyl chloride, 113, 128
1-(2,3,5-*Tri-O*-acetyl-β-D-ribofuranosyl) pyrazolo (3,4-*d*) pyrimidin-4-one, 123, 124
3-(2,3,5-*Tri-O*-acetyl-β-D-ribofuranosyl) pyrazolo (1,5-*a*)-1,3,5-triazin-4-one, 298
N_3-*S*-Triazine nucleoside, 222
Triazine nucleosides, 2,2-anhydro-, 203

as - Triazine 1-oxides, 306
Triazinc ring system, unsymmetrical, 200
Triazines, symmetrical, 206, 287
1,2,4-Triazole(s), 124, 269–274
1,2,4-Triazole-3-carbosamide, 275, 302, 305
1,2,4-Triazole-3-carboxamide nucleoside, 270–271, 277
1,2,3-Triazole-4-carboxamide ribonucleosides, 278, 280
1,2,4-Triazole-3-carboxylate, 270–271
Triazole nucleosides, 276
1,2,3-Triazole nucleoside, 267, 278–279
1,2,4-Triazole nucleoside, 273
1,2,4-Triazole nucleosides, 267–278
Triazolopyrimidine, 37
1,2,4-Triaolo (1,5-*a*) pyrimidine, 288
3H-(1,2,3) Triazolo (4,5-*d*) pyrimidine-5-amine, 7-(benzylthio)-3-β-D-ribofuranosyl-, 42
1,2,3 Triazolo (4,5-*d*) pyrimidine- (1,2,3) thiadiazolo (5,4-*d*) pyrimidine: furnaose-pyranose rearrngement, 42
7H-1,2,3-Triazole (4,5-*d*) pyrimidine-7-thione, 3,6-dihydro-3-β-D-ribofuranosyl-, 42, 45
Triazolopyrimidinones, 244
s-Triazolo (1,3-*a*) pyrimidinone, 245
s-Triazolo (1,5-*a*) pyrimidinone, 245
s-Triazolo (1,5-*a*) pyrimidin-7-one, 305
s-Triazolo (1,5-*a*) pyrimidin-7-one, 5-chloro-3-(β-D-ribofuranosyl)-, 244
s-Triazolo (1,5-*a*) pyrimid-7-one, 244
s-Triazolo (4,3-*a*) pyrimid-5-one, 246
2,3,5-*Tri-O*-benzoyl-D-arabinofuranosyl bromide, 48
Tribenzoyl ribofuranosyl acetate, 221
Tribenzoyl ribofuranosyl bromide, 221
2,3,5-*Tri-O*-benzoyl-D-ribofuranosyl-1-*O*-acetate, 222
2,3,5-*Tri-O*-benzoyl-β-D-ribofuranosyl azide, 278
2,3,5-*Tri-O*-benzoyl-D-ribofuranosyl-bromide, 22
2,3,5-*Tri-O*-benzoyl-β-D-ribofuranosyl-cyanide, 361, 367
3-(2,3,5-*Tri-O*-benzoyl-β-D-ribofuranosyl)-5-ethoxycarbonyl-isoxazole, 365
2-(2,3,5-*Tri-O*-benzoyl-β-D-ribofuranosyl)-4-methyl-thiazole, 382
2,3,5-*Tri-O*-benzoyl-β-D-ribofuranosyl thiocarboxamide, 381–384
2,3,5-*Tri-O*-benzyl-D-arabinosyl chloride, 293–294
2',3',5'-*Tri-O*-benzylarabinofuranosyl pyrazole-D-nucleoside intermediates, 315
2,3,5-*Tri-O*-benzyl-D-ribofuranosyl bromide, 374
2,3,5-*Tri-O*-benzyl-D-ribose, 371, 374
5-Trifluoromethyl-2'-deoxyuridine, 5'-amino analogs, 101
Triester approach to oligonucleoside synthesis, 355
Trimethylsilyl (TMS) group, 239
Trimethylsilyl heterocycle, 245, 248
O-Trimethylsilyl-5-methyl-1,2,4-triazolo (1,5-*a*) pyrimidin-7-one, 290
5'-Triphosphate of AIU (AIdUTP), 85–95
2,'3,'5'-*Tri-O-tert*-butyldimethylsilyl-N-benzoyl- . guanosine, 348
N^2, 3',5'-*Tri*-tetrahydropyranyl-2'-mesylarabion-furanosylguanine, 395
5'-Trityl FUDR, 147
5-*O*-Trityl-2,3-*O*-isopropylidene-β-D-ribofuranosyl chloride, 416
p-Tritylphenysulfonylethyl (TPSE), 26
p-Tritylphenylthioethyl (TPTE), 26
dTTP, 98, 103, 105
Tubercidin, 127, 314

Tumor cells, 65
Tyrosine suppressor tRNA gene of *E. coli*, 1–36

U

sh⁵UDP, 62
sm⁵UDP, 63
Ultraviolet optical rotatory dispersion spectra, 312
UMP, 87
Unblocking nucleosides, 321
Unnatural nucleosides, 2′,3′-*O*-isospropylidene derivatives, 294, 404
Uracil or 2′-deoxyuridine, 56, 57, 112, 137, 139–140, 216, 221, 259, 398–400
Uracil nucleosides, formation, 137, 258–259
Uracil nucleotides, 57
Uracils, silylated, 251, 254, 257, 261
Uracil, 5-substituted derivatives, 142
Uridine, 329, 331, 332, 397–400
α-Uridine, effect of hydrogenation on chemical shift, 316
Uridine, deoxy-, 99, 101, 103
Uridine, 5-ethynyl-2′-deoxy-, 140–141
Uridine, dideoxy-, (AIU), 100, 106
Uridine, thymidine, 101
UV spectroscopy, 26–27

V

Vaccinia virus, 98, 103, 199, 212
Venom, snake, 77–78
Vero cell, 104
Viruses, 56, 97–99, 101, 103, 211–212
Virus-infected systems, 99
Vinblastine and thymidine, 149
Vincristine, 189

W

"Wanzlick reagent," in aldehyde chemistry, 361
Watson–Crick pairing, 70
"Wobble" pairing, 70

X

X-ray crystallography, 60

Y

Yeast hexokinase, 89